Impact Studies

Series Editor: Christian Koeberl

Springer
Berlin
Heidelberg
New York
Barcelona
Hong Kong
London
Milan
Paris
Tokyo

E. Buffetaut · C. Koeberl (Eds.)

Geological and Biological Effects of Impact Events

with 85 Figures and 23 Tables

ESF IMPACT

Springer

DR. ERIC BUFFETAUT
CNRS
16 cour du Liegat
75013 Paris
France

PROFESSOR CHRISTIAN KOEBERL
Institute of Geochemistry
University of Vienna
Althanstrasse 14
1090 Vienna
Austria

ISBN-13:978-3-642-63960-9 Springer-Verlag Berlin Heidelberg New York

Library of Congress Cataloging-in-Publication Data applied for

Die Deutsche Bibliothek – CIP Einheitsaufnahme
Geological and biological effects of impact events : ESF IMPACT ; with 23 tables / Eric Buffetaut and Christian Koeberl (ed.). - Berlin ; Heidelberg ; New York ; Barcelona ; Hong Kong ; London ; Milan ; Paris ; Tokyo : Springer, 2001
ISBN-13:978-3-642-63960-9 e-ISBN-13:978-3-642-59388-8
DOI: 10.1007/978-3-642-59388-8

Springer-Verlag is a member of BertelsmannSpringer Science+Business Media GmbH

http://www.springer.de

Camera ready by authors
Cover design: E. Kirchner, Heidelberg

Printed on acid-free paper SPIN 10843395 32/3130/as 5 4 3 2 1 0

Preface

This book is the first volume of a new interdisciplinary series on "Impact Studies". The volumes of this series aim to include all aspects of research related to impact cratering – geology, geophysics, paleontology, geochemistry, mineralogy, petrology, planetolgy, etc. Future volumes will include monographs, field guides, conference proceedings, etc. All contributions in this book were peer-reviewed to ensure high scientific quality.

The thirteen papers in the present volume result from a workshop of the European Science Foundation (ESF) IMPACT programme ("Response of the Earth System to Impact Processes"). This programme is an interdisciplinary effort aimed at understanding impact processes and their effects on the Earth System, including environmental, biological, and geological changes, and consequences for the biodiversity of ecosystems. The goals of the programme, and details about our activities, can be found on the web at <http://pssri.open.ac.uk/ESF/>. The IMPACT programme has currently 15 member nations from all over Europe. The activities of the programme range from workshops to specific topics regarding impact cratering, short courses on impact stratigraphy, shock metamorphism, etc., mobility grants for students and young researchers, development of teaching aids, and publications.

The third IMPACT workshop was held in Quillan, in the foothills of the French Pyrenees, in September 1999. The theme chosen for the workshop was "Geological and biological evidence for global catastrophes", and the papers in this volume reflect the diversity of approaches that can be used to investigate the complex causal chains linking a catastrophic event to its ultimate environmental and biological consequences. It is now widely accepted that large impact events can have a considerable influence on the global environment of the Earth and on the biosphere. However, finding a chronological coincidence between an extraterrestrial impact and an extinction episode in the fossil record is only a first step in the elucidation of what may actually have happened at that particular time. The general question to be asked (and, ideally, answered) is: what are the causal links between the physical impact phenomenon and the extinctions of species or groups of species revealed by palaeontology? Unravelling those links can only be a multidisciplinary endeavour, including the fields of (among others) astrophysics, geophysics, geochemistry, stratigraphy, mineralogy, climate modelling, and palaeobiology. Only in this way can we expect to reliably identify global catastrophes on the basis of the (often difficult to interpret) geological and biological evidence they have left behind.

A taste of both the diversity of approaches and the common goal of impact researches is provided by the papers in this volume. Contributions cover a wide time span, ranging from the Late Devonian mass extinction to the Miocene Ries/ Steinheim impact (which did not cause any mass extinction, confirming the existence of a threshold in impactor size below which no global effects can be expected) and to the Tunguska event of 1908. Not unexpectedly, the K/T boundary is discussed in several papers, both of a descriptive and of a more general nature, but papers on the crises at the Permian/Triassic, Triassic/Jurassic, and Jurassic/Cretaceous boundaries, as well as the Late Eocene event, are also included. Finally, a group of more theoretical papers dealing with impact models rounds up the volume.

All in all, the papers collected here both reflect the contents of what was presented and discussed during the three days of the Quillan workshop, and suggest directions for future researches and future meetings. It is not often that physicists, astrophysicists, geochemists, stratigraphers, and palaeontologists can meet to discuss matters of really common interest, and one of the merits of impact research is that it makes such meetings not only possible, but necessary. We hope this collection of papers will be useful both as a source of information and as an incentive for further joint work and confrontation of results and ideas.

Eric Buffetaut
CNRS
Paris, France

Christian Koeberl
University of Vienna
Vienna, Austria

April 2001

Acknowledgments

The editors are grateful to the ESF IMPACT programme for making both the Quillan workshop and the preparation of this volume possible. We thank Janet Dryden (Open University, Milton Keynes, UK) and Dieter Mader (Universität Wien, Austria) for their essential contribution to the practical preparation of this volume, and all the authors and referees for their assistance. Special thanks to Jean Le Loeuff (Musée des Dinosaures, Espéraza) for efficiently organising the Quillan workshop in surroundings the participants are not likely to forget.

Contents

List of Contributors

Gilberto A. Albertão
PETROBRAS
Av. Elias Agostinho, 665
Ponta da Imbetiba, CEP 27913-350 Macaé, R.J.
Brasil
(albertao@ep-bc.petrobras.com.br)

Natalia N. Artemieva
Institute for Dynamics of Geospheres
Leninsky pr., 38, bld.6
Moscow, 117979
Russia
(nata_art@mtu-net.ru)

Madelaine Böhme
Bavarian State Collection for Palaeontology and Historical Geology
Richard-Wagner-Str. 10
D-80333 Munich
Germany
(m.boehme@lrz.uni-muenchen.de)

Henk Brinkhuis
Laboratory of Palaeobotany and Palynology
University of Utrecht
Budapestlaan 4
3584 CD, Utrecht
The Netherlands
(h.brinkhuis@bio.uv.nl)

Jean-Georges Casier
Department Paleontology
Belgium Royal Institute of Natural Sciences
Vautier Street, 29
B-1000, Brussels
Belgium
(casier.pal@kbinirsnb.be)

Lionel Cavin
GIS PalSédCo (Toulouse-Espéraza)
Musée des Dinosaures
11260 Espéraza
France
(lionel.cavin@dinosauria.org)

Philippe Claeys
Department of Geology
Vrije Universiteit Brussel
Pleinlaan 2
B-1050 Brussels
Belgium
(phclaeys@vub.ac.be)

Henning Dypvik
Department of Geology
University of Oslo
P.O.Box 1047 Blindern
N-0316 Oslo
Norway
(henning.dypvik@geologi.uio.no).

Hans-Joachim Gregor
Naturmuseum
Im Thäle 3
D-86152 Augsburg
Germany
(H.-J.Gregor@t-online.de)

Kurt Heissig
Bavarian State Collection for Palaeontology and Historical Geology
Richard-Wagner-Str. 10
D-80333 Munich
Germany
(k.heissig@lrz.uni-muenchen.de)

Heinz Huber
Institute of Geochemistry
University of Vienna
Althanstrasse 14
A-1090 Vienna
Austria
(heinzhuber@hotmail.com)

Wolfgang Kiessling
Department of Geophysical Sciences
University of Chicago
5734 S Ellis Avenue
Chicago IL 60637
USA
(kiessl@geosci.uchicago.edu)

David T. King, Jr.
Department of Geology
Auburn University
Auburn, AL 36849-5305
USA
(kingdat@mail.auburn.edu)

Christian Koeberl
Institute of Geochemistry
University of Vienna
Althanstrasse 14
A-1090 Vienna
Austria
(christian.koeberl@univie.ac.at)

Gerhard Kudielka
Institute of Geochemistry
University of Vienna
Althanstrasse 14
A-1090 Vienna
Austria
(a8925919@unet.univie.ac.at)

J.W. de Leeuw
Department of Geochemistry
Institute of Earth Sciences
Utrecht University
P.O. Box 80.021, 3508 TA, Utrecht
The Netherlands
and
Department of Marine Biogeochemistry and Toxicology
Netherlands Institute for Sea Research (NIOZ)
P.O. Box 59, 1790 AB, Den Burg, Texel
The Netherlands

Irène Lefevre
Laboratoire des Sciences du Climat et de l'Environnement (LSCE)
Bât. 12, Domaine du C.N.R.S.
Avenue de la Terrasse
91198 Gif-sur-Yvette cedex
France
(irene.lefevre@lsce.cnrs-gif.fr)

Francis Lethiers
Laboratory of Micropaleontology
Paris VI University
4 Place Jussieu
F-75252 Paris Cedex 05
France
(lethiers@ccr.jussieu.fr)

Cindy V. Looy
Laboratory of Palaeobotany and Palynology
University of Utrecht
Budapestlaan 4
3584 CD, Utrecht
The Netherlands

Paulo P. Martins Jr.
Fundação CETEC
Av. J. C. da Silveira 2000, Horto
CEP 31170-000 Belo Horizonte
and
Universidade Federal de Ouro Preto
Escola de Minas, DEGEO, Campus
CEP 35400-000 Ouro Preto, M.G.
Brasil
(pmartin@cetec.br)

Alessandro Montanari
Osservatorio Geologico di Coldigioco
I-62020 Frontale di Apiro (MC)
Italy
(sandro.ogc@fastnet.it)

Atle Mørk
SINTEF Petroleum Research
N-7465 Trondheim
Norway
(atle.mork@iku.sintef.no).

Jason Newton
University of California Santa Cruz
Department of Earth Sciences
Santa Cruz, CA 95064
USA
(jnewton@es.ucsc.edu)

Lucille W. Petruny
Department of Geology
Auburn University
Auburn, AL 36849-5305
USA

Olivier Pierrard
Laboratoire des Sciences du Climat et de l'Environnement (LSCE)
Bât. 12, Domaine du C.N.R.S.
Avenue de la Terrasse
91198 Gif-sur-Yvette cedex
France
(olivier.pierrard@opri.fr)

Wolf Uwe Reimold
Impact Cratering Research Group
Department of Geology
University of the Witwatersrand
Johannesburg 2050
South Africa
(065wur@cosmos.wits.ac.za)

Eric Robin
Laboratoire des Sciences du Climat et de l'Environnement (LSCE)
Bât. 12, Domaine du C.N.R.S.
Avenue de la Terrasse
91198 Gif-sur-Yvette cedex
France
(robin@ede.cfr.cnrs-gif.fr)

Robert Rocchia
Laboratoire des Sciences du Climat et de l'Environnement (LSCE)
Bât. 12, Domaine du C.N.R.S.
Avenue de la Terrasse
91198 Gif-sur-Yvette cedex
France
(Robert.Rocchia@lsce.cnrs-gif.fr)

Mark A. Sephton
Planetary and Space Sciences Research Institute
Open University
Milton Keynes, MK7 6AA
Buckinghamshire
United Kingdom
(M.A.Sephton@open.ac.uk)

Valery V. Shuvalov
Institute for Dynamics of Geospheres
Leninsky pr., 38, bld.6
Moscow, 117979
Russia
(valery@vshuvalov.mccme.ru)

Morten Smelror
Geological Survey of Norway
N-7491 Trondheim
Norway
(morten.smelror@ngu.no)

Jan Smit
Institute of Earth Sciences
Vrije University
de Boelelaan 1085
1081HV Amsterdam
The Netherlands
(smit@geo.vu.nl)

Vladimir V. Svetsov
Institute for Dynamics of Geospheres
Leninsky pr., 38, bld.6
Moscow, 117979
Russia
(svetsov@idg1.chph.ras.ru).

Ruben J. Veefkind
Department of Geochemistry
Institute of Earth Sciences
Utrecht University
P.O. Box 80.021, 3508 TA, Utrecht
The Netherlands
and
Department of Marine Biogeochemistry and Toxicology
Netherlands Institute for Sea Research (NIOZ)
P.O. Box 59, 1790 AB, Den Burg, Texel
Netherlands

Henk Visscher
Laboratory of Palaeobotany and Palynology
University of Utrecht
Budapestlaan 4
3584 CD, Utrecht
The Netherlands

Ostracods Prove that the Frasnian/Famennian Boundary Mass Extinction was a Major and Abrupt Crisis

Jean-Georges Casier[1] and Francis Lethiers[2]

[1]Department Paleontology, Belgium Royal Institute of Natural Sciences, Vautier Street, 29, B-1000, Brussels, Belgium. (casier.pal@kbinirsnb.be)
[2]Paris VI University, Laboratory of Micropaleontology, 4 Place Jussieu, F-75252 Paris Cedex 05, France. (lethiers@ccr.jussieu.fr)

Abstract. For the study of the Late Devonian mass extinction, probably the second largest during the Phanerozoic, more than 40,000 ostracods have been extracted from samples collected close to the Frasnian/Famennian boundary in principal reference sections all over the world. Their study shows that about 75% of all marine species disappeared abruptly close to this boundary.

1 Generalities

The Late Devonian mass extinction is one of the big five (Raup and Sepkoski 1982; Sepkoski 1990), and probably the second largest occurring during the Phanerozoic (Casier and Lethiers 1998a; Copper 1998). Twenty-one percent of all animal families and 50 percent of all animal genera were wiped out in the marine realm during this event (Sepkoski 1982) and this extinction has been followed by one of the longest period of poor diversity known after such a crisis. Last but not least: the most extensive reef development this planet has ever seen disappeared definitively during this event (Copper 1994). This reefal activity extended almost without discontinuity from western Canada to Australia and covered ten times the areal extent of reefal ecosystems seen in modern oceans (ibid.). Reefal, peri-reefal and shallow-water benthic organisms were decimated during the Frasnian/Famennian boundary mass extinction, especially corals, stromatoporoids, trilobites and brachiopods.

Several hypotheses have been proposed to explain the Late Devonian mass extinction: a bolid impact (McLaren 1970, 1982; Sandberg et al. 1988; Claeys et al. 1992), an anoxic event - the Kellwasser event of Walliser (1984) and House (1985) or the longer Kellwasser crisis of Schindler (1990a), a cooling of the biosphere (Caputo 1985; Lethiers and Raymond 1991), an eustatic sea-level variation (Johnson et al. 1985; Casier and Devleeschouwer 1995; Girard and Feist 1997), a change in oceanic circulation patterns (Copper 1986). A combination of several of these factors, for example the co-occurrences of sea-level fluctuations, anoxia and global climatic changes (Joachimski and Buggisch 1993), is probable and might be under control of the global Late Devonian tectonic activity which has been recently reviewed by Racki (1998). This scenario is not incompatible with the occurrence of an impact or of several impacts of asteroids. But until now, evidences of cataclysmic events in the Late Devonian are scarce (Claeys et al. 1996).

2
Ostracods as a Tool

Widespread controversies subsist regarding the extinction rates, the duration and the cause of the Late Devonian mass extinction. For McLaren (1970) and Sandberg et al. (1988) the extinction is quasi-instantaneous at the Frasnian/Famennian boundary[1], but gradual for Schindler (1990a) and McGhee (1996), over several conodont zones.

Ostracods are generally very abundant, diversified and present in all types of environments in the Late Devonian and their study shows that they provide the most precise temporal representation of how rapid and devastator the Late Devonian mass extinction was.

For the study of the Late Devonian mass extinction, more than 40,000 ostracods have been extracted from samples collected close to the Frasnian/Famennian boundary in Belgium, France, Germany, Poland, Algeria, Canada, Nevada, Utah and China. As ostracods are very ecologically sensitive crustaceans, their study has provided valuable informations about the causes and the aftermath of this event, summarized in Lethiers and Casier (1999a). The goal of this paper is to review the principal results concerning the magnitude and the duration of the event.

3
Extinction of Ostracods in Shallow Settings

In **southern Belgium**, the type area for the definition of the Frasnian and Famennian stages, we have investigated the former Frasnian/Famennian boundary stratotype at Senzeille (Casier 1989, 1992) and the Sinsin (Casier and Devleeschouwer 1995), Lambermont, and Hony sections. In this region, it is well

[1] The Frasnian/Famennian boundary was selected by the Subcommission on Devonian Stratigraphy, in 1991, to coincide with the Lower Palmatolepis triangularis Zone in the conodont biostratigraphy (Klapper et al. 1994).

known that the reefal ecosystems disappeared and reappeared several times during the Frasnian in relation with sea-level fluctuations. The influence of the bathymetry and of the rate of sea-level change on the history of these buildups have been recently reviewed by Boulvain and Herbosch (1994). But the *definitive* disappearance of reefal and perireefal ecosystems, characterized by ostracods belonging to the Eifelian ecotype, seems related to the influx of oxygen depleted water into shallow marine environments (Casier 1987). These hypoxic conditions are responsible in the south-western part of the Dinant Basin, for the deposition of the Matagne black shales Formation which contains ostracods (Entomozoacea and Cypridinacea) belonging to the Myodocopid ecotype alone. But owing to the important thickness of the Matagne black shales Formation and also because the exact position of the Frasnian/Famennian boundary is controverted (Casier 1992; Bultynck and Martin 1995) in this region, the determination of the extinction rate of ostracod species belonging to the Eifelian ecotype and of the precise timing of this event are impossible. In addition, the study of ostracods has shown an important regression in the base of the Famennian. Ostracods in the base of the Famennian of all sections investigated in Belgium are poorly diversified attesting of restricted water conditions. For example, in the Sinsin section, 98 percent of the ostracod fauna in the Early Famennian are composed only of five species and in some samples 95 percent of ostracods belong to one species. Last but not least, nested ostracods are frequent in Sinsin and Lambermont. This "saucer stack" arrangement is favoured by the monospecificity and is observed today in shallow environments such as shores of a lake or of a sebkha. Owing to this important marine regression, it is not possible also to determine either the percentage or the duration of the extinction of ostracods in Belgium.

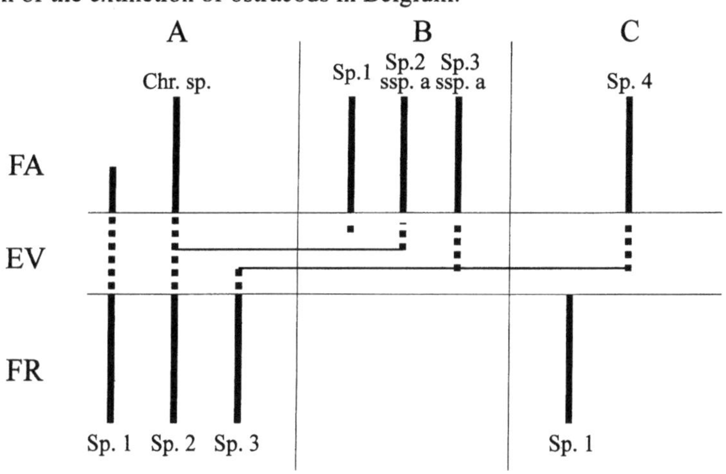

Fig. 1. Ranges of ostracod species during the Frasnian (= FR) - Famennian (= FA) event (= EV) in 3 paleobiogeographic provinces (A, B, C). An unscathed surviving species (Sp. 1) in the province A may be a false new arriving species in the province B and a false disappearing species in the province C. Sp. 2 is a chronocline surviving species. Subspeciations (Sp. 2, ssp. a and Sp. 3, ssp. a) and speciations (Sp. 4) are always allopatric. (After Lethiers and Casier 1999b, c).

In **Poland** where we have investigated the Kostomloty (Casier et al. 2000) and the Psiegorky (study in progress) sections in the Holy Cross Mountains, the apparently absence of hypoxic water conditions in the Late Frasnian is more favourable for the determination of the extinction rate. In the Kostomloty quarry which exposes a monotonous and apparently uninterrupted series of shallow-water limestones, 68 ostracod species belonging to the Eifelian ecotype are recorded close to the Frasnian/Famennian boundary. In this section, the ostracod fauna is very rich and diversified in the Frasnian, but scarce and poorly diversified close to the boundary and in the base of the Famennian. Only 6 (8?) species out of 53 occurring in the Late Frasnian, pass through the boundary. The study of these ostracods shows a regressive trend from a marine environment below wave base in the Late Frasnian to semi-restricted water conditions in the Early Famennian (Casier et al. 2000), confirmed by the sedimentological study (ibid.). For this reason the extinction of ostracods close to the Frasnian/Famennian is overestimated in this quarry. But false disappearances and false appearances of ostracod species are the rule at the Frasnian/Famennian boundary (Lethiers and Casier 1999b) (Fig. 1). These are linked either to local paleoecological factors or to the mobility of species during the Frasnian/Famennian event (ibid.). In the Kostomloty quarry, 5 false disappearances and one false appearance are numbered and consequently a total of 12 (14?) species survived the Frasnian/Famennian event. Thus, we can estimate that the real rate of extinction is consequently about 75 percent at Kostomloty.

4 Extinction of Ostracods in Deeper Settings

The best reference section in **America** for the Frasnian/Famennian boundary is the Devils Gate Pass section, located close to the mining town of Eureka, in Nevada. This section has been studied in detail by Sandberg and colleagues (1988) for the lithology, the sedimentology and conodonts. The Frasnian/Famennian boundary is located in the upper member of the Devils Gate Limestone which consists of deep-water slope deposits, mainly debris-flow and turbiditic limestones which may have occurred rapidly, so that the distribution of conodonts is not even slightly affected (ibid.) and this is necessarily also valid for ostracods. The study of ostracods from the Devils Gate Pass section has formed the subject of several papers (Casier et al. 1996, Casier and Lethiers 1997, 1998b,c). In the Late Frasnian *linguiformis* conodont Zone, ostracods are very abundant and diversified and 70 species have been identified. They belong to the Eifelian ecotype and are indicative of a well oxygenated marine environment below wave base. But close to the Frasnian/Famennian boundary, ostracods are scarce undoubtedly in relation to an increase of the sedimentation rate. According to Sandberg et al. (1988), the 10 m of shales, calcareous mudstones, and argillaceous limestones containing large flows of carbonate-rich siltstones overlapping the Frasnian/Famennian boundary at Devils Gate have been deposited rapidly. This has been confirmed by the discovery of very disseminated silicified ostracod carapaces in the Frasnian part of these turbiditic beds.

The ostracod distribution in the Early Famennian at Devils Gate is greatly influenced by regional sedimentological factors principally induced by a drastic sea-level eustatic fall. In the base of the Famennian, abundant ostracods characteristic of shallow marine environments, mixed with reworked Frasnian conodonts typical of a deeper water environment, give evidence for a dramatic environmental change, the collapse of the platform margin.

Among the 70 species occurring in the Late Frasnian at Devils Gate, only 12 survived the mass extinction. But 4 false disappearances are numbered at Devils Gate and, thus, a total of 16 species out of 70 survived the Frasnian/Famennian event. Consequently, the extinction rate is over 75 percent in the Devils Gate pass section and comparable to that observed in the Kostomloty section.

Our most important control point is of course the Global Stratotype Section and Point (GSSP) for the Frasnian/Famennian boundary recently designated in the Upper Quarry of Coumiac in the Montagne Noire, **Southern France**. The Coumiac section exposes hemi-pelagic facies that are ideal for conodont and cephalopod biozonations and consequently provides optimal biostratigraphic precision (Klapper et al. 1994). In the Upper Quarry at Coumiac, a continuous limestone boundary is marked by a grey coarse-grained dolomitic limestone (Becker et al. 1989). This bed is believed to be equivalent to the German Upper Kellwasser Limestone, a black bituminous limestone deposited under hypoxic conditions in the Kellwasser Valley in the Harz Mountains.

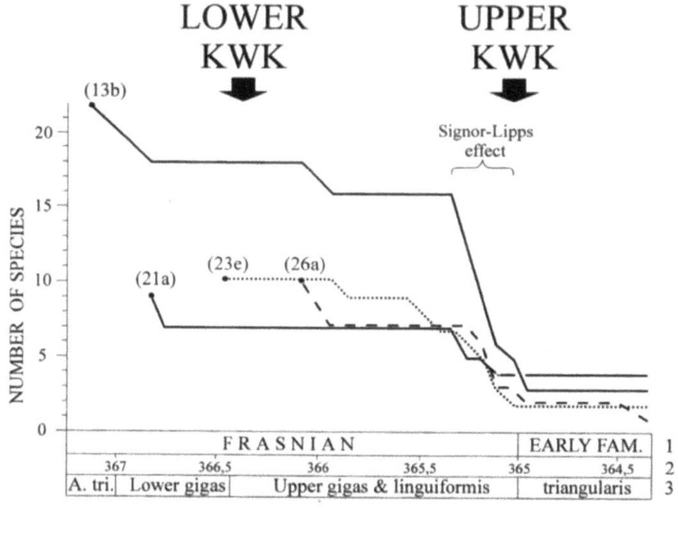

Fig. 2. Survivorship curves of successive ostracod faunas from several Frasnian beds (numbers in brackets) in the Coumiac GSSP. LKW = lower Kellwasser horizon; UKW = upper Kellwasser horizon. 1. stratigraphic scale; 2. geochronologic calibration used by Schindler (1990a,b); 3. conodont zonation. Note that no ostracod disappearances are recorded in relation with the Lower Kellwasser event. (After Lethiers and Casier 1996b.)

Ostracods present in the Coumiac GSSP have also formed the subject of several papers (Lethiers and Casier 1995, 1996a,b, 1999b). Forty-nine ostracod species from the latest Frasnian disappeared at the top or just below the bed equivalent to the Upper Kellwasser Limestone (Fig. 2) and only 17 passed through the boundary and survived the hypoxic event. Consequently, the extinction rate in the Coumiac section is similar to the extinction rate observed in Nevada and Poland.

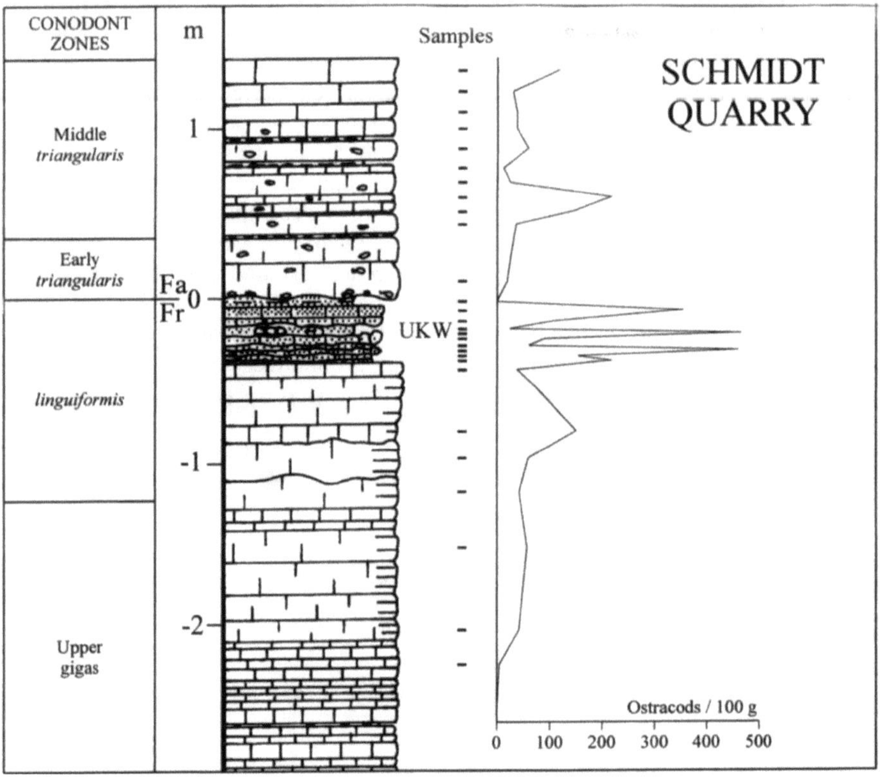

Fig. 3. Lithological column of the Late Frasnian and Early Famennian in the Schmidt quarry parastratotype (after Schindler 1990) and amount of ostracods per 100 g of sample (after Casier and Lethiers 1998a). Fr = Frasnian; Fa = Famennian; UKW = Upper Kellwasser horizon.

In **Germany** we have investigated the Schmidt quarry near Bad Wildungen, in the Kellerwald (Casier and Lethiers 1998a; Casier et al. 1999). This section was an alternative candidate selected some years ago by the Subcommission of Devonian Stratigraphy for the choice of the Frasnian/Famennian boundary GSSP (Klapper et al. 1994). The Schmidt quarry exposes an hemi-pelagic light-grey limestone sequence, interrupted by two horizons of black or dark-grey interstratified shales and limestones, the Lower and Upper Kellwasser horizons (Fig. 3). These are considered as deposited during periods of basin stagnation and anoxic bottom conditions proved by the absence of benthic life (Buggisch 1972; Groos-Uffenorde and Schindler 1990). But the study of ostracods in this section has demonstrated, on the contrary, that the oxygen content of the bottom waters during the deposition of the Upper Kellwasser horizon was very variable: sometimes bottom waters were sufficiently oxygenated and ostracods of the Eifelian and Myodocopid ecotypes occurred together, sometimes bottom waters were deficient in oxygen and ostracods of the Myodocopid ecotype were alone present. Only in the last 5 cm of the Upper Kellwasser horizon, bottom waters were anoxic and ostracods were not able to survive. The most remarkable change for conodonts occurs also above these 5 cm-thick dark carbonaceous shales (Sandberg et al. 1988), which do not contain any specimen. Thus, there is no doubt that this last bed corresponds to the acme of the biological crisis.

Of the 60 benthonic species occurring in the Late Frasnian of the Schmidt quarry, only 7 have continued into the Famennian. But 5 false disappearances and appearances of ostracod species are numbered in the Schmidt quarry and consequently a total of 12 species survived the Frasnian/Famennian event. Thus, 80 percent of the ostracod fauna has disappeared in the Schmidt quarry.

5 Conclusions and Discussions

The study of ostracods shows that about 75 percent, and sometimes more, of all marine neritic ostracod species disappeared abruptly during the F/F boundary mass extinction. This percentage of extinction seems stable whatever the depth of the platform and the geographic localisation and this plead for a global event. The study shows also that, owing to a drastic regression in the base of the Famennian, the rate of extinction is over-estimated in very shallow setting, because in this case regional eustatic factors predominate.

Until now, no other events recorded in the Late Devonian seems to be so catastrophic for ostracods. Ostracod disappearances have been recorded in relation to the Lower Kellwasser event neither in the Coumiac quarry (Fig. 2) nor in Belgium. Current study of ostracods close to the Devonian/Carboniferous boundary in the Montagne Noire, southern France, shows also that the impact of the Hangenberg event on ostracods was moderate and can be explained by sedimentological factors.

Our best control points are the Coumiac GSSP and the Schmidt quarry Parastratotype in deeper settings. In these sections the paleoecological study of ostracods shows few environmental changes on both sides of the Frasnian/Famennian boundary. In the Coumiac quarry, disappearances of ostracods began about 400,000 years before the boundary according to the chronologic calibration of Sandberg and Ziegler (1996). But at Coumiac the "Signor-Lipps" effect (Fig. 2) has probably emphasized the duration of the extinction. When species are rare or become rare exactly before an event, they can give artificial precocious disappearances in biostratigraphical charts (Signor and Lipps 1982). Thus, the maximal and real duration of the extinction is certainly inferior to 400,000 years. In the Schmidt quarry, the extinction seems more drastic and took place in the last 5 cm of the Upper Kellwasser horizon (Fig. 3). This thickness is in accord with a duration of some thousand years for the mass extinction, as estimated by Sandberg et al. (1988).

Acknowledgement

This work has been supported by the "Crisevole" and ESA 7073 programs of the CNRS. We thank Alain Herbosch of the Free University of Brussels and Raimund Feist of the University of Montpellier for constructive reviews of our manuscript.

References

Becker R, Feist R, Flajs G, House M., Klapper G (1989) Frasnian-Famennian extinction events in the Devonian at Coumiac, southern France. Compt Rend Acad Sci Paris 309 sér II: 259-266

Boulvain F, Herbosch A (1996) Anatomie des monticules micritiques du Frasnien belge et contexte eustatique. Bull Soc Géol France 3: 391-398

Buggisch W (1972) Zur Geologie und Geochemie der Kellwasserkalke und ihrer begleitenden Sedimente (Unteres Oberdevon). Abhandlungen des Hessischen Landesamtes für Bodenforschung 62: 1-68

Bultynck P, Martin F (1995) Assessment of an old stratotype: the Frasnian/Famennian boundary at Senzeilles, Southern Belgium. Bull Inst Roy Sci Nat Belgique, Sci de la Terre 65: 5-34

Caputo M (1985) Late Devonian glaciation in South America. Palaeogeogr Palaeoclimatol Palaeoecol 51: 291-317

Casier J-G (1987) Etude biostratigraphique et paléoécologique des Ostracodes du récif de marbre rouge du Hautmont à Vodelée (partie supérieure du Frasnien, Bassin de Dinant, Belgique). Rev Paléobio 6 (2): 193-204

Casier J-G (1989) Paléoécologie des Ostracodes au niveau de la limite des étages Frasnien et Famennien, à Senzeilles. Bull Inst Roy Sci Nat Belgique, Sci de la Terre 59: 79-93

Casier J-G (1992) Description et étude des Ostracodes de deux tranchées traversant la limite historique Frasnien-Famennien dans la localité-type. Bull Inst Roy Sci Nat Belgique, Sci de la Terre 62: 109-119

Casier J-G, Devleeschouwer X (1995) Arguments (Ostracodes) pour une régression culminant à proximité de la limite Frasnien - Famennien, à Sinsin (Bord sud du Bassin de Dinant, Belgique). Bull Inst Roy Sci Nat Belgique, Sci de la Terre 65: 51-68

Casier J-G, Lethiers F (1997) Les Ostracodes survivants à l'extinction du Dévonien Supérieur dans la coupe du col de Devils Gate (Nevada, U.S.A). Geobios 30, 6: 811-821
Casier J-G, Lethiers F (1998a) Ostracod Late Devonian mass extinction: the Schmidt quarry parastratotype (Kellerwald, Germany). Compt Rend Acad Sci Paris, Earth and Planet Sci 326: 71-78
Casier J-G, Lethiers F (1998b) Les Ostracodes du Frasnien terminal (Zone à linguiformis des Conodontes) de la coupe du col de Devils Gate (Nevada, USA). Bull Inst roy Sci nat Belgique, Sci de la Terre 68: 77-95
Casier J-G, Lethiers F (1998c) The recovery of the ostracod fauna after the Late Devonian mass extinction: the Devils Gate Pass section example (Nevada, USA). Compt Rend Acad Sci Paris, Earth and Planet Sci 327: 501-507
Casier J-G, Lethiers F, Claeys P (1996) Ostracod evidence for an abrupt mass extinction at the Frasnian/Famennian boundary (Devils Gate, Nevada, USA). Compt Rend Acad Sci Paris 322 sér IIa: 415-422
Casier J-G, Lethiers F, Baudin F (1999) Ostracod, organic matter and anoxic events associated with the Frasnian/Famennian boundary in the Schmidt quarry Parastratotype section (Kellerwald, Germany). Geobios 32 (6): 869-881
Casier J-G, Devleeschouwer X, Lethiers F, Preat A, Racki G (2001) Ostracods and sedimentology of the Frasnian-Famennian boundary in the Kostomloty section (Holy Cross Mountains, Poland) in relation with the Late Devonian mass extinction. Bull Inst Roy Sci Nat Belgique, Sci de la Terre 70 (submitted)
Claeys P, Casier J-G, Margolis S (1992) Microtektites and mass extinction: Evidence for a Late Devonian asteroid impact. Science 257: 1102-1104
Claeys P, Kyte F, Herbosch A, Casier J-G (1996) Frasnian-Famennian boundary: mass extinction, anoxic oceans, microtektite layers, but not much iridium? In: Ryder G, Fastovsky D, Gartner S (eds) The Cretaceous-Tertiary Event and Other Catastrophes in Earth History. Geol Soc of Amer Spec Paper 307: 491-504
Copper P (1986) Frasnian/Famennian mass extinction and cold-water oceans. Geology 14: 835-839
Copper P (1994) Ancient reef ecosytem expansion and collapse. Coral Reefs 13: 3-11
Copper P (1998) Evaluating the Frasnian-Famennian mass extinction: Comparing brachiopod faunas. Acta Palaeontol Polonica 43, 2: 137-154
Girard C, Feist R (1997) Eustatic trends in conodont diversity across the Frasnian-Famennian boundary in the stratotype area, Montagne Noire, Southern France. Lethaia 29: 329-337
Groos-Uffenorde H, Schindler E (1990) The effect of global events on entomozoacean Ostracoda In Whatley R, Maybury C (eds) Ostracoda and Global Events. British Micropal Soc Pub Ser, Chapman and Hall, pp 101-112.
House M (1985) Correlation of mid-Palaeozoic ammonoid evolutionary events with global sedimentary perturbations. Nature 313: 17-22
Joachimski M, Buggisch W (1993) Anoxic events in the late Frasnian - Causes of the Frasnian-Famennian faunal crisis? Geology 21: 675-678
Johnson J, Klapper G, Sandberg C (1985) Devonian eustatic fluctuations in Euramerica. Bull Geol Soc America 96: 567-587
Klapper G, Feist R, Becker T, House M (1994) Definition of the Frasnian/Famennian stage boundary. Episodes 16, 4: 433-441
Lethiers F, Casier J-G (1995) Les Ostracodes du Frasnien terminal ("Kellwasser" supérieur) de Coumiac (Montagne Noire, France). Rev Micropal 38, 1: 63-77
Lethiers F, Casier J-G (1996a) Les Ostracodes qui disparaissent avec l'événement Frasnien/Famennien au limitotype de Coumiac (Montagne Noire, France). Bull Inst Roy Sci Nat Belgique, Sci de la Terre 66: 73-91
Lethiers F, Casier J-G (1996b) Les Ostracodes survivants à l'événement F/F dans le limitotype de Coumiac (Montagne Noire, France). Ann Soc Géol Belgique 117 (1994), 1: 137-153

Lethiers F, Casier JG (1999a) Autopsie d'une extinction biologique. Un exemple: La crise de la limite Frasnien-Famennien (364 Ma). Comptes Rendus de l'Académie des Sciences, Earth and Planet Sci 329: 303-315

Lethiers F, Casier J-G (1999b) Les Ostracodes du Famennien inférieur au stratotype de Coumiac (Montagne Noire, France): la reconquête post-événementielle. Bull Inst Roy Sci Nat Belgique, Sci de la Terre 69: 47-66

Lethiers F, Casier J-G (1999c) Taxons "Lazares" et taxons "Elvis": à utiliser avec circonspection. Geobios 32 (5): 727-731

Lethiers F, Raymond D (1991) Les crises du Dévonien Supérieur par l'étude des faunes d'Ostracodes dans leur cadre paléogéographique. Palaeogeogr Palaeoclimatol Palaeoecol 88, 1-2: 133-146

McGhee G (1996) The Late Devonian Mass Extinction. The Frasnian/Famennian crisis. Columbia University Press, New York, 378 pp

McLaren D (1970) Time, life and boundaries. J Palaeontol 44: 801-815

McLaren D (1982) Frasnian-Famennian extinctions. In: Silver L, Schultz P (eds) Geological implications of impact of large asteroids and comets on the Earth. Geol Soc Amer Spec Pap, 190: 477-484

Racki G (1998) Frasnian-Famennian biotic crisis: undervalued tectonic control? Palaeogeogr Palaeoclimatol Palaeoecol 141: 177-198

Raup D, Sepkoski J (1982) Mass extinction in the marine fossil record. Science 215: 1501-1503

Sandberg C, Ziegler W (1996) Devonian conodont biochronology in geologic time calibration. Senckenbergiana lethaea 76, 1/2: 259-265

Sandberg C, Ziegler W, Dreesen R, Butler J (1988) Late Frasnian mass extinction: Conodont event stratigraphy, global changes, and possible cause. Cour Forschungsinst Senckenberg 102: 267-297

Schindler E (1990a) Die Kellwasser-Krise (hohe Frasne-Stufe, Ober-Devon). Göttinger Arb zur Geol und Paläontol 46, 115 pp

Schindler E (1990b) The Late Frasnian (Upper Devonian) Kellwasser crisis In Kauffman E, Walliser O (eds) Extinction Events in Earth History. Lecture Notes in Earth Sciences 30: 151-159

Sepkoski J (1982) Mass extinctions in the Phanerozoic oceans: A review. In: Silver L, Schultz P (eds) Geological implications of impact of large asteroids and comets on the Earth. Geol Soc Amer Spec Pap 190: 283-289

Sepkoski J (1990) The taxonomic structure of periodic mass extinction. In: Sharpton VL, Ward PD (eds) Global catastrophes in Earth history; an interdisciplinary conference on impacts, volcanism, and mass mortality. Geol Soc Amer Spec Pap 247: 33-44

Signor P, Lipps J (1982) Sampling bias, gradual extinction patterns and catastrophes in the fossil record. In: Silver L, Schultz P (eds) Geological implications of impact of large asteroids and comets on the Earth. Geol Soc Amer Spec Pap 190: 291-296

Walliser O (1984) Geologic processes and global events. Terra Cognita 4: 17-20

Lateral Variations in End-Permian Organic Matter in Northern Italy

Mark A. Sephton[1,2*], Ruben J. Veefkind[1,2], Cindy V. Looy[3], Henk Visscher[3], Henk Brinkhuis[3], and J.W. de Leeuw[1,2]

[1]Department of Geochemistry, Institute of Earth Sciences, Utrecht University, P.O. Box 80.021, 3508 TA, Utrecht, Netherlands.
[2]Department of Marine Biogeochemistry and Toxicology, Netherlands Institute for Sea Research (NIOZ), P.O. Box 59, 1790 AB, Den Burg, Texel, Netherlands.
[3]Laboratory of Palaeobotany and Palynology, University of Utrecht, Budapestlaan 4, 3584 CD, Utrecht, Netherlands.
[*]Corresponding author. Present address: Planetary and Space Sciences Research Institute, Open University, Milton Keynes, Buckinghamshire, MK7 6AA, United Kingdom. (M.A.Sephton@open.ac.uk)

Abstract. The end-Permian mass extinction is the most profound biotic disturbance in the Phanerozoic and is accompanied by a dramatic change in the carbon chemistry of the oceans. The cause of these disturbances remains controversial. One valuable source of information for the end-Permian event is that contained within sedimentary organic matter. We have used organic geochemical methods to investigate end-Permian organic matter in the Dolomites of northern Italy. Samples of the same marl, which marks the junction between the Bellerophon Formation and Werfen Formation, from five different sections were analysed to establish lateral variations in organic matter deposited at the end of the Permian. As the basal unit of the Werfen Formation (the Tesero Oolite Horizon) is thought to have been deposited almost synchronously throughout the Southern Alps, the marl that lies immediately beneath it may also show some synchroneity within the area. Organic geochemical data were compared to mineralogical data from the same samples. The amounts of organic matter vary with distance from the palaeocoastline but similar organic molecules are found in each sample. Both the inorganic and organic constituents point to a widespread and rapid introduction of land-derived material into the western Tethys Ocean in the Late Permian. It is possible that the features observed in the marl are the result of environmental changes occurring at the end of the Period. Acidifying emissions from the massive volcanism of the Siberian Traps may have caused a terrestrial ecosystem collapse, a loss of soil deposits and an increase in weathering rates. Future organic geochemical work will benefit from the identification of relatively low maturity organic matter in the sections in the west of the study area.

1 Introduction

The close of the Permian, approximately 250 million years ago, represents the most far-reaching and unsparing extinction of Phanerozoic time (Raup and Sepkoski 1982). Profound changes occurred in the marine and terrestrial biospheres with the loss of 90% of marine species (Raup 1979) and 70% of land vertebrate species (Maxwell 1992). In addition, Late Permian sediments record remarkable shifts in carbon isotope ratios, testifying to significant changes in oceanic and atmospheric chemistry (Magaritz et al. 1988; Holser et al. 1989; Baud et al. 1989).

Traditionally these biological and chemical changes were thought to have occurred over a period of several million years (Holser et al. 1989). However, recent radiometric dating methods have constrained their duration to less than 165,000 years (Bowring et al. 1998), making the end-Permian disturbances appear all the more dramatic.

The close association of the biological and chemical phenomena suggests that they have a common cause. A number of mechanisms have been proposed to account for the end-Permian event, including a period of extreme volcanism (Renne et al. 1995), the impact of an extraterrestrial object (Xu et al. 1985), the catastrophic overturn (Knoll et al. 1996) or transgression (Wignall and Hallam 1992) of a stagnant ocean and the rapid decomposition of the gas hydrate reservoir (Erwin 1993).

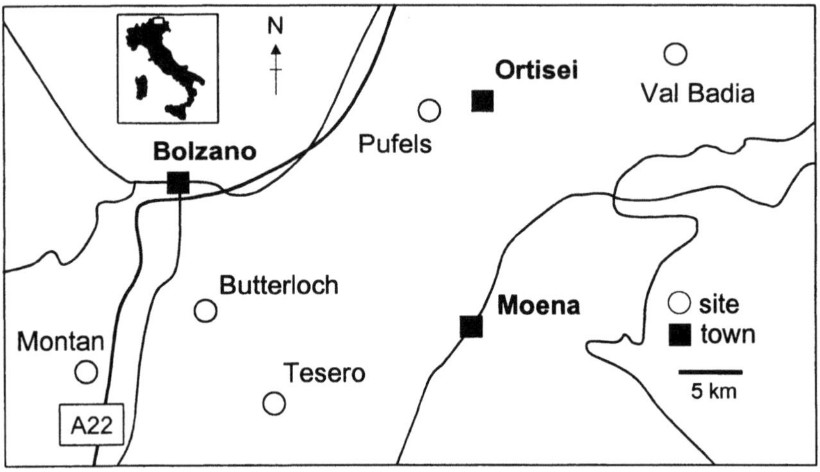

Figure 1. Location map for the studied P-Tr sections in the Dolomites of northern Italy.

A number of approaches can be followed to further understand the end-Permian disturbances. One valuable source of information is provided by the organic constituents in sediments. These components include organic compounds that are characteristic of known biochemicals produced by specific organisms from well defined habitats. Such 'molecular fossils' help in the reconstruction of palaeoenvironments. With respect to the end-Permian, these organic geochemical methods are particularly appropriate, as efforts to gain a better appreciation of the event are hindered by relatively poor body fossil preservation (Erwin 1993).

Most organic geochemical studies aim to establish vertical changes in sedimentary organic matter to reveal how palaeoenvironments have varied with time. Yet, determining the lateral variation in organic matter in a single bed is also important for a number of reasons. These variations may assist in the reconstruction of the palaeoenvironment and indicate just how representative an individual section is of the wider setting. Lateral differences can also indicate the relative extent of post-depositional alteration, such as metamorphism, experienced by a number of sections. This information can be used to identify the locations at which the organic information source is the least degraded and, therefore, reveal which sections may be suitable for more detailed investigation.

During our studies of five Permian-Triassic (P-Tr) sections in the Dolomites of northern Italy, we investigated an organic-rich marl which is laterally continuous throughout the area. The marl represents the junction between the Bellerophon and Werfen Formations (Cirilli et al. 1998; Sephton et al. 2001). Here, we present data and interpretations that constitute the first attempt to assess the lateral variations in end-Permian organic matter in the region.

2 Experimental

2.1 Sample Locations and Stratigraphical Setting

Five P-Tr sections in the Dolomites of northern Italy were sampled for this study. The sampled locations cover an area of approximately 500 km^2 (Figure 1). Descriptions of the P-Tr sections of northern Italy can be found in Bosellini (1964), Broglio-Loriga et al. (1986, 1988), Noé (1987), Brandner (1988), Wignall and Hallam (1992) and Massari et al. (1994), but a brief summary of the stratigraphy follows.

The general stratigraphy of the P-Tr sediments in the area is illustrated in Figure 2. The latest Permian deposits in the area are represented by the Bellerophon Formation, which consists of limestones, dolomites and evaporites deposited in a shallow marine inner shelf or lagoon in the western Tethys Ocean. The Bellerophon Formation is overlain by the Tesero Oolite Horizon, which is the laterally widespread basal unit of the Werfen Formation. The Tesero Oolite Horizon is an oolitic grainstone thought to have been deposited almost

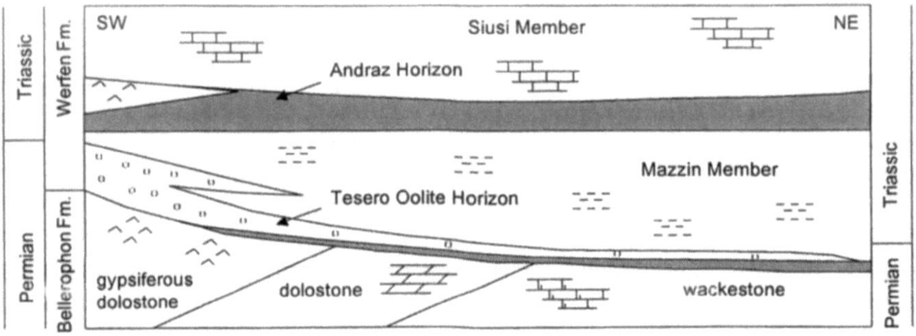

Figure 2. Stratigraphical setting of the P-Tr sediments in northern Italy (after Wignall and Hallam 1992). The proximity to the basin margin (from present-day south west to north east) decreases as follows: Butterloch < Montan and Tesero < Pufels and Val Badia.

Figure 3. A close-up of the end-Permian marl. This marl lies below the Tesero horizon, which is considered to have been deposited synchronously throughout the southern Alps.

synchronously throughout the southern Alps during rapid a marine transgression (Massari et al. 1994). The Tesero Oolite Horizon is overlain by the fine grained marly limestones of the Mazzin Member, which were deposited in a distal deep water setting.

Historically, the contact between the Bellerophon Formation and Werfen Formation has been used to indicate the boundary between the Permian and the Triassic (Posenato 1988) and this definition is in harmony with recent research from Val Badia, which indicates that the biostratigraphic and chemostratigraphic indicators of the crisis occur within a few tens of centimetres from the sedimentological changes at the Bellerophon/Werfen boundary (Cirilli et al. 1998, Sephton et al. 2001). However, the first appearance of the conodont *H. Parvus* is currently recommended to mark the onset of the Triassic (Yin et al. 1996). In the Italian Dolomites this places the P-Tr boundary in the base of the Mazzin Member. One of the implications of this definition is that the crisis occurred before, not at, the P-Tr boundary.

The sampled bed is a marly unit that marks the junction between the Bellerophon Formation and Werfen Formation (Figure 3). This sedimentary rock was chosen as a focus for this study because it is rich in organic matter, conspicuous in all sections and, based on palynological and chemical evidence, represents the end Permian crisis (Cirilli et al. 1998, Sephton et al. 2001). The marl lies immediately beneath the Tesero Oolite Horizon, which is a unit thought to have been deposited synchronously throughout the southern Alps and the transition between the two units is conformable according to biostratigraphical, sedimentological (Massari et al. 1994 and refs. therein), isotopic (Magaritz et al. 1988) and palaeomagnetic (Scholger et al. 2000) data. Hence, to some extent, the marl may provide a laterally synchronous record of the organic inputs to the western Tethys at the end of the Permian.

2.2 Bulk Geochemical Analyses

For determination of the total organic carbon content (TOC) the samples were treated twice with 0.1 N hydrochloric acid to remove the carbonate. The residue was then rinsed with distilled water and dried at 60 °C overnight. TOC was calculated from the volume of CO_2 produced by combustion at 900 °C.

2.3 X-Ray Diffraction

Powdered samples were pressed into pellets and analysed with a purpose-built goniometer equipped with a Cu XRD tube, computer steered variable divergence and anti-scatter slits and a Si(Li) solid state detector (Kevex). The generator setting was 40 kV and 40 mA. Measurements were made from 1 to 40° 2θ in steps of 0.02°. During the analyses, a relative humidity of 50% was maintained by flushing the sample container with N_2 gas from a humidity chamber.

2.4 Extraction and Fractionation

Crushed rock samples were subjected to soxhlet extraction with dichloromethane/methanol (93:7 v/v). The extracts were concentrated by rotary evaporation and separated into maltene and asphaltene fractions by repeated (x3) precipitation in heptane. Internal standards [6,6-d_2-3-methylhenicosane; 2,3-dimethyl-5-(1',1'-d_2-hexadecyl) thiophene; 2-methyl-2-(4,8,12-trimethyltridecyl) chroman and 2,3-dimethyl-5-(1',1'-d_2-hexadecyl) thiolane] were added to the maltenes before fractionation by Al_2O_3 column chromatography to produce apolar (hexane/dichloromethane 9:1 v/v) and polar (dichloromethane/methanol 1:1 v/v) fractions.

2.5 Gas Chromatography-Mass Spectrometry

Compound detection and identification was performed by gas chromatography-mass spectrometry (GC-MS) using a Hewlett Packard 5890 gas chromatograph interfaced to a VG Autospec Ultima mass spectrometer operated at 70 eV with a mass range m/z 40-800 and a cycle time of 1.8 seconds (resolution 1000). Analyses were by on-column injection onto a CP Sil-5 capillary column (25 m x 0.32 mm x 0.2 μm) with helium as a carrier gas. The GC oven was programmed from 70 °C to 110 °C at 20 °C min^{-1} and then from 110 °C to 320 °C at 4 °C min^{-1}. The final temperature was held for 15 min.

2.6 Gas Chromatography-Isotope Ratio Mass Spectrometry

Individual molecules, representative of the main aliphatic and aromatic species present in the apolar extracts had their carbon isotopic compositions determined by gas chromatography- isotope ratio mass spectrometry (GC-IRMS). The gas chromatograph (Hewlett Packard 5890) was operated as for GC-MS analyses above. Isotopic values were calculated by integrating the mass 44, 45 and 46 ion currents of the peaks produced by combustion of chromatographically separated compounds and of CO_2 spikes generated by admitting CO_2 of known ^{13}C content at regular intervals into the mass spectrometer. All obtained carbon isotopic ratios are expressed in the usual δ notation relative to the international PDB standard as described below:

$$\delta^{13}C\ ‰ = [(^{13}C/^{12}C)_{sample}/(^{13}C/^{12}C)_{PDB} - 1] \times 10^3$$

3 Results and Discussion

3.1 Total Organic Carbon

The total organic carbon content (TOC) of the marl in the different sections are listed in Table 1. Generally the TOC increases towards the east of the area. That is, away from the palaeocoastline which was situated towards the west and northwest (Magaritz et al. 1988). The TOC values indicate that the more distal environment was characterised by high autochthonous primary productivity or allochthonous input and/or storage and preservation of organic matter.

Table 1. Total organic carbon content (TOC) of the marl at different locations.

Sample	TOC (wt%)
Butterloch	0.02
Montan	0.01
Tesero	0.01
Pufels	0.51
Val Badia	2.56

3.2 Mineralogy

The X-ray diffraction results in Table 2 reveal the relative abundance of inorganic components in the different marl samples. All samples contain a significant proportion of land-derived clastic material. Quartz, feldspar, and mica are common terrigenous minerals. Kaolinite is a weathering product of feldspar and chlorite and mixed layer clays can be produced by the weathering of mica. The presence of easily weathered minerals such as feldspar and mica suggests that the transport and burial of this land-derived material was relatively rapid. Calcite is a common marine mineral derived from the skeletal remains of marine organisms and is a widespread diagenetic cement. In the more distal Pufels and Val Badia sections, the terrigenous clastics become less abundant and calcite begins to dominate (Table 2).

It is noticeable that the relative proportions of minerals can be quite different for sections that are close together geographically. For example, Montan and Tesero are approximately 15 kilometres apart but the marl from the former location is dolomitic whilst the marl from the latter is dominated by calcite. It is possible that the results presented in Table 2 have been influenced by variations in both palaeogeography and diagenetic alteration.

Table 2. X-ray diffraction results.

	Butterloch	Montan	Tesero	Pufels	Val Badia
calcite			●●●●	●●●●	●●●●
dolomite	●●●●	●●			
quartz	●●	●●●●	●●●	●●	●
feldspar	●●	●●	●●	●	
mica	●	●●●	●●	●●	●
kaolinite		●●●	●	●	
chlorite	●		●	●	
mixed layer clays	●	●●	●●	●	●
gypsum		●			

3.3 Apolar Fractions

Figure 4 shows the total ion chromatograms (TIC) from GC-MS analyses of the apolar fractions from the marl in the different sections. The aliphatic hydrocarbons content of the apolar fractions mainly consist of a homologous series of n-alkanes in the range C_{15} to C_{31} and the isoprenoidal hydrocarbons pristane and phytane. The aromatic content of the apolar fractions comprises several types of compound including polyaromatic hydrocarbons, aromatic ethers and aromatic ketones.

3.4 Origin of the Organic Matter

Visscher and Brugman (1986) studied the latest Permian organic matter in the southern Alps and discovered that towards the top of the Bellerophon Formation the marls are increasingly characterised by the abundant debris of land plants, thought to have been transported into the marine depositional environment by river action. This interpretation is consistent with our X-ray diffraction data which indicate that there has been a significant input of land-derived inorganic material to the sediment. Hence, it is reasonable to consider that a large part of the organic matter in the studied marl may also be land-derived.

Sephton et al. (1999) studied the Val Badia marl using both optical and chemical methods to reveal a terrigenous origin for the majority of the organic matter. However, compound specific carbon isotopic measurements for representative molecules for the aliphatic and aromatic organic matter at Val Badia indicate that the two fractions are isotopically distinct and, therefore, have different sources. Methyl dibenzofuran, an oxygen-containing aromatic compound has a $\delta^{13}C$ value of -29.3 ‰ whereas the aliphatic C_{19}-C_{29} n-alkanes exhibit average $\delta^{13}C$ values of -27.3 ‰.

Previous work has investigated the origin of the aliphatic material at Val Badia using flash pyrolysis-gas chromatography mass spectrometry (Sephton et al. 2001). The aliphatic pyrolysis products from the Val Badia kerogen show a

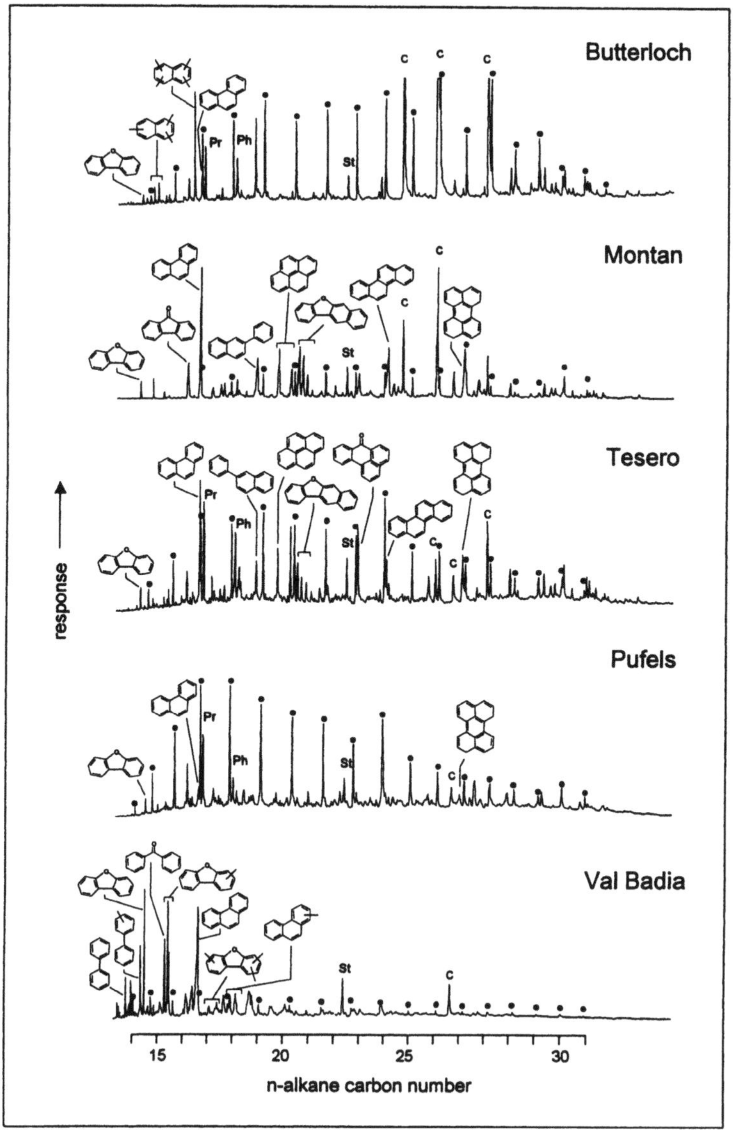

Figure 4. Total ion chromatograms from GC-MS analyses of the apolar fractions of solvent extracts of the end-Permian marl from the different sections. Assignments are as follows: (●) n-alkanes, (Pr) pristane, (Ph) phytane, (St) internal standard, (C) contaminant.

remarkable similarity to those from fossil leaf cuticles of *Ortiseia*, a common conifer genus in the Upper Permian of the western Dolomites. Furthermore, the vitrinite reflectance optical maturity parameter (mean vitrinite reflectance values at Val Badia: 0.62%) indicates that the aliphatic organic matter is thermally mature and is breaking down to produce n-alkanes. Hence, the n-alkanes at Val Badia originate from the leaf cuticles of land plants. Whether the n-alkanes in the other sections have a similar origin is, as yet, uncertain.

The provenance of the aromatic and oxygen-containing aromatic compounds at Val Badia has also been studied (Sephton et al. 1999). High resolution gas chromatography-mass spectrometry confirmed a dominance of aromatic ethers such as dibenzofuran and its alkylated homologues. These compounds are produced by the dehydration and condensation of polysaccharides such as cellulose (Pastorova et al. 1994). The aromatic ketones may also be produced by this process (cf. Kilzer and Brodio 1965). A possible source for these molecules is redeposited soil organic matter. The high abundances of these compounds seen in the marl are not common in the geological record and, therefore, the molecules in all sections are likely to have a mutual origin.

Polysaccharides are easily degraded by microbial action. The presence of polysaccharide markers in the sediment suggests that the original land-derived organic matter was either stored in oxygen-poor conditions, quickly transported and buried, or both, before the polysaccharides could be completely degraded. Therefore, both the inorganic and organic constituents of the Late Permian marl imply that significant amounts of land-derived materials may have been rapidly introduced into the marine depositional setting.

3.5
Significance of End-Permian Environmental Change

It is useful to consider the inorganic and organic data in the context of environmental change occurring at the end of the Permian. The close of the Permian was accompanied by widespread volcanism and estimates of the original size of the Siberian Traps suggest that this end-Permian volcanic province may have been the largest flood basalt episode of the Phanerozoic (e.g., Renne et al. 1995). As a result of a change in atmospheric composition due to the gaseous emissions of the Siberian eruptions, the terrestrial biosphere may have been subjected to the destabilising effects of acid rain (Eshet et al. 1995; Visscher et al. 1996). Evidence of a terrestrial ecosystem collapse at this time is provided by an unparalleled abundance of fungal remains representing the dieback of gymnosperms (Visscher et al. 1996). A crisis in the terrestrial ecosystem may have left the soil deposits surrounding the western Tethys Ocean (Figure 5) vulnerable to erosion and redeposition. This interpretation is in harmony with changes in stream morphology during the end-Permian event, evidence for which is recorded in the sedimentary record of the Karoo Basin of South Africa (Ward et al. 2000). It appears that a rapid and widespread change from meandering to braided river systems may be the result of a catastrophic and global terrestrial ecosystem collapse causing a loss of rooted plant life and a marked increase in sediment yield. Acid rain may have also increased weathering rates of inorganic materials.

This scenario could explain why significant amounts of land-derived materials were transferred to the Tethys Ocean towards the close of the Permian.

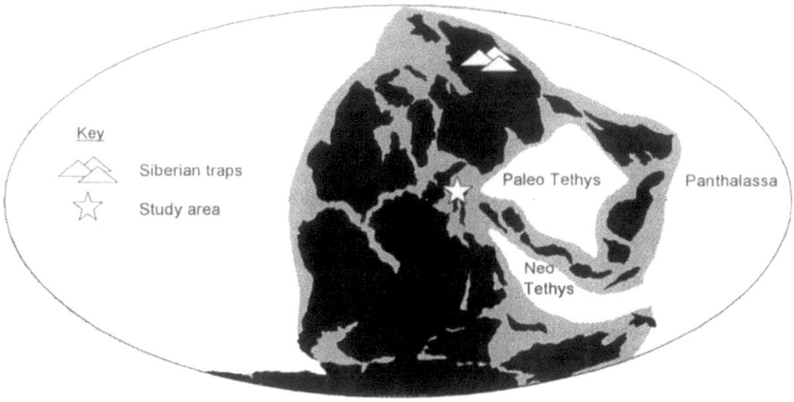

Figure 5. Palaeogeographical reconstruction illustrating the setting of the study area in the Late Permian (after Scotese and Langford 1995).

3.6 Thermal Alteration of the Organic Matter

Further organic geochemistry-based work may provide additional constraints on the causes and effects of the end-Permian environmental catastrophe. Such work will be most successful if focused on the best possible sections. Following burial, the information recorded in organic matter can be lost. Labile structures are degraded by heat and pressure. Generally the most useful information is found in thermally immature sediments where the biological heritage of the organic molecule is most easily recognised. For this reason it is important to establish the relative maturities of the P-Tr sections in the study area.

During maturation, certain compounds can be converted to others by temperature-induced reactions. Hopanes are one example of this type of compound and the ratios of particular hopanes can be used as maturity parameters (Figure 6). At Val Badia, values for the hopane maturity parameters imply that the organic matter is thermally mature, a result consistent with various published optical and chemical maturity indicators for this section (Sephton et al. 1999). This agreement between the various data illustrates the utility of hopane maturity parameters as indicators of thermal maturation levels in these sedimentary rocks. The hopane maturity parameters for the Late Permian marl indicate that the degree of thermal maturation experienced by the organic matter generally increases from the west to the east of the study area. Consequently, future searches for labile molecular fossils would be most fruitful in the west of the region. It is, however, unfortunate that the western P-Tr sections appear to have the lowest overall abundances of organic carbon.

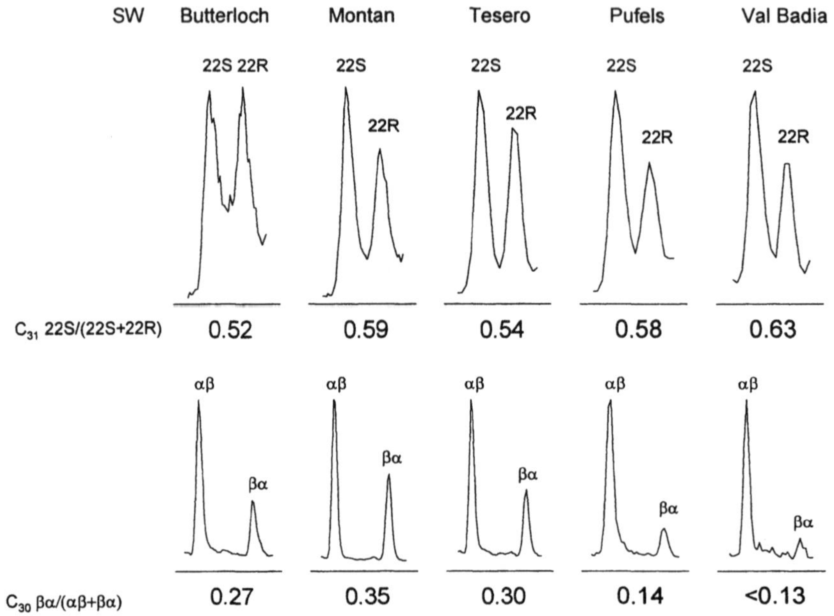

Figure 6. Hopane maturity parameters for the various end-Permian marl samples. C_{31}22S/(C_{31}22S+ C_{31}22R) increases with thermal maturity; $C_{30}\beta\alpha/(C_{30}\alpha\beta+ C_{30}\beta\alpha)$ decreases with thermal maturity.

4 Conclusions

1. In the Dolomites of northern Italy, the organic content of the youngest Bellerophon marl is dominated by land-derived organic matter.
2. The mineralogy of the marl layer reveals a significant input of an immature assemblage of terrigenous clastics. The organic constituents of the marl contain markers of polysaccharides which are usually degraded by microbial action if not quickly removed and buried. These data indicate that land-derived material was quickly transported into the marine depositional environment at the end of the Permian.
3. The rapid influx of land-derived material may have been a result of a terrestrial ecosystem collapse caused by the acidifying emissions from the extreme volcanism of the Siberian Traps. The loss of vegetation which bound soil

deposits together would have left the soil vulnerable to erosion and redeposition.

4. Future organic geochemical studies will be most fruitful at sections with a high organic carbon content and low thermal maturity. Unfortunately, while organic matter is most abundant in the east of the study area, the most immature organic matter is present in the west.

Acknowledgements

These investigations were, in part, supported by the Netherlands Geosciences Foundation (GOA) with financial aid from the Netherlands Organization for Scientific Research (NWO). This contribution is NIOZ number 3446.

References

Baud A, Magaritz M, Holser, WT (1989) Permian-Triassic of the Tethys: Carbon isotope studies. Geologische Rundschau 78: 649-677

Bowring SA, Erwin DH, Jin YG, Martin MW, Davidek K, Wang W (1998) U/Pb zircon geochronology and tempo of the end-Permian mass extinction. Science 280: 1039-1045

Bosellini A (1964) Stratigrafia, petrographia e sedimentologia delle facies carbonatiche al limite Permiano-Trias nelle Dolomiti Occidentali. Memorie del Museo di Storia Naturale della Venzia Tridentina 15: 59-110

Brandner R (1988) The Permian-Triassic boundary in the Dolomites (Southern Alps, Italy), Sant' Antonio section. In Rare Events in Geology, IGCP Project 199, Abstract Vol, Vienna, p. 49-56

Broglio Loriga C, Neri C, Pasini M, Posenato R (1988) Marine fossil assemblages from Upper Permian to lowermost Triassic in the western Dolomites (Italy). Mem Soc Geol It 34: 5-44

Broglio Loriga C, Conti MA, Farabegoli E, Fontana D, Mariotti N, Massari F, Neri C, Nicosia U, Pasini M, Parri MC, Pittau P, Posenato R, Venturini C, Viel G (1986) Upper Permian sequence and P/T boundary in the area between Carnia and the Adige Valley. In Italian IGCP 203 Group (ed). Field conference on Permian and Permian-Triassic boundary in the South-Alpine segment of the western Tethys, Bresica, June 1986, Tipolitografia Commerciale Pavese, Pavia, 23-28 pp

Cirilli S, Pirini Radrizzani C, Ponton M, Radrizzani S (1998) Stratigraphical and palaeoenvironmental analysis of the Permian-Triassic transition in the Badia Valley (Southern Alps, Italy). Palaeogeog Palaeoclim Palaeoecol 138: 85-113

Erwin DH (1993) The Great Palaeozoic Crisis: Life and Death in the Permian. Cambridge University Press, New York, 327pp

Eshet Y, Rampino MR, Visscher H (1995) Fungal event and palynological record of ecological crisis and recovery across the Permian-Triassic boundary. Geology 23: 967-970

Holser WT, Schönlaub HP, Attrep Jr M, Boeckelmann K, Klein P et al. (1989) A unique geochemical record at the Permian/Triassic boundary. Nature 337: 39-44

Kilzer FJ, Brodio A (1965) Speculations on the nature of cellulose pyrolysis. Pyrodynamics 2: 151-163

Knoll AH, Bambach RK, Canfield DE, Grotzinger JP (1996) Comparative Earth history and the Late Permian mass extinction. Science 273: 452-457

Magaritz M, Bar R, Baud A, Holser WT (1988) The carbon-isotope shift at the Permian/Triassic boundary in the southern Alps is gradual. Nature 331: 337-339

Massari F, Neri C, Pittau P, Fontana D, Stefani C (1994) Sedimentology, palynostratigraphy and sequence stratigraphy of a continental to shallow-marine rift-related succession: Upper Permian of the eastern Southern Alps (Italy). Mem Sci Geol (Padova) 46: 119-243

Maxwell WD (1992) Permian and early Triassic extinction of non-marine tetrapods. Palaeontology 35: 571-583

Noé S (1987) Facies and paleogeography of the marine Upper Permian and of the Permian-Triassic boundary of the southern Alps (Bellerophon Formation, Tesero Horizon). Facies 16: 89-142

Pastorova I, Botto RE, Arisz W, Boon J (1994) Cellulose char structure: a combined analytical Py-GC-MS, FTIR and NMR study. Carbohydrate Research 262: 27-47

Posenato R (1988) The Permian/Triassic boundary in the western Dolomites, Italy. Review and proposal. Ann Univ di Ferrara Scienze della Terra 1: 30-51

Raup DM (1979) Size of the Permo-Triassic bottleneck and its evolutionary implications. Science 206: 217-218

Raup DM, Sepkoski JJ (1982) Mass extinctions in the marine fossil record. Science 215: 1501-1503

Renne PR, Zichao Z, Richards MA, Black MT, Basu AR (1995) Synchrony and causal relations between Permian-Triassic boundary crises and Siberian flood volcanism. Science 269: 1413-1416

Scholger R, Mauritsch HJ, Brandner R (2000) Permian-Triassic boundary magnetostratigraphy from the Southern Alps (Italy). Earth Planet Sci Lett 176: 495-508

Scotese CR, Langford RP (1995) Pangea and the paleogeography of the Permian, In Scholle PA, Peryt TM, Ulmer-Scholle DS (eds) The Permian of Northern Pangea. Springer-Verlag, Berlin, pp 3-19

Sephton MA, Looy CV, Veefkind RJ, Visscher H, Brinkhuis H, de Leeuw JW (1999) Cyclic diaryl ethers in a Late Permian sediment. Org Geochem 30: 267-273

Sephton MA, Looy CV, Veefkind RJ, Brinkhuis H, de Leeuw JW, Visscher H (2001) A synchronous record of $\delta^{13}C$ shifts in the oceans and atmosphere at the end of the Permian, In: Koeberl C, MacLeod K (eds) Catastrophic Events and Mass Extinctions: Impacts and Beyond, Geological Society of America. Special Paper, 356, in press

Visscher H, Brugman WA (1986) The Permian-Triassic boundary in the southern Alps: a palynological approach. Mem Soc Geol It 34: 121-128

Visscher H, Brinkhuis H, Dilcher DL, Elsik WK, Eshet Y, Looy CV, Rampino MR, Traverse A (1996) The terminal Paleozoic fungal event: Evidence of terrestrial ecosystem destabilization and collapse. Proc Natl Acad Sci USA 93: 2155-2158

Ward PD, Montgomery DR, Smith R (2000) Altered river morphology in South Africa related to the Permian-Triassic extinction. Science 289: 1740-1743

Wignall PB, Hallam A (1992) Anoxia as a cause of the Permian/Triassic extinction: facies evidence from northern Italy and the western United States. Palaeogeogr Palaeoclimatol Palaeoecol 93: 21-46

Xu Dao-Yi, Ma S, Chia Z-F, Mao X-Y, Sun Y-Y, Zhang Q, Yang ZZ (1985) Abundance variation of iridium and trace elements at the Permian/Triassic boundary at Shangsi in China. Nature 314: 154-156

Yin H, Sweet WC, Glenister BF, Kotlyar G, Kozur H, Newell ND, Sheng J, Yang Z, Zakharov YD (1996) Recommendation of the Meishan section as global stratotype section and point for basal boundary of Triassic System. Newsletters on Stratigraphy 34: 81-108

Stable-Isotope and Trace Element Stratigraphy of the Jurassic/Cretaceous Boundary, Bosso River Gorge, Italy

Gerhard Kudielka[1], Christian Koeberl[1,2], Alessandro Montanari[2], Jason Newton[1,3] and Wolf Uwe Reimold[4]

[1]Institute of Geochemistry, University of Vienna, Althanstrasse 14, A-1090 Vienna, Austria. (christian.koeberl@univie.ac.at)
[2]Osservatorio Geologico di Coldigioco, I-62020 Frontale di Apiro (MC), Italy.
[3]Department of Earth Sciences, University of California, Santa Cruz, CA 95064, USA.
[4]Impact Cratering Research Group, Department of Geology, Univ. of the Witwatersrand, Johannesburg 2050, South Africa.

Abstract. Traditionally, the Jurassic/Cretaceous (J/K) boundary has not been well defined. Recently, a large impact structure in South Africa, the Morokweng impact structure, was found to have a radiometric age (145 Ma) indistinguishable from the stratigraphic age of the J/K boundary. Thus, it is conceivable that at the boundary, if indeed coeval with this impact structure, a record of the environmental changes associated with the impact event might exist. This could be manifested in the form of stratigraphic and geochemical features, such as sudden stable-isotope shifts, but also by the presence of impact markers, such as siderophile element anomalies. Prompted by the age similarity between Morokweng and the J/K boundary, we searched for possible impact signatures in a set of samples spanning about 40 m across the J/K boundary at the Bosso River Gorge in the Marche region of northern Italy. Various stratigraphic levels have been proposed for the boundary: by calcareous nannofossils (at 329 m), calpionellids (at 334.1 m), and magnetostratigraphy (at 318 m, in this stratigraphic interval). Carbon and oxygen stable-isotope analyses were performed on 110 carbonate samples, and trace element contents were determined in corresponding decarbonated samples. In the lower part of the sampled section, $\delta^{13}C$ and $\delta^{18}O$ values show only limited variations between 1.3‰ and 1.6‰ and a gradual shift from –1.2‰ to –2.0‰, respectively; at 333.5 m both values decrease significantly by 0.3‰ and 1.0‰, respectively. Most trace element abundances increase slightly across the lower third of the stratigraphic interval, onto which highly variable values are superimposed, probably as a result of variable amounts of a silicate or clay fraction. Very close to the boundary as defined by calcareous nannofossils (at 329 m), the abundances of Fe, Co, and Ni, and, to a lesser degree, Cr, show a marked enrichment compared to other samples. Another minor anomaly is present at 333.5 m for Ir and Cr. Thus, at best, the present data might be interpreted to hint at – but not to confirm - the presence of an impactoclastic layer at the Bosso River Gorge.

1 Introduction

One of the most important driving forces for impact research in the past decades was the study of rocks from the Cretaceous-Tertiary (K/T) boundary. Research into the cause of the mass extinction at the end of the Cretaceous took a new turn when Alvarez et al. (1980) found that concentrations of Ir and other siderophile elements in the thin clay layer that marks the K/T boundary are considerably enriched compared to those found in Cretaceous and Tertiary rocks of this stratigraphic sequence. These significant enrichments (up to a factor of 30), and the characteristic interelement ratios, were interpreted by Alvarez et al. (1980) to be the result of a large asteroid or comet impact, which also caused extreme environmental stress on a global scale. An important result of K/T boundary studies is the demonstration of how the various types of distal ejecta can be correlated with proximal ejecta and, eventually, an impact structure. The determination of the enrichment in the platinum group elements, a characteristic of extraterrestrial material, in the sediments from the K/T boundary provided the first direct evidence for a large impact event at the end of the Cretaceous. This led, in turn, to the theory that the mass extinction of the end of the Cretaceous was caused by a large asteroid or comet impact (for historical remarks, see Ryder 1996). Studies of the K/T boundary event provided data that helped with the understanding of the origin and significance of other, possibly impact-related, boundaries and exotic layers in the stratigraphic record (cf. papers in Ryder et al. 1996; Montanari and Koeberl 2000).

From studies of the K/T record, seven stratigraphic features have been proposed that may result from the global effects of a large impact: (1) global, (near)synchronous mass extinction in the fossil record; (2) negative shift in $\delta^{13}C$ $\geq 2‰$ and (3) in $\delta^{18}O$ ~2‰; (4) positive shift in $\delta^{34}S$, (5) a decrease of marine biogenic $CaCO_3$ production (Rampino and Haggerty 1996), (6) a siderophile-element anomaly (McLaren and Goodfellow 1990) and (7) presence of shocked minerals (see Montanari and Koeberl 2000, and references therein). (1)–(5) may represent mass extinction effects that could be indications for an impact event, whereas (6) and (7) verify an impact event. The six features (2–6) could be used as criteria to search for the possible correlation between a mass extinction and an impact event in the geological record.

2 The Jurassic-Cretaceous (J/K) Boundary

The biostratigraphy of the sedimentary record close to the J/K boundary has been investigated in detail at several locations in Italy, France, Spain, Slovakia, and at DSDP (deep-sea drilling program) sites. In particular, nannofossils are of critical interest, as their abundance and diversity increased rapidly from the early Tithonian to the Early Cretaceous. However, the only biostratigraphic marker rather widely accepted is Calpionellids (Bralower et al. 1989, Lowrie and Channell 1983), and the first occurrence of *Calpionella elliptica* appears to be a useful biostratigraphic event (Channell and Grandesso 1987). There is still no unambiguous and unanimously accepted definition of the boundary, as no major geological or evolutionary event, which would have left globally noticeable features, seems to have occurred during latest Jurassic or earliest Cretaceous times (Ogg and Lowrie 1986; Remane 1991). Stratigraphic positioning is rather uncertain, and previous estimates for the numerical age of the J/K boundary have varied between 133 Ma and 144 Ma, depending on which timescale is used (Rampino and Haggerty 1996, Odin 1994; Gradstein et al. 1994). Several regional working definitions for the boundary based upon the first occurrences of specific nannofossils exist, but as the faunal realms change with latitude, the biostratigraphic-based geochronological boundaries remain incompatible. It is possible to correlate the differing definitions to one another via magnetostratigraphy, since polarity events are globally isochronous, but the establishment of the J/K boundary in terms of magnetostratigraphy is still a matter of debate (Housa et al. 1999) as well. Lowrie and Channel (1983) positioned the boundary at the base of magnetochron CM17r; however, as the biostratigraphy used at this section has been updated, Ogg and Lowrie (1986) placed the boundary at the base of magnetochron CM18r. Kent and Gradstein (1985) provided an integrated magnetostratigraphic, geochronometric and biostratigraphic record as a working hypothesis, but, due to the lack of a globally recognizable event marking the boundary, no universally accepted definition exists yet (Lowrie and Channell 1983, Ogg et al. 1991).

Raup and Sepkoski (1986), in their statistical investigation of extinction rates, described the Late Jurassic mass extinction as one of the eight major extinctions during the past 250 Ma. This was corroborated by Rampino and Haggerty (1996) and by Sepkoski (1992), who indicated a 12.7% mass extinction of marine genera. Benton (1985, 1995) noticed a slightly elevated extinction rate, combined with a depressed origination rate, for continental organisms at that time. Podlaha et al. (1998) recognized a maximum of seawater temperature at 143 ± 1 Ma. Several timescales of the Mesozoic that correlate actual ages with biostratigraphic events exist. Of those, the one by Gradstein et al. (1994) seems to be widely accepted. These authors suggested an interpolated age for the J/K boundary of 144.2 ± 2.6 Ma.

Rampino and Haggerty (1996), among others, summarized arguments of the discussion whether there might be a correlation between impacts and extinctions in general, not just at the K/T boundary. Despite the relatively minor extinctions present at the J/K boundary, some indications of impact signatures were reported (cf. Rampino and Haggerty 1996), but no large coeval impact structure (or an impactoclastic layer) were known at the time. This changed a few years ago with the discovery of the Morokweng structure.

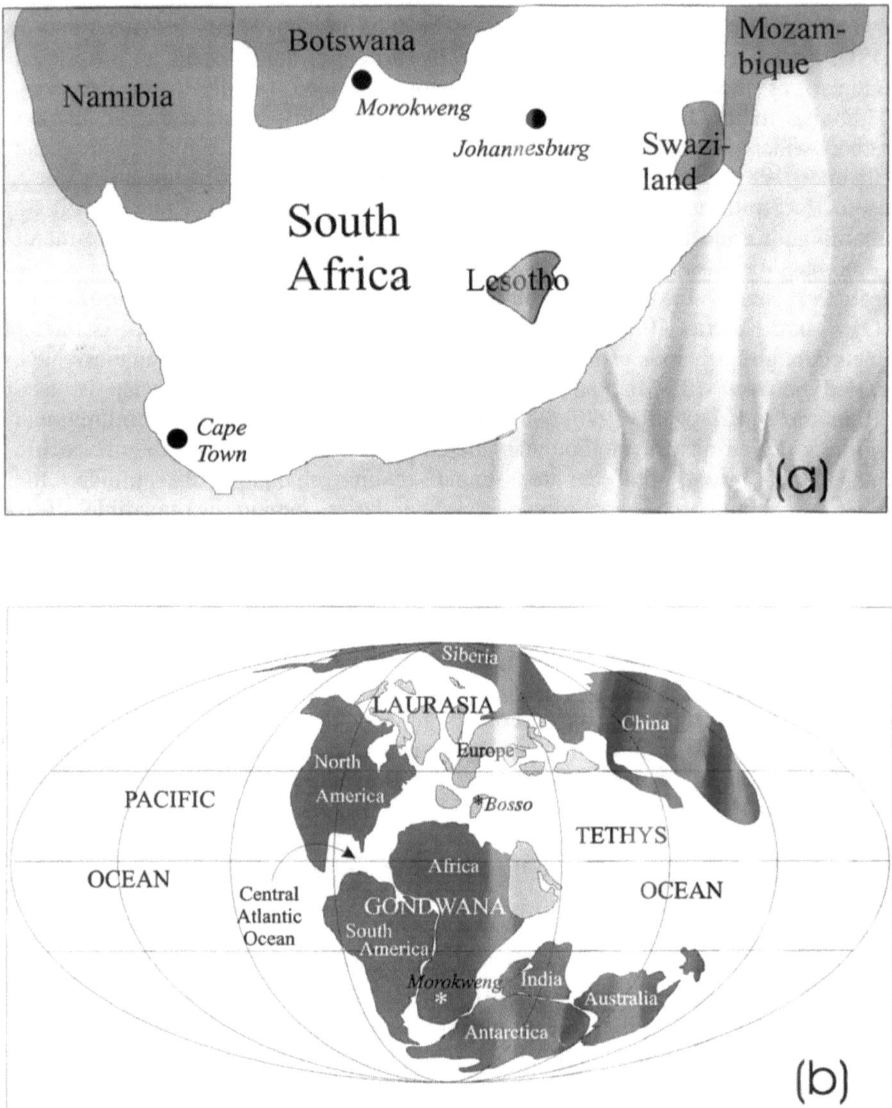

Fig. 1. (a) Geographical location of the Morokweng Impact Structure and **(b)** Continental reconstruction for the Late Jurassic (ca. 152 Ma), after Scotese (1991, 1997). The dark- and light-grey patterns represent exposed continental areas and those covered by the Tethys Ocean, respectively.

3 The Morokweng Impact Structure

The existence of a large, almost circular, subsurface impact structure in the area around Morokweng, South Africa (centered at 23° 32' E and 26° 31' S) (Fig. 1a), was inferred on the basis of gravity and magnetic investigations (Corner et al. 1997). Geophysical data showed the presence of an anomaly with about 70 to 80 km diameter, possibly surrounded by much larger arcuate features of 200 and 340 km diameter. However, current drill core data (e.g., Reimold et al. 2000) seem to favor a diameter of about 80 km. Melt rocks were recovered from drill cores near the center of the structure and indicated the presence of a thick layer of impact melt rock with high abundances of Cr, Ni, Co, and the PGEs, consistent with up to 5% of a meteoritic (chondritic) component (Koeberl et al. 1997). SHRIMP ion probe dating of zircons extracted from the impact melt rocks yielded a U–Th–Pb age of 144.7 ± 1.9 Ma (Koeberl et al. 1997). This age is indistinguishable from the currently accepted age of about 145 Ma for the J/K-boundary mentioned above (Gradstein et al. 1994). A reconstruction of the paleogeography at 152 Ma is shown in Fig. 1b (Scotese 1991, 1997). At that time southern Europe was still covered by the Tethys Ocean.

4 Is there a Link between the Morokweng impact and the J/K boundary?

The time around the J/K boundary is generally known to have been a geologically active period (see, e.g., Hynes 1990, Renne et al. 1992, Vergara et al. 1995). Hart et al. (1997) linked at least some of this enhanced activity to the Morokweng impact event. These authors suggested that additional stress from the impact fallout contributed to the already very unstable geological conditions, which, in turn, could have triggered additional geological activity, such as rifting, volcanism and tectonic movement. Thus, environmental conditions changed significantly and finally led to a (minor) mass extinction.

Based upon the (more or less) established link between a major impact event and a global mass extinction at the end of the Cretaceous Period, and given the coincidence between the age of the Morokweng impact structure and the J/K boundary, we have begun a search for distal impactoclastic layers linked to the Morokweng and/or J/K boundary event. Worldwide disturbances caused by an impact should have globally left distinct stratigraphic anomalies. Evidence of any of the stratigraphic features mentioned above anywhere in the northern hemisphere may indicate a global effect of an impact event, probably Morokweng, and hence might assist with a possible definition of the J/K boundary. The Bosso River Gorge, Italy, is an ideal place for such stratigraphic research, as this locality is well exposed, easily accessible, and appears to be stratigraphically continuous (cf. Montanari and Koeberl 2000).

5
The Bosso River Gorge: Geological Background and Location

The Bosso River section is located in the Marche Region in the northeastern Apennines of Italy, near Cagli, some 70 km west of Ancona, in a valley where the river has carved into the north-west striking folded mountains. Figure 2 shows the location and the simplified geology of this area.

The fold-and-thrust belt of the Umbria-Marche Apennines comprises a thick marine, sedimentary sequence, which can be subdivided into two main parts (cf. Montanari and Koeberl 2000): (1) a carbonate sequence (ca. 3 km thick), composed of limestones and marls, with occasional chert, extending from the Upper Triassic to the Lower Miocene, and (2) a Middle Miocene to Recent siliciclastic sequence (ca. 3 km thick), composed of synorogenic flysch (marls and sandstones), hemipelagic clays and silt, Messinian evaporites and black shales, and molasse deposits.

The Upper Triassic to uppermost Jurassic portion of the carbonate sequence is composed of a series of formations that are laterally discontinuous, because they represent a long period during which a vast carbonate platform, resting on the so-called Adria promontory of Africa, subsided differentially, creating a complex horst and graben topography. While thin, incomplete sequences of shallow water limestones were deposited on the top of the horst, deep water limestones and marls filled the adjacent graben structure.

At the very beginning of the Cretaceous, the complex topography of the Jurassic epeiric basin of the Umbria-Marche region was almost completely leveled by the deposition of the homogeneous, deep-water, pelagic limestones of the Maiolica Formation, at a time when the differential subsidence and block faulting regime of the Jurassic were followed by regional subsidence of the Adriatic crust. Thus, it is in the graben facies of the carbonate sequence that a complete and continuous transition across the J/K boundary can be found, which corresponds roughly with the boundary between the Diaspri and the Maiolica formations.

The stratigraphy of the Bosso J/K boundary has been relatively well studied in a roadcut exposure in the Bosso River Gorge by a number of researchers (Corfield et al. 1991, Cecca et al. 1994, Housa et al. 1999). This section is located in the northwestern part of the Marche region, along the provincial road that leads from the town of Cagli to the village of Pianello (Fig 2: 12° 34' 44" W, 44° 33' 30" N). A stratigraphic synthesis of the Bosso River Gorge section is shown in Fig. 3. This exposure is about 300 m wide and consists of well-stratified white limestone in beds of 5 to 10 cm thickness. The outcrop can be conveniently accessed along a road over its entire width (Fig. 4).

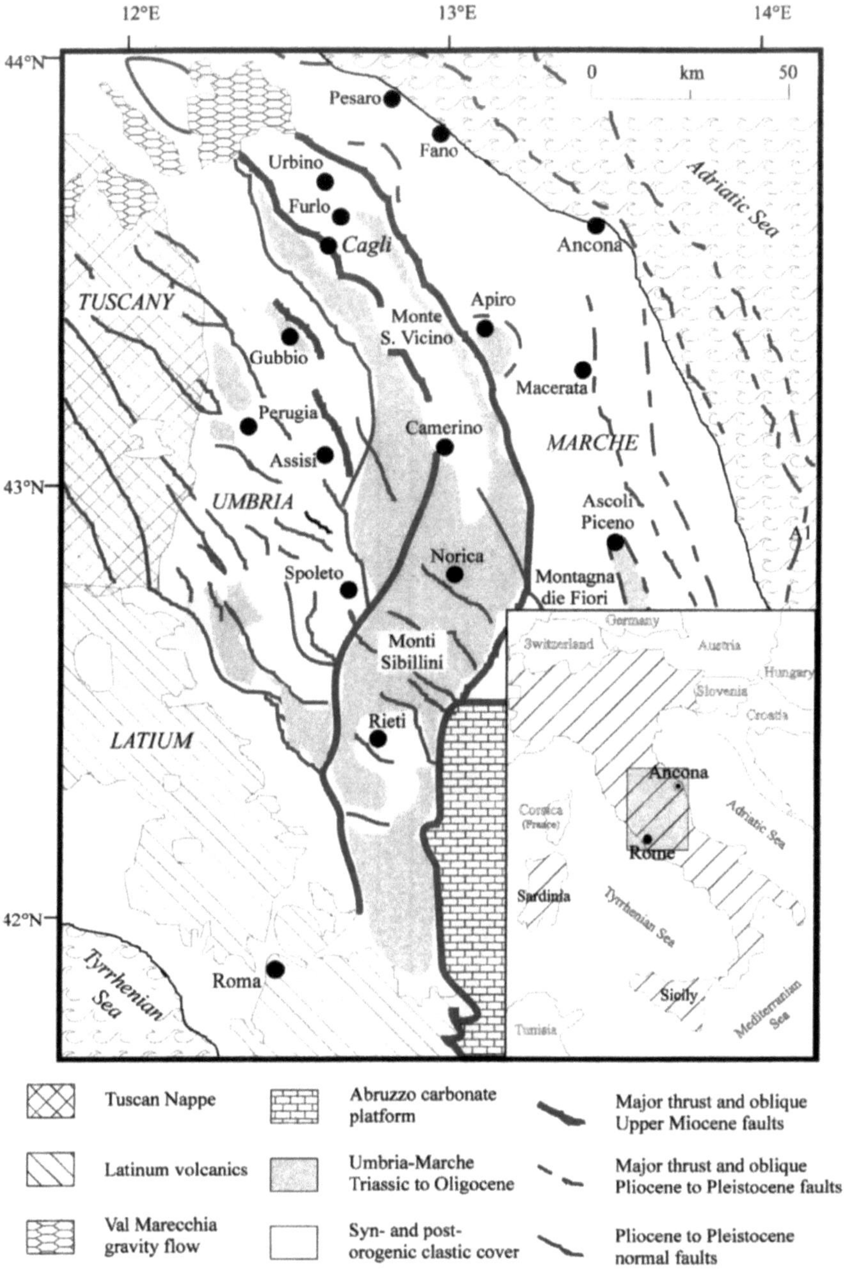

Fig. 2. Simplified geological map of the northeastern Appenines. Taken from Montanari and Koeberl (2000). The Bosso section is located in the Umbria-Marche region close to Cagli, some 70 km west of Ancona.

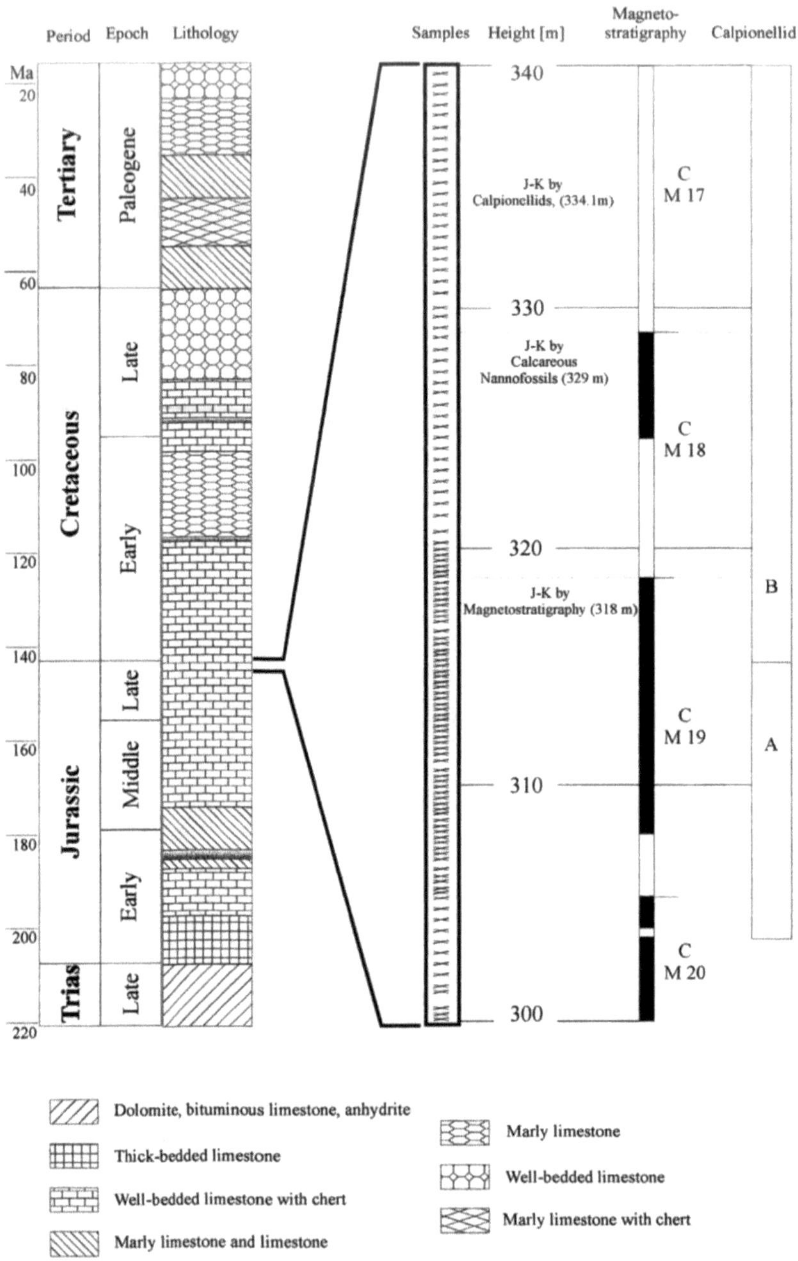

Fig. 3. Stratigraphic succession of the Umbria-Marche region. The magnified section on the right side shows the stratigraphic location of the samples and the several definitions of the J/K boundary based upon biostratigraphy as well as on magnetostratigraphy. Modified after Ogg and Lowrie (1986), Cresta et al. (1989), and Montanari and Koeberl (2000)

Fig. 4. Outcrop photograph of part of the sampled area (at stratigraphic location 330 m), showing the well-stratified structure of the lower Maiolica Formation. This section is about 300 m thick and consists mainly of limestone beds of 5 to 10 cm thickness.

In this section the J/K boundary has been defined stratigraphically by three different methods: twice in terms of biostratigraphy on the basis of calcareous nannofossils (first occurrence of *Nannoconus steimmannii steimmannii* at 329 m, Bralower et al. 1989) and calpionellids (first occurrence of *Calpionella elliptica* at 334.1 m, Cresta et al. 1989), as well as by magnetostratigraphy (at the base of magnetochron CM18r at 318 m, Ogg and Lowrie 1986).

6 Methodology

The Bosso River section was sampled at high-resolution (Fig. 4) and each sample was analyzed using two methods, isotope-ratio mass spectrometry (IRMS) and instrumental neutron activation analysis (INAA). Two batches with a total of 110 fresh bulk samples (avoiding surface samples) were taken in the form of large chips at intervals of 25 to 30 cm, following the meter count of Cresta et al. (1989). The first batch covers the section from 300 to 330.1 m and the second one the range from 329.5 to 339.5 m. The overlap helped to verify the correspondence between the two batches (see results section). The samples were relatively pure limestone with some gray calcite veins that were manually removed. The samples were then subdivided for two analytical methods using procedures described below.

For the stable isotope measurements, the (handpicked) clean limestone samples were powdered in a ball mill (alumina ceramics), and 200 ± 50 μg were used for isotope ratio mass spectrometry (IRMS).

Table 1. Sample yields after HCl treatment

Strat Loc [m]	Initial quantity [g]	Yield [g]	Yield [%]
329.5	11.375	0.092	0.80
330.2	14.562	0.122	0.84
330.5	11.815	0.072	0.61
331.0	10.543	0.064	0.61
332.0	16.338	0.139	0.85
333.0	9.321	0.094	1.00
333.5	12.342	0.108	0.87
334.0	10.365	0.115	1.11
335.0	10.094	0.127	1.26
335.5	10.052	0.110	1.09
335.8	9.913	0.120	1.21
336.1	11.784	0.133	1.13
337.0	9.991	0.091	0.92
337.4	13.379	0.099	0.74
337.8	9.863	0.095	0.97
338.0	9.749	0.104	1.07
338.6	12.330	0.120	0.98
339.0	12.217	0.107	0.88
339.4	10.978	0.121	1.10

The preparation of samples for trace element determinations by instrumental neutron activation analysis (INAA) was more elaborate. Approximately 15 g of the coarsely crushed limestone was treated with 0.25 N HCl to remove the carbonate component as CO_2. The HCl was added in 250 ml portions and the residue then filtered. This procedure was repeated until no more gas was evolved; about 2 litres of acid were used for each sample. The residues were washed three times with de-ionized water and then dried at 80 °C for two days. Quantitative analysis indicated that, on average, ~99 wt% of the samples were composed of carbonate, i.e., the residue comprised only about 1 wt% of the original sample mass (Table 1). The first batch of samples was prepared at the Department of Geology, University of the Witwatersrand, Johannesburg, South Africa, and the second one at the Institute of Geochemistry, University of Vienna. For samples in the first batch we obtained two fractions, a light one, consisting of suspended matter, and a heavy one that rapidly precipitated. Considering the minor yield of the "heavy" fraction (ca. 1 mg), in the second batch fractions were not separated.

6.1
Isotope Ratio Mass Spectrometry – Experimental Details

The isotopic ratios of carbon and oxygen were determined by isotopic ratio mass spectrometry (IRMS). As the instrument has only recently been installed at the Institute of Geochemistry (University of Vienna), no description of our analytical procedures has yet been published. Thus, we are providing here, for the first time, detailed information on our IRMS procedures, experimental parameters, standards, accuracy, and precision.

The automated liberation of the analyte gas (CO_2) by phosphoric acid digestion of multiple carbonate samples is a fairly new method and yields precise results even for small sample quantities (Ball et al. 1996). The analyte gas is prepared in a temperature-controlled system (MultiPrep), which delivers acid and purifies the CO_2 successively for up to 60 samples. The analyte gas is fed into the mass spectrometer via a dual inlet system, whereby sample and reference gas are alternately let into the mass spectrometer via a changeover valve (Fig. 5).

For the gas analyses we used a Micromass Optima electromagnetic field mass spectrometer (MS) with a 90° geometry, an electron impact ion source and a triple Faraday arrangement, which operates with an acceleration voltage of 3705.8 V, a trap current of 600 μA, and a magnet current of 3.74 A. The instrument parameters are set to a sensitivity of 4 ppm (i.e., contribution of mass 44 to mass 45 expressed relative to mass 44) and a mass resolution of 100.

The MultiPrep system consists of a temperature-controlled sample rack, with the temperature of 90 °C kept with a precision of better than ±0.05 °C, and can hold up to 60 sample vials. 200 ± 50 μg of sample are weighed into the vials using an UM3 ultramicro-balance (Mettler) with an accuracy of ±0.2 μg, sealed, and treated with some 100 μl of 103% H_3PO_4 to initiate the release of CO_2 by the reaction $3\,CaCO_3 + 2\,H_3PO_4 \rightarrow Ca_3(PO_4)_2\downarrow + 3\,CO_2\uparrow + 3\,H_2O$. For acid preparation some 250 g P_2O_5 (97% purity) were dissolved under constant stirring in 500 ml of 85 wt% H_3PO_4 until the solution reached a density of 1.9 g ml^{-1}. The acid is stored in an evacuated desiccator.

The evolved CO_2 gas passes through an electronically cooled water trap (–25 °C) to remove moisture from the reaction, and is accumulated as a solid in a liquid nitrogen-cooled cold finger (CF I at –196 °C; Fig. 5). After the reaction is complete (20 minutes), the sample gas is purged by heating CF I to –70 °C and collecting the gas in another cold-finger (CF II) at –196 °C to remove all non-gaseous contaminants above –70 °C. The gas is then injected into the mass spectrometer for C and O stable-isotope ratio analysis via the dual inlet system. The Micromass Optima instrument contains three Faraday cups for detection. This fixed-detector method implies a loss of flexibility, as the MS can not scan over a certain range of masses but it allows the simultaneous measurement of several individual isotopes with a very high precision and good efficiency (Brand 1996).

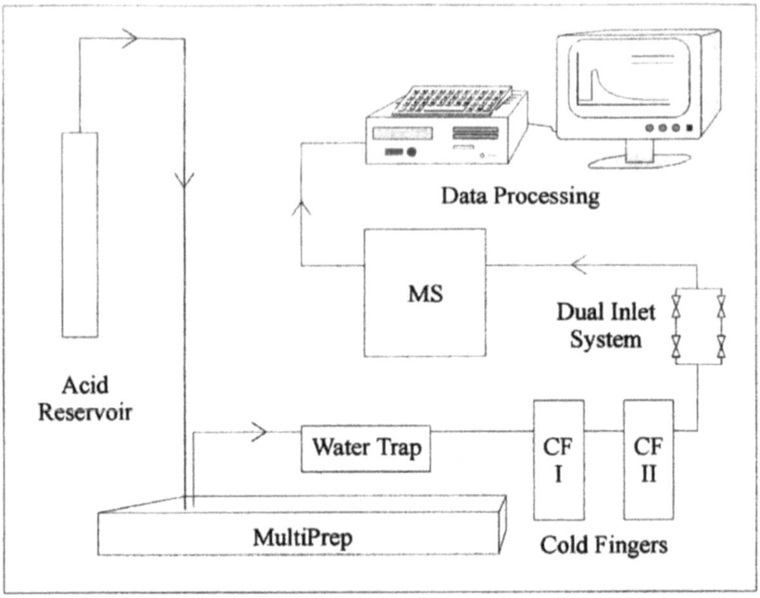

Fig. 5. Schematic set-up of the Micromass Optima isotope ratio mass spectrometer (IRMS): 103% H_3PO_4 is pumped from the acid reservoir onto the carbonate sample which is in a vial in the MultiPrep device, which is held at precisely 90 °C. Cold finger CF I (cooled with liquid N_2) collects the evolving CO_2. After the reaction, CF I is heated up to –70 °C for purification and the analyte gas is transferred into cold finger CF II, which is cooled with liquid N_2, as well. Subsequently CF II is rapidly heated up and the CO_2 is transferred into the MS via a dual inlet system for C and O stable-isotope measurements.

6.2 Standard, Precision, Reproducibility, Working Conditions

The isotopic ratio is defined as:

$$\delta^{h}E = \frac{\left(\frac{^{h}E}{^{l}E}\right)_{Sam} - \left(\frac{^{h}E}{^{l}E}\right)_{Std}}{\left(\frac{^{h}E}{^{l}E}\right)_{Std}} \times 1000‰$$

where ^{h}E represents the heavy isotope of an element E and ^{l}E the light one. The indices Sam and Std represent sample and standard, respectively.

$\delta^{18}O$–values are given vs. PDB (*P*ee *D*ee *B*elemnite, Cretaceous belemnite form the Pee Dee formation) in low-temperature carbonate studies, otherwise they are related to SMOW (*S*tandard *M*ean *O*cean *W*ater). The $\delta^{13}C$-values are referred to the Vienna PDB (VPDB) standard. Several secondary standards have been introduced as the original Pee Dee standard has been exhausted (e.g., Coplen et al. 1983; Gonfiantini et al. 1995). For standardization and accuracy measurements, we used the NBS19 marble ($CaCO_3$) standard (Friedman et al. 1982). The isotope ratios of the NBS19 standard are: $\delta^{13}C_{VPDB}$ = 1.95‰ and $\delta^{18}O_{VPDB}$ = -2.20‰ (cf. Gonfiantini et al. 1995).

Table 2 shows a set of data obtained in our laboratory for repeated measurements of the NBS19 standard throughout the sample measurement to determine accuracy and precision. The data show precisions of 1.950‰ ± 0.014 ‰ for $\delta^{13}C$ and –2.201‰ ± 0.024‰ for $\delta^{18}O$ and accuracy values of 0.03‰ for $\delta^{13}C$ and 0.06‰ for $\delta^{18}O$. The results of the isotopic measurements on the Bosso River samples are given in Table 3. All results are averages of at least two analyses.

Table 2. $\delta^{13}C$ and $\delta^{18}O$ for NBS 19

Sample	$\delta^{13}C$	$\delta^{18}O$
	1.957	–2.193
	1.957	–2.204
	1.906	–2.264
	1.956	–2.220
NBS 19	1.948	–2.170
	1.958	–2.178
	1.948	–2.220
	1.952	–2.183
	1.961	–2.170
Average	1.949	–2.200
Absolute standard deviation	0.016	0.030
Relative standard deviation	0.856	1.388

Reference values according to Gonfiantini et al. (1995): $\delta^{13}C_{PDB}$ = 1.95‰, $\delta^{18}O_{PDB}$ = –2.20‰

6.3 Instrumental Neutron Activation Analysis

Trace element analyses were carried out by instrumental neutron activation analysis (INAA) at the Institute of Geochemistry, University of Vienna, using routine procedures as described by Koeberl (1993). In short, about 50 mg (up to 150 mg) of each sample residue after decarbonation, depending on the available total sample quantity, was used for analysis. The samples were irradiated for 8 hours at a neutron flux of about $2 \cdot 10^{12}$ n cm^{-2} s^{-1}. For the analysis of the heavy fractions obtained during acid treatment, only very small quantities in the range between 0.7 mg and 7 mg were available. These samples were irradiated twice as long at the same neutron flux.

We used four standards for the calculation of element contents and for quality control: granite *AC-E* (CRPG-Nancy) for the evaluation of Na, K, Fe, Zn, As, Rb, Zr, Sb, La, Ce, Nd, Sm, Lu, Eu, Gd, Tb, Yb, Lu, Hf, Ta, Th, U; granite *G2* (U.S.

Geological Survey) for the evaluation of Na, K, Sc, Cr, Fe, Co, Zn, Rb, Zr, Ba, La, Ce, Nd, Sm, Eu, Gd, Tb, Tm, Yb, Lu, Hf, Ta, Th, U; Allende carbonaceous chondrite *ALL* (Smithsonian Institution, Washington) for the evaluation of Sc, Cr, Fe, Co, Ni, Zn, Ir, Au, and ore standard *WMG-1* (CANMET) for the evaluation of Au and Ir. Three counting series were carried out, with the first one starting 4 days after irradiation. The second counting period started 10 days, and the third one four weeks after irradiation. For further details on standards, instrumentation, accuracy, and precision, see Koeberl (1993).

7 Results

7.1 Isotope Ratio Mass Spectrometry

The results are presented in Table 3. The letters A and B indicate affiliation to the first and the second batch, respectively. Fig. 6 shows the stable isotope ratios plotted vs. stratigraphic height; we superimposed a 6th order polynomial trend-line to indicate the general trends more clearly.

The $\delta^{13}C$ and $\delta^{18}O$ data basically show corresponding zig-zag trends, with the carbon curve being the smoother one, having smaller variations superimposed. Both isotope ratio curves show a gradual decline in the first part of the section, up to 315 m. Over this interval, the $\delta^{13}C$ values decline by 0.3‰ to 1.3‰, and the $\delta^{18}O$ values decline by 1.3‰ to –2.6‰. The subsequent moderate incline of the isotope ratios back to almost the initial values continues up to 333 m. In these first parts of the sequence both data series show minima at 306.35 m and 315.75 m, with an additional minimum for $\delta^{18}O$ at 312 m. The final part of the section (after 333 m) reveals a significant drop in the $\delta^{13}C$ value by 0.3‰ over three meters, and in the $\delta^{18}O$ by 1‰ over one meter. For the remaining part of the sampled section, the trends differ slightly: the carbon isotope ratios stabilize close to 1.3‰, whereas the oxygen values increase again right after the drop.

Thus, the stable isotope ratio trends for both carbon and oxygen remain rather uneventful over most of the sampled sequence, but show some notable changes at 333 m.

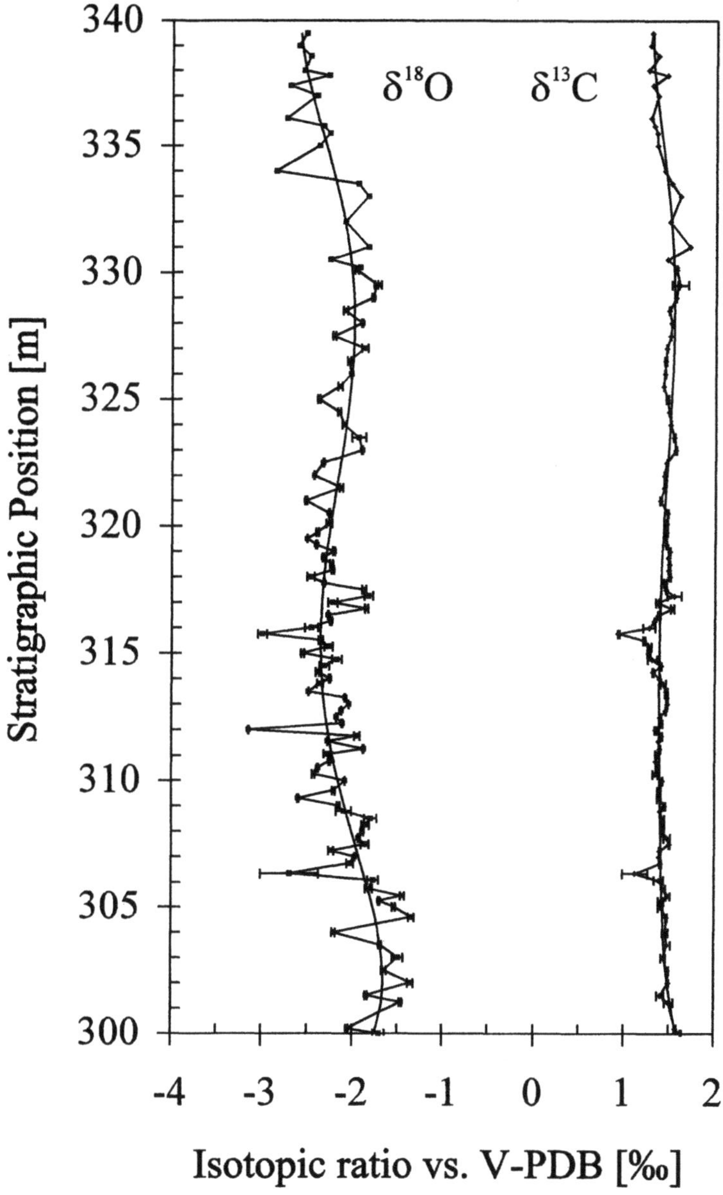

Fig. 6. $\delta^{13}C_{PDB}$ and $\delta^{18}O_{PDB}$ vs. stratigraphic position (height). A 6th order polynomial trend-line is superimposed to indicate the general trends more clearly. In the lower part the $\delta^{13}C_{PDB}$ data show little variation between 1.3‰ and 1.6‰ and the $\delta^{18}O_{PDB}$ – data indicate a slight and gradual shift from –1.2‰ to –2.0‰. At 333.5 m a significant drop and a subsequent increase is noticeable for both isotope ratios.

Table 3. Carbon and oxygen isotope data for samples from the Bosso River Gorge, Italy

Strat Loc [m]	$\delta^{13}C_{PDB}$ [‰]	$\delta^{18}O_{PDB}$[‰]	Strat Loc [m]	$\delta^{13}C_{PDB}$[‰]	$\delta^{18}O_{PDB}$[‰]
300.00	1.64 ±0.01	-1.70 ±0.06	316.50	1.37 ±0.01	-2.28 ±0.00
300.20	1.58 ±0.01	-2.05 ±0.01	316.75	1.53 ±0.02	-1.86 ±0.02
300.50	1.73 ±0.09	0.74 ±1.14	317.00	1.36 ±0.02	-2.23 ±0.05
301.25	1.50 ±0.05	-1.46 ±0.01	317.25	1.55 ±0.08	-1.83 ±0.05
301.52	1.41 ±0.04	-1.84 ±0.01	317.50	1.46 ±0.01	-1.88 ±0.02
302.00	1.50 ±0.00	-1.35 ±0.03	317.75	1.45 ±0.01	-2.32 ±0.00
302.50	1.49 ±0.01	-1.64 ±0.03	318.00	1.50 ±0.01	-2.47 ±0.04
303.00	1.44 ±0.03	-1.49 ±0.06	318.25	1.49 ±0.01	-2.23 ±0.01
303.50	1.48 ±0.04	-1.69 ±0.01	318.50	1.50 ±0.00	-2.24 ±0.02
304.00	1.46 ±0.03	-2.21 ±0.02	318.75	1.49 ±0.00	-2.33 ±0.01
304.60	1.48 ±0.01	-1.34 ±0.03	319.00	1.50 ±0.00	-2.22 ±0.01
305.00	1.41 ±0.03	-1.54 ±0.02	319.25	1.46 ±0.00	-2.41 ±0.01
305.25	1.41 ±0.03	-1.70 ±0.01	319.50	1.45 ±0.00	-2.51 ±0.00
305.45	1.49 ±0.02	-1.45 ±0.02	319.75	1.46 ±0.01	-2.40 ±0.01
305.75	1.44 ±0.02	-1.82 ±0.04	320.10	1.46 ±0.01	-2.27 ±0.03
306.10	1.39 ±0.05	-1.77 ±0.06	320.50	1.48 ±0.00	-2.27 ±0.00
306.35	1.13 ±0.14	-2.69 ±0.32	321.00	1.39 ±0.01	-2.53 ±0.01
306.75	1.41 ±0.01	-2.03 ±0.04	321.50	1.43 ±0.01	-2.14 ±0.02
307.00	1.39 ±0.00	-1.97 ±0.00	322.00	1.43 ±0.00	-2.44 ±0.00
307.25	1.39 ±0.01	-2.24 ±0.03	322.50	1.46 ±0.00	-2.34 ±0.01
307.50	1.50 ±0.01	-1.86 ±0.04	323.00	1.56 ±0.01	-1.91 ±0.01
307.75	1.48 ±0.04	-1.94 ±0.00	323.50	1.54 ±0.01	-1.94 ±0.08
308.00	1.43 ±0.01	-1.89 ±0.02	324.00	1.50 ±0.01	-2.11 ±0.02
308.25	1.43 ±0.02	-1.86 ±0.04	324.50	1.49 ±0.00	-2.16 ±0.02
308.50	1.42 ±0.02	-1.80 ±0.07	325.00	1.47 ±0.01	-2.38 ±0.01
308.80	1.40 ±0.00	-2.10 ±0.09	325.50	1.43 ±0.00	-2.15 ±0.02
309.00	1.45 ±0.01	-2.16 ±0.01	326.00	1.44 ±0.01	-2.03 ±0.00
309.30	1.39 ±0.02	-2.60 ±0.01	326.50	1.44 ±0.01	-2.05 ±0.02
309.60	1.39 ±0.02	-2.21 ±0.02	327.00	1.46 ±0.00	-1.88 ±0.04
310.00	1.42 ±0.00	-2.09 ±0.00	327.50	1.50 ±0.00	-2.22 ±0.02
310.25	1.34 ±0.03	-2.43 ±0.02	328.00	1.51 ±0.00	-1.91 ±0.01
310.50	1.37 ±0.02	-2.38 ±0.01	328.50	1.49 ±0.00	-2.09 ±0.03
310.75	1.35 ±0.00	-2.26 ±0.01	329.00	1.56 ±0.00	-1.79 ±0.01
311.05	1.36 ±0.02	-2.28 ±0.04	329.50 A	1.58 ±0.00	-1.76 ±0.02
311.25	1.39 ±0.01	-1.88 ±0.01	329.50 B	1.60 ±0.09	-1.74 ±0.04
311.55	1.38 ±0.00	-2.28 ±0.00	330.10	1.56 ±0.01	-1.97 ±0.02
311.75	1.41 ±0.01	-1.95 ±0.03	330.20	1.56 ±0.03	-1.94 ±0.00
312.00	1.34 ±0.01	-3.15 ±0.00	330.50	1.46 ±0.05	-2.26 ±0.10
312.25	1.41 ±0.01	-2.12 ±0.01	331.00	1.71 ±0.06	-1.84 ±0.08
312.50	1.40 ±0.01	-2.18 ±0.00	332.00	1.49 ±0.03	-2.10 ±0.05
312.75	1.46 ±0.00	-2.13 ±0.00	333.00	1.61 ±0.04	-1.84 ±0.08
313.00	1.48 ±0.00	-2.04 ±0.00	333.50	1.51 ±0.02	-1.95 ±0.02
313.25	1.47 ±0.01	-2.09 ±0.00	334.00	1.43 ±0.06	-2.86 ±0.05

Table 3. Continued

Strat Loc [m]	$\delta^{13}C_{PDB}$ [‰]	$\delta^{18}O_{PDB}$[‰]	Strat Loc [m]	$\delta^{13}C_{PDB}$[‰]	$\delta^{18}O_{PDB}$[‰]
313.50	1.47 ±0.02	-2.49 ±0.00	335.00	1.35 ±0.03	-2.39 ±0.03
313.80	1.43 ±0.03	-2.37 ±0.03	335.50	1.34 ±0.05	-2.27 ±0.02
314.00	1.39 ±0.01	-2.26 ±0.01	335.80	1.31 ±0.02	-2.34 ±0.03
314.25	1.31 ±0.01	-2.40 ±0.02	336.10	1.28 ±0.01	-2.74 ±0.07
314.50	1.41 ±0.01	-2.32 ±0.05	337.00	1.35 ±0.05	-2.41 ±0.04
314.75	1.30 ±0.05	-2.18 ±0.05	337.40	1.30 ±0.02	-2.71 ±0.01
315.00	1.27 ±0.02	-2.56 ±0.02	337.80	1.46 ±0.08	-2.28 ±0.07
315.25	1.28 ±0.02	-2.27 ±0.05	338.00	1.25 ±0.01	-2.56 ±0.07
315.50	1.22 ±0.01	-2.36 ±0.02	338.60	1.35 ±0.04	-2.49 ±0.03
315.75	0.93 ±0.01	-2.99 ±0.05	339.00	1.27 ±0.05	-2.62 ±0.01
316.00	1.27 ±0.07	-2.46 ±0.07	339.50	1.28 ±0.01	-2.53 ±0.05
316.25	1.33 ±0.00	-2.25 ±0.01			

7.2 Neutron Activation Analysis

The abundances of the measured elements and the ratios K/U, La/Th, Th/U, Ni/Ce, La_{CN}/Yb_{CN} and Eu_{CN}/Eu_{CN}^* ($Eu_{CN}^* = \sqrt{Sm_{CN} \cdot Gd_{CN}}$) are listed in Table 4 and the results for the heavy fractions are given in Table 5. Some selected trace elements (Na, Sc, Cr, Fe, Co, Ni, As, Rb, Ba, La, Ce, Nd, Sm, Yb, Hf, and Th) are plotted in a logarithmic diagram vs. stratigraphic height (Fig. 7). In general, relatively high abundances of elements such as Zn, As, Sb, and Au are observed in practically all samples. As the analyzed samples represent only the non-carbonate fraction of the rocks (i.e., silicate minerals and accessory minerals, such as sulfides) it is very likely that the high values of these elements result from trace amounts of opaque minerals, such as pyrite, chalcopyrite, etc. A mineralogical study of the residues (including a search for shocked quartz) has not yet been done. The results obtained for the rare earth elements (REE) have been recalculated to chondrite-normalized abundances (with normalization factors after Taylor and McLennan 1985) and are shown in Fig. 8. In general, the results show high abundances of the light REE with a steeply decreasing slope, typically a negative Eu anomaly, and a flat heavy REE distribution pattern. This pattern is typical for upper continental crust and likely results from the accessory silicate minerals. In the lower part of the sampled section, some minor Ce-anomalies were observed, with the sample at 315.5 m having the most noticeable one. Compared to the average REE content, the sample at 300.5 m is enriched in the REEs by about a factor of two, whereas those at 309.3 m and 312 m are extremely depleted (by a factor of 10 to 20).

Table 4. Trace element abundances of HCl-leached residues from the Bosso River Gorge, Italy. All data in ppm, except as noted. Analyses by Instrumental Neutron Activation Analysis (INAA).

Strat Loc [m]	300.2	300.5	301.25	301.52	302.0	302.5	303.0	303.5	304.0	304.6
Na [wt%]	0.25	0.34	0.20	0.18	0.15	0.21	0.19	0.23	0.27	0.24
K [wt%]	2.50	4.96	2.34	2.23	1.98	2.81	2.52	3.46	3.38	3.33
Sc	7.38	16.3	6.53	5.88	5.13	8.25	6.23	9.38	7.93	17.6
Cr	38.3	94.7	34.7	39.3	32.0	84.6	39.8	57.4	55.3	56.2
Fe [wt%]	1.28	2.49	0.99	1.04	0.94	1.56	1.23	1.67	1.46	1.30
Co	7.24	12.7	7.71	5.88	5.51	8.99	7.16	9.15	8.43	9.84
Ni	30	65	34	38	17	47	51	72	42	11
Zn	35	60	19	30	16	35	52	34	37	53
As	0.97	7.72	1.61	1.27	0.96	1.68	3.74	5.91	3.92	2.33
Rb	97.4	215	75.6	81.5	70.6	102	87	130	110	106
Zr	100	150	65	55	50	50	50	60	95	75
Sb	0.69	1.86	0.40	0.54	0.47	0.69	0.55	0.97	0.88	0.93
Cs	12.7	45.8	15.1	10.5	8.95	14.3	12.1	14.5	12.3	14.4
Ba	98	330	150	173	90	153	99	132	124	217
La	10.2	23.7	8.35	8.57	6.66	13.1	8.42	12.9	11.0	11.6
Ce	17.2	39.0	12.9	14.0	10.3	15.8	13.5	8.49	18.3	17.8
Nd	7.00	13.1	4.29	6.64	5.36	6.76	6.13	18.0	9.19	5.65
Sm	1.13	2.85	0.91	0.98	0.67	1.16	0.93	1.26	1.23	1.20
Eu	0.22	0.55	0.13	0.19	0.14	0.18	0.17	0.22	0.23	0.24
Gd	1.02	2.24	0.58	0.88	0.58	0.78	0.70	1.15	<1.1	1.21
Tb	0.21	0.43	0.12	0.16	0.12	0.16	0.14	0.23	0.23	0.21
Tm	0.16	0.30	0.14	0.12	0.11	0.16	0.11	0.17	0.17	0.13
Yb	1.15	2.64	1.07	0.99	0.8	1.11	1.07	1.40	1.54	1.21
Lu	0.18	0.39	0.18	0.16	0.14	0.17	0.16	0.23	0.23	0.19
Hf	2.20	4.83	1.84	1.85	1.46	2.06	1.78	2.94	3.08	2.75
Ta	0.53	1.29	0.41	0.54	0.26	0.55	0.45	0.66	0.83	0.74
Ir [ppb]	< 2.3	< 2.2	0.2	< 0.6	< 1.8	< 1.3	< 1	< 1.6	< 2	< 1
Au [ppb]	6	6	133	11	13	11	6	11	43	63
Th	3.42	9.12	2.69	2.49	1.76	3.03	2.42	3.38	3.04	6.35
U	1.45	3.42	1.50	1.05	0.53	1.13	1.25	2.52	1.81	2.61
K/U (x 10^4)	1.72	1.45	1.56	2.12	3.74	2.49	2.02	1.37	1.87	1.28
La/Th	2.98	2.60	3.10	3.44	3.78	4.32	3.48	3.82	3.62	1.83
Th/U	2.36	2.67	1.79	2.37	3.32	2.68	1.94	1.34	1.68	2.43
Ni/Ce	1.74	1.67	2.64	2.71	1.65	2.97	3.78	8.48	2.30	0.62
La_{CN}/Yb_{CN}	5.99	6.06	5.28	5.85	5.59	7.97	5.32	6.22	4.83	6.48
Eu/Eu*	0.63	0.67	0.54	0.62	0.59	0.58	0.64	0.56	0.63	0.61

Table 4. Continued

Strat Loc [m]	305.0	305.25	305.75	306.1	306.35	306.75	307.0	307.25	307.5	307.75
Na [wt%]	0.21	0.20	0.25	0.22	0.27	0.26	0.18	0.26	0.25	0.19
K [wt%]	2.91	2.79	3.27	3.96	3.25	3.30	3.30	3.82	4.61	4.03
Sc	17.9	6.37	6.96	9.14	6.61	7.84	7.65	9.27	9.59	8.85
Cr	49.4	45.6	50.1	63.5	47.6	54.5	55.3	55.6	61.7	57.7
Fe [wt%]	1.26	1.33	1.32	1.68	1.27	2.18	1.78	2.21	2.88	2.67
Co	7.89	8.23	8.02	10.3	7.84	13.2	11.8	16.1	27.0	17.8
Ni	52	25	12	33	34	44	38	86	22	104
Zn	43	27	27	37	41	30	48	69	94.8	78
As	2.20	1.41	2.09	4.13	4.06	29.2	29.4	6.06	141	144
Rb	93.2	89.5	97.1	127	9.96	107	111	111	132	127
Zr	950	50	60	110	65	65	60	95	95	70
Sb	0.59	0.60	0.74	0.92	0.77	1.04	1.03	1.36	13.2	1.77
Cs	11.6	11.4	12.3	15.8	11.3	13.6	13.4	10.5	13.2	15.5
Ba	165	193	135	317	168	199	165	150	278	155
La	9.39	7.80	8.43	14.3	8.00	10.8	10.8	10.3	14.5	15.7
Ce	14.7	12.9	14.5	18.9	11.6	17.7	17.9	15.3	22.4	20.8
Nd	5.94	5.72	4.91	9.01	4.48	5.42	8.07	5.59	10.0	9.72
Sm	0.98	0.81	0.94	1.32	0.82	1.13	1.13	1.07	1.61	1.41
Eu	0.16	0.18	0.19	0.26	0.18	0.27	0.23	0.24	0.28	0.28
Gd	0.91	0.24	1.23	1.32	0.91	1.47	1.43	0.30	1.42	1.26
Tb	0.15	0.11	0.21	0.25	0.16	0.28	0.25	0.25	0.28	0.24
Tm	0.13	0.13	0.14	0.19	0.17	0.17	0.19	0.16	0.24	0.20
Yb	1.24	1.08	1.27	1.41	1.27	1.35	1.32	1.41	1.57	1.61
Lu	0.21	0.17	0.20	0.22	0.20	0.21	0.21	0.22	0.27	0.26
Hf	2.94	1.98	2.38	2.8	2.43	2.55	2.43	2.69	3.05	2.8
Ta	0.52	0.57	0.62	0.77	0.57	0.68	0.64	0.65	0.85	0.73
Ir [ppb]	< 2	< 0.8	< 1	< 1.8	< 1.7	< 0.3	< 1.4	< 1.1	< 2.5	< 2.7
Au [ppb]	50	31	17	42	41	90	67	33	153	19
Th	5.94	2.24	2.43	3.37	2.31	2.96	2.80	3.17	3.51	3.88
U	1.70	2.08	2.14	2.17	1.28	1.53	1.64	1.92	2.79	1.87
$K/U (x 10^4)$	1.71	1.34	1.53	1.82	2.54	2.16	2.01	1.99	1.65	2.16
La/Th	1.58	3.48	3.47	4.24	3.46	3.65	3.86	3.25	4.13	4.05
Th/U	3.49	1.08	1.14	1.55	1.80	1.93	1.71	1.65	1.26	2.07
Ni/Ce	3.54	1.94	0.83	1.75	2.93	2.49	2.12	5.62	0.98	5.00
La_{CN}/Yb_{CN}	5.12	4.89	4.65	6.85	4.26	5.41	5.53	4.93	6.24	6.59
Eu/Eu*	0.52	1.25	0.54	0.60	0.64	0.64	0.55	1.30	0.57	0.64

Table 4. Continued

Strat Loc [m]	308.0	308.25	308.5	308.8	309.0	309.05	309.3	309.6	310.0	310.25
Na [wt%]	0.21	0.20	0.18	0.22	0.22	0.25	0.05	0.27	0.08	0.26
K [wt%]	4.15	3.98	4.27	4.22	3.87	5.33	0.21	4.84	1.35	0.37
Sc	10.5	9.43	10.1	13.9	12.9	17.9	0.53	9.82	9.45	15.1
Cr	57.4	60.1	62.7	62.7	62.3	104	3.45	71.2	65.4	65.1
Fe [wt%]	2.07	2.07	2.40	2.03	2.27	3.83	0.11	2.44	1.96	2.12
Co	15.5	16.1	20.1	14.4	14.7	27.3	1.07	16.3	16.9	16.7
Ni	30	94	104	61	31	120	4	57	82	101
Zn	63	56	72	67	89	182	3	73	54	86
As	72.4	81.8	94.0	76.7	105	625	11.2	145	3.89	304
Rb	127	138	132	13.0	124	182	4.36	141	125	122
Zr	170	115	50	75	70	145	5	70	55	80
Sb	1.38	1.28	1.48	1.50	1.39	4.95	0.30	2.45	0.31	3.61
Cs	11.7	14.6	12.6	16.2	16.2	31.6	0.79	15.1	15.8	14.8
Ba	188	254	115	252	153	284	201	217	164	192
La	13.5	12.5	13.2	13.5	12.6	21.5	0.66	14.3	11.7	15.4
Ce	20.2	20.5	18.7	21.0	17.2	32.1	0.93	22.0	20.5	24.4
Nd	8.90	9.18	6.54	10.3	8.73	12.1	0.56	11.3	6.95	9.14
Sm	1.41	1.42	1.30	1.40	1.73	2.20	0.08	1.28	1.22	1.47
Eu	0.31	0.26	0.28	0.28	0.37	0.36	0.02	0.24	0.25	0.33
Gd	<3.5	1.13	1.20	1.23	2.02	2.19	0.11	1.20	1.10	1.53
Tb	0.25	0.20	0.23	0.24	0.31	0.39	0.02	0.28	0.21	0.28
Tm	0.17	0.16	0.18	0.22	0.21	0.28	0.01	0.23	0.16	0.21
Yb	1.57	1.40	1.67	1.63	1.59	2.47	0.11	1.10	0.93	1.55
Lu	0.28	0.22	0.27	0.26	0.28	0.40	0.02	0.31	0.16	0.26
Hf	3.06	2.83	2.73	2.87	2.73	4.87	0.17	3.23	2.56	2.67
Ta	0.62	0.75	0.61	0.70	0.57	1.22	0.02	0.73	0.65	0.71
Ir [ppb]	< 5.7	< 1.4	< 0.9	< 2.1	< 1.9	< 1.7	< 0.3	< 1.4	< 1.4	< 0.6
Au [ppb]	68	50	22	30	28	23	4	30	13	40
Th	3.59	3.57	3.23	4.21	5.61	7.80	0.20	3.21	3.38	4.32
U	1.86	2.77	1.81	1.59	1.84	2.30	0.12	1.81	0.64	1.87
K/U (x 10^4)	2.23	1.44	2.36	2.65	2.10	2.32	1.75	2.67	2.11	0.20
La/Th	3.76	3.50	4.09	3.21	2.25	2.76	3.30	4.45	3.46	3.56
Th/U	1.93	1.29	1.78	2.65	3.05	3.39	1.67	1.77	5.28	2.31
Ni/Ce	1.49	4.59	5.56	2.90	1.80	3.74	4.30	2.59	4.00	4.14
La_{CN}/Yb_{CN}	5.81	6.03	5.34	5.60	5.36	5.88	4.09	5.72	7.18	6.71
Eu/Eu*	n.d.	0.63	0.69	0.65	0.60	0.50	0.65	0.59	0.63	0.61

Table 4. Continued

Strat Loc [m]	310.5	311.05	311.25	311.55	311.75	312.0	312.25	312.5	312.75	313.0
Na [wt%]	0.25	0.22	0.25	0.22	0.24	0.17	0.24	0.24	0.23	0.11
K [wt%]	4.28	4.87	4.61	5.08	4.64	0.59	4.80	4.39	4.57	2.29
Sc	11.1	14.3	11.4	9.82	10.1	1.33	11.9	11.0	9.80	9.37
Cr	61.4	71.5	66.7	81.0	65.4	8.70	63.2	65.4	81.0	66.2
Fe [wt%]	2.10	2.91	2.41	2.53	2.19	0.22	2.41	2.26	2.40	2.55
Co	13.9	16.0	14.7	14.0	13.6	1.93	15.3	12.3	13.8	14.5
Ni	78	90	100	76	82	11	11	43	30	95
Zn	67	114	71	71	68	8	73	52	47	71
As	251	813	213	197	188	18.2	243	185	112	177
Rb	114	130	128	158	126	15.7	129	142	143	127
Zr	80	65	85	95	100	15	95	105	70	70
Sb	2.61	6.00	2.24	2.65	2.74	0.26	3.37	1.9	<0.53	1.58
Cs	14.1	15.8	16.9	19.3	14.8	1.72	78.3	16.8	14.4	17.9
Ba	153	169	167	233	195	45	182	231	155	174
La	13.8	12.3	10.4	13.9	12.8	1.69	11.8	13.9	13.4	9.13
Ce	19.1	19.7	23.5	23.9	21.3	2.71	17.8	21.9	23.8	20.2
Nd	8.48	7.01	8.08	10.8	9.44	1.24	4.63	8.88	9.44	5.84
Sm	1.31	1.31	1.27	1.4	1.36	0.18	1.17	1.45	1.55	0.88
Eu	0.28	0.28	0.32	0.20	0.20	0.04	0.19	0.22	0.33	0.19
Gd	1.34	1.71	1.60	1.22	1.20	0.27	1.12	0.96	1.94	1.13
Tb	0.23	0.28	0.30	0.26	0.27	0.05	0.20	0.30	0.36	0.24
Tm	0.21	0.23	0.19	0.23	0.21	0.03	0.20	0.24	0.22	0.18
Yb	1.46	1.60	1.35	1.81	1.57	0.23	1.49	1.63	1.62	1.34
Lu	0.23	0.27	0.21	0.31	0.25	0.11	0.24	0.28	0.26	0.23
Hf	2.55	2.67	2.76	3.25	3.01	0.41	2.67	3.01	2.79	2.69
Ta	0.53	0.62	0.73	0.75	0.77	0.09	0.57	0.73	0.76	0.66
Ir [ppb]	< 1.5	< 0.2	< 1.4	< 0.4	< 0.5	< 0.2	< 0.8	< 1.3	< 1	< 1
Au [ppb]	38	83	61	38	9	5	9	1	26	1
Th	4.45	3.35	3.8	3.49	3.22	0.47	3.52	3.69	3.51	3.05
U	1.47	1.58	2.04	1.59	2.17	0.08	1.56	1.98	1.63	0.85
K/U (x 10^4)	2.91	3.08	2.26	3.19	2.14	7.38	3.08	2.22	2.80	2.69
La/Th	3.10	3.67	2.74	3.98	3.98	3.60	3.35	3.77	3.82	2.99
Th/U	3.03	2.12	1.86	2.19	1.48	5.88	2.26	1.86	2.15	3.59
Ni/Ce	4.08	4.57	4.26	3.18	3.85	4.06	0.62	1.96	1.26	4.70
La_{CN}/Yb_{CN}	6.38	5.20	5.21	5.19	5.51	4.95	5.35	5.76	5.59	4.61
Eu/Eu*	0.65	0.57	0.69	0.47	0.48	0.56	0.51	0.45	0.58	0.58

Table 4. Continued

Strat Loc [m]	313.25	313.5	313.8	314.0	314.25	314.5	314.75	315.0	315.25	315.5
Na [wt%]	0.24	0.07	0.24	0.30	0.51	0.24	0.24	0.24	0.21	0.22
K [wt%]	4.60	0.08	4.50	4.86	10.9	5.46	4.66	4.96	4.39	5.62
Sc	12.5	0.41	9.24	9.97	13.6	10.2	10.7	8.71	12.8	9.47
Cr	78.3	2.98	65.5	63.6	70.4	85.7	70.0	70.1	63.1	102
Fe [wt%]	2.55	0.08	2.09	2.07	2.34	2.5	2.55	2.21	2.10	2.55
Co	16.9	1.63	13.7	22.6	15.8	13.2	11.9	11.4	13.9	16.1
Ni	102	4	36	116	90	93	108	36	71	114
Zn	93	5	51	46	56	51	83.9	46	54	62
As	381	35.1	321	439	472	172	144	277	315	500
Rb	144	3.60	133	130	142	172	146	137	129	161
Zr	95	5	75	95	60	95	100	100	95	100
Sb	4.47	0.37	2.99	4.77	4.37	2.56	6.1	3.28	3.58	5.31
Cs	19.2	0.54	14.1	17.1	18.6	20.2	15.4	13.6	13.2	14.2
Ba	222	47	187	163	161	239	153	185	186	290
La	15.2	1.34	13.0	13.3	20.5	15.4	14.5	14.2	13.3	15.2
Ce	24.6	1.15	22.1	18.7	19.3	24.3	22.7	25.5	20.7	28.9
Nd	10.8	1.45	9.11	5.21	5.96	11.5	10.8	10.1	13.3	8.51
Sm	1.56	0.27	1.36	1.30	1.25	1.61	1.41	1.46	1.47	1.55
Eu	0.28	0.07	0.27	0.27	0.28	0.33	0.27	0.30	0.25	0.29
Gd	1.54	0.22	0.29	1.90	1.34	2.16	1.66	1.50	1.40	1.50
Tb	0.34	0.04	0.21	0.35	0.28	0.40	0.30	0.26	0.30	0.27
Tm	0.21	0.20	0.16	0.22	0.19	0.24	0.21	0.19	0.23	0.26
Yb	1.83	0.12	1.58	1.60	1.5	1.96	1.93	1.66	1.54	1.78
Lu	0.32	0.02	0.27	0.26	0.24	0.33	0.30	0.26	0.27	0.29
Hf	3.18	0.13	2.90	2.75	2.9	3.53	3.07	3.09	2.99	3.08
Ta	0.73	0.02	0.65	0.62	0.71	1.00	0.93	0.70	0.64	0.85
Ir [ppb]	< 0.8	< 0.5	< 1.2	< 1	< 2.2	< 2.3	< 1.3	< 1	< 1	< 1.7
Au [ppb]	15	5	3	9	11	31	18	30	24	20
Th	4.04	0.27	3.25	3.03	3.83	3.71	3.73	3.23	3.97	3.78
U	1.45	0.11	2.05	1.21	3.49	1.83	1.20	1.76	1.41	2.35
K/U (x 10^4)	3.17	0.73	2.20	4.02	3.12	2.98	3.88	2.82	3.11	2.39
La/Th	3.76	4.96	4.00	4.39	5.35	4.15	3.89	4.40	3.35	4.02
Th/U	2.79	2.45	1.59	2.50	1.10	2.03	3.11	1.84	2.82	1.61
Ni/Ce	4.15	3.48	1.63	6.20	4.66	3.83	4.76	1.41	3.43	3.94
La_{CN}/Yb_{CN}	5.61	7.60	5.56	5.62	9.23	5.31	5.66	5.78	5.84	5.77
Eu/Eu*	0.55	0.87	1.31	0.52	0.66	0.54	0.54	0.62	0.53	0.58

Table 4. Continued

Strat Loc [m]	315.75	316.0	316.25	316.5	316.75	317.0	317.25	317.5	317.75	318.0
Na [wt%]	0.18	0.17	0.25	0.26	0.21	0.22	0.19	0.23	0.22	0.21
K [wt%]	3.72	3.62	4.99	4.79	4.47	4.88	3.82	4.25	4.65	5.06
Sc	10.6	7.58	9.73	10.6	15.1	10.6	32.9	8.78	14.6	15.8
Cr	65.9	50.5	78.9	77.3	63.9	91.5	67.9	69.0	70.7	73.9
Fe [wt%]	1.78	1.78	1.94	2.09	2.29	2.54	5.76	1.97	2.14	2.26
Co	12.5	11.6	10.6	13.6	13.0	18.3	16.3	13.8	12.7	12.9
Ni	47	46	9	45	78	75	99	70	96	96
Zn	35	38	55	47	58	181	143	46	54	23
As	27.7	66.1	48.9	25.2	39.3	47.4	203	25.0	16.4	14.2
Rb	120	99.1	148	141	136	126	111	129	137	135
Zr	65	105	70	130	120	60	115	90	115	95
Sb	< 0.04	0.18	0.19	0.12	0.21	0.16	< 0.06	0.13	0.11	0.10
Cs	15.9	9.29	18.5	17.9	17.3	13.4	13.8	15.2	15.6	13.8
Ba	140	125	247	227	224	161	120	224	192	139
La	13.1	9.83	17.8	16.6	14.2	13.1	14.4	12.9	14.3	10.6
Ce	20.1	16.1	28.6	29.4	23.7	22.1	23.3	19.9	24.8	18.4
Nd	7.73	8.25	8,68	12.1	7.54	7.52	8.62	9.57	7.08	7.76
Sm	1.25	1.2	1.65	1.90	1.38	1.63	1.36	1.33	1.52	1.22
Eu	0.26	0.14	0.27	0.35	0.28	0.17	0.23	0.24	0.29	0.16
Gd	1.26	1.20	1.77	2.01	1.52	1.22	1.18	1.34	1.80	1.25
Tb	0.23	0.21	0.36	0.34	0.32	0.25	0.20	0.26	0.32	0.26
Tm	0.19	0.15	0.24	0.23	0.20	0.15	0.15	0.24	0.18	0.21
Yb	1.24	0.99	1.63	1.61	1.41	1.13	1.10	1.69	1.18	1.46
Lu	0.21	0.17	0.24	0.26	0.22	0.21	0.19	0.26	0.19	0.22
Hf	2.67	2.36	3.57	3.30	2.75	2.85	3.37	2.58	2.84	2.43
Ta	0.67	0.56	0.87	0.80	0.78	0.72	0.80	0.80	0.90	0.66
Ir [ppb]	< 0.7	< 2.8	< 2.9	< 2.4	< 1.3	< 4.3	< 0.4	< 0.9	< 1.9	< 2.8
Au [ppb]	27	12	16	58	13	73	38	< 17	24	22
Th	3.21	2.98	4.81	4.31	4.38	3.5	7.82	3.39	3.17	4.03
U	0.30	0.46	0.33	0.23	< 0.35	0.21	0.18	0.11	0.31	0.26
K/U (x 10^4)	12.4	7.87	15.1	20.8	n.d.	23.2	21.2	38.6	15.0	19.5
La/Th	4.08	3.30	3.70	3.85	3.24	3.74	1.84	3.81	4.51	2.63
Th/U	10.7	6.48	14.6	18.7	n.d.	16.7	43.4	30.8	10.2	15.5
Ni/Ce	2.34	2.86	0.31	1.53	3.29	3.39	4.25	3.52	3.87	5.22
La_{CN}/Yb_{CN}	7.14	6.71	7.38	6.89	6.80	7.83	8.84	5.16	8.18	4.90
Eu/Eu*	0.63	0.36	0.48	0.55	0.59	0.37	0.55	0.55	0.54	0.40

Table 4. Continued

Strat Loc [m]	318.25	318.5	318.75	319.0	319.25	319.5	319.75	320.1	320.5	321.0
Na [wt%]	0.18	0.20	0.24	0.20	0.26	0.27	0.21	0.22	0.23	0.29
K [wt%]	4.26	4.60	5.34	5.34	4.86	4.63	4.39	4.83	4.95	4.76
Sc	14.4	17.5	9.76	11.8	11.3	10.96	14.6	12.6	13.3	10.0
Cr	69.7	76.3	74.5	64.9	88.0	86.7	72.6	74.9	87.2	92.1
Fe [wt%]	2.00	2.09	2.17	3.66	2.47	2.64	2.55	2.37	2.30	2.09
Co	13.5	10.7	16	16	14.4	15.9	17.1	13.5	13.3	18.2
Ni	35	13	89	111	74	109	79	77	92	61
Zn	38	< 12	48	75	61	88	77	66	53	66
As	10.8	22.4	7.88	167	2.18	5.80	7.96	6.82	5.53	6.79
Rb	136	133	145	135	158	143	155	145	164	133
Zr	75	45	95	130	100	110	105	95	115	60
Sb	0.05	0.02	0.07	2.23	0.05	0.06	0.05	0.04	0.04	0.04
Cs	14.8	17.7	18.0	19.6	22.1	21.4	17.7	17.4	19.7	14.4
Ba	198	220	181	226	230	227	226	225	238	210
La	18.0	16.2	13.8	14.7	12.3	15.3	14.9	12.8	14.8	14.9
Ce	25.6	29.2	23.4	23.1	21.4	25.1	26.2	22.5	27.1	22.4
Nd	10.9	6.71	8.48	7.64	12.2	10.4	13.3	7.87	12.0	6.71
Sm	1.45	1.63	1.68	1.54	1.54	1.85	1.63	1.53	1.69	1.42
Eu	0.26	0.26	0.27	3.23	0.26	0.34	0.31	0.27	0.32	0.27
Gd	1.61	1.96	1.54	1.76	1.82	1.73	2.02	1.47	1.96	1.55
Tb	0.36	0.32	0.31	0.27	0.38	0.32	0.41	0.31	0.33	0.26
Tm	0.26	0.19	0.21	0.18	0.26	0.23	0.24	0.20	0.23	0.18
Yb	1.81	1.52	1.58	1.38	1.73	1.62	1.78	1.52	1.64	1.35
Lu	0.27	0.26	0.27	0.25	0.26	0.25	0.29	0.25	0.26	0.23
Hf	3.13	3.41	3.09	2.80	3.47	3.46	3.15	3.18	3.31	3.27
Ta	0.59	0.70	0.85	0.68	0.90	0.88	0.92	0.87	0.95	0.79
Ir [ppb]	< 3.1	< 1.8	0.6	< 2.3	< 0.5	< 0.5	< 2.9	< 0.3	< 0.7	< 1
Au [ppb]	17	10	23	96	22	19	16	38	26	36
Th	4.49	5.98	3.66	4.27	3.68	4.53	5.05	4.28	4.81	3.57
U	0.16	0.29	0.33	1.73	0.21	0.28	0.55	0.36	0.5	0.24
K/U (x 10^4)	26.6	15.9	16.2	3.09	23.14	16.50	8.00	13.4	9.90	19.8
La/Th	4.01	2.71	3.77	3.44	3.34	3.38	2.95	2.99	3.08	4.17
Th/U	28.1	20.6	11.1	2.47	17.5	16.2	9.18	11.89	9.62	14.9
Ni/Ce	1.37	0.45	3.80	4.81	3.46	4.34	3.02	3.42	3.39	2.72
La_{CN}/Yb_{CN}	6.72	7.20	5.90	7.20	4.80	6.38	5.65	5.69	6.10	7.46
Eu/Eu*	0.52	0.44	0.51	0.43	0.47	0.58	0.52	0.55	0.54	0.56

Table 4. Continued

Strat Loc [m]	321.5	322.0	322.5	323.0	323.5	324.0	324.5	325.0	325.5	326.0
Na [wt%]	0.27	0.26	0.11	0.21	0.20	0.17	0.19	0.19	0.18	0.19
K [wt%]	6.00	4.88	1.58	5.26	4.37	3.90	5.03	5.24	5.23	4.50
Sc	15.4	28.6	4.63	12.5	14.0	33.4	12.7	18.8	13.2	24.0
Cr	110	89.1	24.4	80.0	78.7	72.7	87.1	71.1	82.8	56.2
Fe [wt%]	2.69	3.43	0.62	2.32	2.14	4.38	2.22	2.62	1.98	5.30
Co	14.9	25.3	4.01	17.4	15.1	28.6	11.6	24.8	13.3	38.3
Ni	110	136	67	106	50	130	35	140	77	266
Zn	139	2050	9	51	1875	202	2663	64	48	95
As	24.9	31.1	6.58	9.76	10.1	92.0	127	183	26.8	610
Rb	103	145	42.2	152	131	121	137	142	139	96.2
Zr	105	120	50	125	70	70	70	90	100	90
Sb	0.08	0.14	0.03	0.12	0.03	< 0.05	n.d.	1.55	0.87	7.02
Cs	13.5	15.8	3.72	18.6	15.8	11.3	16.7	0.17	16.8	13.4
Ba	133	228	71	191	168	91	191	175	162	202
La	23.2	18.3	6.17	14.9	11.4	12.0	14.5	17.8	14.3	15.8
Ce	28.0	28.9	7.01	25.1	19.5	19.8	23.8	25.4	24.4	16.2
Nd	13.5	11.9	2.29	11.7	7.61	7.27	9.56	8.51	13.4	5.29
Sm	1.87	1.94	0.50	2.02	1.52	1.32	1.72	1.73	1.58	1.16
Eu	0.28	0.36	0.06	0.23	0.24	0.19	0.18	0.22	0.15	0.19
Gd	2.71	1.84	< 1.0	0.94	1.28	1.15	0.73	1.20	0.82	0.96
Tb	0.47	0.34	0.04	0.23	0.18	0.21	0.19	0.26	0.16	0.19
Tm	0.26	0.24	0.09	0.18	0.18	0.16	0.18	0.19	0.19	0.14
Yb	1.26	1.80	0.64	1.48	1.27	1.28	1.69	1.53	1.58	1.16
Lu	0.34	0.29	0.08	0.23	0.19	0.22	0.35	0.28	0.25	0.20
Hf	3.52	3.52	0.95	3.30	2.94	2.94	3.23	3.31	3.01	2.64
Ta	1.07	0.89	0.23	105	0.78	0.66	0.62	0.74	0.72	0.58
Ir [ppb]	< 5.9	1.27	< 2.6	< 0.5	0.8	< 2.5	< 3.2	<1	< 1.6	1.1
Au [ppb]	112	97	76	91	268	349	60	85	92	72
Th	5.3	9.28	1.32	4.00	4.05	6.59	4.4	4.29	4.39	4.14
U	0.28	0.27	0.05	0.32	0.26	0.31	2.52	2.52	2.35	1.51
K/U (x 10^4)	21.4	18.1	31.6	16.4	16.8	12.6	2.00	2.08	2.23	2.98
La/Th	4.38	1.97	4.67	3.73	2.81	1.82	3.30	4.15	3.26	3.82
Th/U	18.9	34.4	26.4	12.5	15.6	21.3	1.75	1.70	1.87	2.74
Ni/Ce	3.93	4.71	9.56	4.22	2.56	6.57	1.47	5.51	3.16	16.42
La_{CN}/Yb_{CN}	5.27	6.87	8.16	6.80	6.07	6.34	n.d.	7.86	6.12	9.20
Eu/Eu*	0.38	0.58	n.d.	0.51	0.53	0.47	n.d.	0.47	0.40	0.55

Table 4. Continued

Strat Loc [m]	326.5	327.0	327.5	328.0	328.5	329.0	329.5 A	329.5 B	330.1	330.2
Na [wt%]	0.24	0.21	0.19	0.16	0.30	0.23	0.04	0.23	0.19	0.21
K [wt%]	5.88	5.05	4.88	4.72	8.58	5.23	0.40	6.20	5.53	6.30
Sc	11.4	10.8	12.0	13.9	14.4	15.0	11.2	10.8	26.5	10.2
Cr	78.0	73.9	77.0	63.9	73.8	87.6	108	82.6	83.4	93.8
Fe [wt%]	2.04	2.02	3.45	2.4	3.12	2.75	9.98	5.10	3.02	3.88
Co	22.4	14.4	31.5	15.4	25.8	23.8	236	47.2	15.9	33.7
Ni	88	57	144	78	118	171	1550	316	141	150
Zn	1550	55	5520	2040	14200	2870	2140	83	121	90
As	49.8	72.5	310	180	266	168	60.6	729	3.37	297
Rb	144	137	145	138	137	167	150	201	147	194
Zr	100	130	130	85	50	150	395	185	75	140
Sb	1.08	0.71	3.56	1.85	3.6	2.09	0.67	5.81	3.48	2.72
Cs	15.9	16.9	15.1	15.3	12.9	16.1	15.4	20.9	15.5	26.4
Ba	217	195	228	181	176	216	225	291	190	320
La	15.9	17.3	18.3	21.5	26.9	36.5	17.5	18.6	24.2	16.6
Ce	23.4	26.5	27.0	23.9	24.2	27.1	23.0	32.3	23.2	28.2
Nd	9.78	10.0	10.1	9.33	8.83	12.1	11.3	15.2	7.53	11.7
Sm	7.64	1.70	1.71	1.53	2.38	1.67	1.31	2.19	1.48	1.88
Eu	0.22	0.24	0.25	0.23	0.22	0.26	0.32	0.32	0.22	0.35
Gd	1.11	<3.64	1.17	1.36	1.07	1.25	2.01	1.56	1.34	1.74
Tb	0.18	0.28	0.24	0.22	0.22	0.29	0.39	0.32	0.26	0.34
Tm	0.17	0.26	0.19	0.19	0.22	0.22	0.22	0.25	0.20	0.28
Yb	1.73	1.78	1.53	1.53	1.62	1.42	1.55	1.87	1.61	2.05
Lu	0.28	0.31	0.27	0.26	0.32	0.26	0.24	0.31	0.25	0.34
Hf	3.48	3.57	3.22	3.08	3.55	3.97	3.00	4.17	4.01	4.25
Ta	0.77	0.78	0.61	0.59	0.63	0.71	0.79	1.28	0.64	1.34
Ir [ppb]	< 0.4	< 2.3	0.9	< 0.04	< 2.2	< 6.3	< 15	1.3	0.6	0.8
Au [ppb]	118	86	98	85	184	294	11.1	53.5	141	32.3
Th	3.69	3.94	4.29	4.01	4.99	4.67	4.33	5.52	7.41	4.51
U	2.67	2.64	2.12	1.94	4.01	3.93	0.24	3.30	2.47	2.59
K/U (x 10^4)	2.20	1.91	2.30	2.43	2.14	1.33	1.67	1.88	2.24	2.43
La/Th	4.31	4.39	4.27	5.36	5.39	7.82	4.04	3.37	3.27	3.68
Th/U	1.38	1.49	2.02	2.07	1.24	1.19	18.0	1.67	3.00	1.74
Ni/Ce	3.76	2.15	5.33	3.26	4.88	6.31	67.39	9.78	6.08	5.32
La_{CN}/Yb_{CN}	6.21	6.57	8.08	9.49	11.2	17.4	7.63	6.72	10.2	5.47
Eu/Eu*	0.50	n.d.	0.54	0.49	0.42	0.55	0.60	0.53	0.48	0.59

Table 4. Continued

Strat Loc [m]	330.5	331.0	332.0	333.0	333.5	334.0	335.0	335.5	335.8	336.1
Na [wt%]	0.31	0.23	0.24	0.20	0.21	0.20	0.20	0.17	0.18	0.18
K [wt%]	7.08	5.80	5.33	6.47	6.94	5.94	6.50	6.52	5.95	6.47
Sc	9.13	9.53	9.59	9.62	11.5	9.19	10.6	9.52	10.1	10.3
Cr	101	93.4	83.0	223	135	89.8	85	114	85.7	89
Fe [wt%]	3.27	2.90	2.99	2.49	3.13	2.37	2.56	2.27	2.45	3.06
Co	24.6	21.3	33.8	17.3	18.5	13.6	16.2	13.1	14.8	28.7
Ni	132	98	126	138	128	92	129	42	112	133
Zn	85	60	75	60	325	55	55	70	45	75
As	194	80.5	247	145	93.9	94.1	91.2	223	109	138
Rb	195	163	169	196	208	170	197	180	186	200
Zr	150	145	140	106	182	154	155	243	150	132
Sb	2.38	1.67	2.35	2.09	1.35	1.27	1.10	2.15	1.94	1.52
Cs	24.6	20.7	18	22.9	24.9	21.6	21.5	17.3	21.3	22.0
Ba	274	341	318	258	296	300	264	234	301	250
La	14.8	16.4	21.5	32.3	18.9	15.5	17.9	23.3	21.0	16.9
Ce	25.7	29.6	37.3	30.8	31.3	25.4	30	30.4	32.8	27.8
Nd	9.42	10.8	15.5	14.3	n.d.	11.4	11.5	13.6	16.1	12.4
Sm	1.61	1.77	2.28	1.8	1.94	1.51	1.74	1.86	2.10	1.78
Eu	0.36	0.33	0.41	0.33	0.34	0.28	0.29	0.31	0.29	0.33
Gd	0.62	1.98	2.41	2.06	2.00	1.92	<1.4	<1.5	2.00	1.99
Tb	0.28	0.38	0.45	0.37	0.34	0.32	0.26	0.33	0.38	0.38
Tm	0.22	0.26	0.29	0.27	0.33	0.29	0.24	0.24	0.31	0.28
Yb	2.02	1.92	2.05	2.08	2.68	2.06	1.97	1.84	2.22	2.13
Lu	0.43	0.34	0.35	0.36	0.51	0.33	0.41	0.31	0.34	0.37
Hf	3.98	4.01	4.36	4.04	4.31	3.72	4.04	3.87	3.99	4.06
Ta	1.20	0.92	0.97	1.02	1.28	1.2	1.35	0.86	1.21	1.23
Ir [ppb]	3.7	<2.3	1.5	2	14	1.5	4.3	2.5	<2.1	0.8
Au [ppb]	28.5	39	57.3	77.3	1169	41.2	1401	197	16.8	19.0
Th	4.08	4.23	5.15	4.08	4.83	3.74	4.68	4.23	4.7	4.15
U	3.22	2.75	3.05	2.27	2.57	2.91	2.63	2.36	2.8	2.62
K/U (x 10^4)	2.20	2.11	1.75	2.85	2.70	2.04	2.47	2.76	2.13	2.47
La/Th	3.63	3.88	4.17	7.92	3.91	4.14	3.82	5.51	4.47	4.07
Th/U	1.27	1.54	1.69	1.80	1.88	1.29	1.78	1.79	1.68	1.58
Ni/Ce	5.14	3.31	3.38	4.48	4.09	3.62	4.30	1.38	3.41	4.78
La_{CN}/Yb_{CN}	4.95	5.77	7.08	10.49	4.76	5.08	6.14	8.56	6.39	5.36
Eu/Eu*	1.10	0.54	0.53	0.52	0.53	0.50	n.d.	n.d.	0.43	0.54

Table 4. Continued

Strat Loc [m]	337.0	337.4	337.8	338.0	338.6	339.0	339.5
Na [wt%]	0.24	0.21	0.22	0.19	0.20	0.24	0.27
K [wt%]	6.37	7.38	7.28	6.34	5.91	6.03	5.78
Sc	9.91	9.8	13.3	10.6	9.33	10.9	9.32
Cr	80.7	88.7	85.6	88.4	86.5	82.3	90.8
Fe [wt%]	2.89	3.01	3.3	3.7	3.15	0.31	2.81
Co	16.4	18.7	21.2	31.5	24.1	24.4	22.4
Ni	100	142	146	148	120	97	124
Zn	60	60	135	100	90	70	85
As	94.9	170	171	243	148	143	134
Rb	180	183	202	210	169	199	167
Zr	140	138	125	220	175	135	125
Sb	1.41	1.66	1.83	2.40	1.36	1.78	1.76
Cs	20.0	19.7	18.8	24.4	20.7	23	19.6
Ba	2.19	241	280	381	233	366	301
La	16.7	17.4	28.2	19.1	17.5	23.4	16.2
Ce	28.6	27.3	33.3	31.0	30.1	39.6	28.4
Nd	10.3	14.2	13.4	12.5	15.4	16.4	12.7
Sm	1.69	1.85	1.98	2.10	1.76	2.53	1.71
Eu	0.30	0.40	0.35	0.38	0.30	0.43	0.30
Gd	1.73	2.27	2.36	1.95	1.58	2.70	1.63
Tb	0.30	0.40	0.38	0.36	0.32	0.45	0.31
Tm	0.22	0.27	0.29	0.26	0.25	0.29	0.24
Yb	1.68	2.05	2.35	1.93	1.82	2.16	1.87
Lu	0.30	0.33	0.42	0.35	0.31	0.39	0.34
Hf	3.74	3.92	4.16	4.08	3.65	4.44	3.79
Ta	0.96	1.13	1.24	1.33	1.04	1.25	1.10
Ir [ppb]	1.2	2.3	<3.3	<2.1	2	<2.1	1.3
Au [ppb]	2.4	232	25.0	12.0	140	22.5	73.3
Th	4.46	4.42	4.89	4.74	4.3	6.01	4.39
U	2.3	2.62	2.13	3.53	2.12	3.05	2.77
K/U (x 10^4)	2.77	2.82	3.42	1.80	2.79	1.98	2.09
La/Th	3.74	3.94	5.77	4.03	4.07	3.89	3.69
Th/U	1.94	1.69	2.30	1.34	2.03	1.97	1.58
Ni/Ce	3.50	5.20	4.38	4.77	3.99	2.45	4.37
La_{CN}/Yb_{CN}	6.72	5.73	8.11	6.69	6.50	7.32	5.85
Eu/Eu*	0.54	0.60	0.49	0.57	0.55	0.50	0.55

Table 5. Trace element abundances of HCl-leached residues from the Bosso River Gorge, Italy, heavy fraction. All data in ppm, except as noted. Analyses by Instrumental Neutron Activation Analysis (INAA).

Strat Loc [m]	319.25	319.5	319.75	320.1	320.5	321.0	321.5	322.0	322.5
Na [wt%]	0.21	0.74	0.78	0.16	0.05	0.07	0.04	0.05	0.07
K [wt%]	0.42	0.25	0.17	0.39	<0.33	0.15	0.19	0.46	0.14
Sc	1.12	1.86	0.33	1.09	0.47	0.28	0.97	0.5	1.03
Cr	41.8	41.3	24.9	12.1	38.0	28.4	69.3	41.1	23.1
Fe [wt%]	52.9	52.9	36.1	24.3	50.4	51.8	53.0	57.4	31.4
Co	104	90.7	84.9	59.2	91.1	177	49.8	480	190
Ni	889	1140	605	455	697	1220	708	1750	1190
Zn	593	749	540	496	694	711	1280	2280	426
As	881	846	958	748	1840	1420	2400	3730	2410
Rb	<34	16.9	2.09	10.6	12.7	20.2	20.9	12.9	36.9
Zr	175	220	85	<175	75	<210	100	105	60
Sb	7.54	9.56	8.46	5.40	7.29	11.6	9.01	16.5	11.8
Cs	1.04	0.86	0.30	0.65	< 1.01	0.46	<0.84	<1.4	0.39
Ba	125	312	122	282	121	135	211	275	122
La	2.21	2.47	1.54	1.25	1.18	1.06	3.12	2.08	16.0
Ce	1.58	2.01	1.39	1.56	1.79	1.21	1.05	1.48	1.59
Nd	1.21	2.87	0.94	7.71	1.28	3.16	6.63	<9.33	1.76
Sm	0.27	0.41	0.14	0.14	0.20	0.14	0.48	0.36	0.19
Eu	0.05	0.06	0.03	0.11	0.09	0.07	0.13	0.10	0.06
Gd	0.34	0.60	0.29	1.48	0.66	0.86	0.54	0.56	0.59
Tb	0.08	0.21	0.09	0.24	0.23	0.13	0.12	0.26	0.13
Tm	0.09	0.06	0.06	0.10	0.13	0.10	0.16	<0.28	<0.15
Yb	0.64	0.38	0.41	0.67	0.36	0.37	1.05	0.67	0.35
Lu	0.08	0.05	0.06	0.03	0.03	0.07	0.17	0.05	0.09
Hf	1.00	0.26	0.52	0.26	<0.58	0.14	0.10	0.21	0.23
Ta	0.19	0.30	0.31	0.06	0.28	0.17	0.28	<0.38	0.37
Ir [ppb]	1.3	<6.2	<5.6	<6.7	<8.0	4.3	7.0	3.5	<5.8
Au [ppb]	63.0	95.0	64.3	120	89.5	157	144	185	67.2
Th	0.56	1.66	0.25	0.21	0.20	0.20	0.30	0.40	0.23
U	0.57	0.51	0.30	1.19	0.67	1.25	0.35	0.97	<1.4
K/U (x 10^4)	0.74	0.49	0.57	0.33	n.d	0.12	0.54	0.47	n.d
La/Th	3.95	1.49	6.16	5.95	5.90	5.30	10.4	5.20	69.6
Th/U	0.98	3.25	0.83	0.18	0.30	0.16	0.86	0.41	n.d.
Ni/Ce	562	566	435	291	389	1000	674	1180	750
La_{CN}/Yb_{CN}	5.99	6.06	5.28	5.85	5.59	7.97	5.32	6.22	4.83
Eu/Eu*	0.63	0.67	0.54	0.62	0.59	0.58	0.64	0.56	0.63

Table 5. Continued.

Strat Loc [m]	**323.0**	**323.5**	**324.0**	**324.5**	**325.0**	**325.5**	**326.0**	**326.5**	**327.0**
Na [wt%]	0.10	0.25	0.06	0.01	0.04	0.01	0.06	0.02	0.04
K [wt%]	0.27	0.55	0.26	<0.18	0.17	<0.19	0.35	0.01	0.27
Sc	1.68	0.63	1.27	0.37	0.47	0.35	0.67	0.30	0.94
Cr	36.4	42.1	67.9	65.9	37.6	32.7	53.0	40.0	40.6
Fe [wt%]	31.0	50.5	55.0	56.6	49.3	52.1	54.8	58.6	33.7
Co	108	110	250	157	667	624	388	618	56.9
Ni	674	829	2010	1320	2240	2140	2140	2400	693
Zn	232	742	2025	1203	1190	637	1563	752	418
As	2380	2300	6200	6480	10900	11400	5920	9670	2410
Rb	15.1	26.4	9.63	9.92	16.5	3.70	17	31.5	55.7
Zr	170	505	<330	110	295	<290	230	145	25
Sb	9.87	17.3	22.3	<0.5	<0.9	<1.2	21.0	39.5	7.30
Cs	<1.12	<4.19	<1.53	0.20	0.82	<1.37	1.41	1.17	0.47
Ba	3500	542	209	244	144	284	718	240	212
La	2.40	8.70	1.66	2.01	2.16	4.14	1.96	1.68	2.18
Ce	2.55	6.62	3.37	1.39	1.08	4.66	2.31	0.96	2.67
Nd	15.1	15.7	3.90	2.13	1.60	4.49	<11.4	<8.1	1.88
Sm	0.19	0.19	0.27	0.23	0.24	0.30	0.25	0.44	0.17
Eu	0.02	1.25	0.15	0.10	0.15	0.15	0.15	0.08	0.06
Gd	1.01	3.58	<4.5	0.53	<3.7	0.81	0.58	0.57	1.47
Tb	0.15	0.87	0.32	0.05	<0.3	0.15	0.43	0.11	0.27
Tm	0.35	0.83	0.17	0.06	<0.3	0.07	0.14	0.05	0.14
Yb	1.29	0.70	0.69	0.35	0.59	0.30	0.55	0.37	0.59
Lu	0.06	0.07	0.06	0.06	0.18	0.04	0.06	0.04	0.03
Hf	0.56	<2.40	0.51	0.10	0.50	0.19	0.38	0.14	0.16
Ta	0.24	0.28	0.53	0.12	0.12	0.49	0.25	0.13	0.23
Ir [ppb]	1.8	9.1	<12	<4.7	<1	3.3	<14	7.4	4.4
Au [ppb]	96.5	818	344	1.04	117	127	586	393	251
Th	0.79	5.62	0.58	0.13	0.29	0.12	0.32	0.45	0.49
U	0.29	2.29	1.08	<1.9	0.32	0.67	0.16	1.05	<1.9
K/U (x 10^4)	0.93	0.24	0.24	n.d.	0.53	n.d.	2.19	0.01	n.d.
La/Th	3.04	1.55	2.86	15.5	7.45	34.5	6.13	3.73	4.45
Th/U	2.72	2.45	0.54	n.d.	0.91	0.18	2.00	0.43	n.d.
Ni/Ce	264	125	596	952	2070	459	926	2500	260
La_{CN}/Yb_{CN}	6.48	5.12	4.89	4.65	6.85	4.26	5.41	5.53	4.93
Eu/Eu*	0.61	0.52	1.25	0.54	0.60	0.64	0.64	0.55	1.30

Table 5. Continued.

Strat Loc [m]	328.0	328.5	329.0	329.5	330.1
Na [wt%]	0.19	0.06	0.19	0.19	0.06
K [wt%]	0.73	<1.39	1.19	<0.68	0.10
Sc	2.40	0.62	0.51	0.29	0.66
Cr	32.7	56.3	71.8	0.93	0.45
Fe [wt%]	38.0	43.0	50.1	56.5	44.2
Co	300	217	417	256	132
Ni	735	890	1760	1200	990
Zn	4240	3130	628	3420	1590
As	9130	15600	17250	16900	15300
Rb	20.1	23.1	56.0	10.9	<115
Zr	255	<320	480	450	<550
Sb	50.7	31.5	40.2	56.8	63.1
Cs	0.53	0.47	<3.2	1.43	1.11
Ba	451	246	233	249	276
La	6.16	2.26	3.98	2.61	5.23
Ce	12.4	5.25	5.97	<5.0	2.80
Nd	4.47	7.79	<16	5.42	2.41
Sm	1.08	0.24	0.14	0.13	0.28
Eu	0.16	0.13	<0.2	0.16	0.25
Gd	<3.8	0.73	10.9	<2.2	2.02
Tb	0.17	0.15	0.11	0.29	0.60
Tm	0.08	0.11	0.81	0.26	0.14
Yb	0.53	0.65	0.82	0.93	0.72
Lu	0.11	0.08	0.01	0.10	0.16
Hf	0.36	0.60	0.92	0.33	0.60
Ta	0.34	<0.4	<1	0.81	0.91
Ir [ppb]	<11	9.7	18	8.5	2.6
Au [ppb]	34.3	344	600	340	269
Th	1.47	0.26	0.75	0.29	0.84
U	10.7	8.50	<8.3	3.87	<11
K/U (x 10^4)	0.07	n.d.	n.d.	n.d.	n.d.
La/Th	4.19	8.69	5.31	9.00	6.23
Th/U	0.14	0.03	n.d.	0.07	n.d.
Ni/Ce	59.3	170	295	n.d.	354
La_{CN}/Yb_{CN}	6.24	6.59	5.81	6.03	5.34
Eu/Eu*	0.57	0.64	n.d.	0.63	0.69

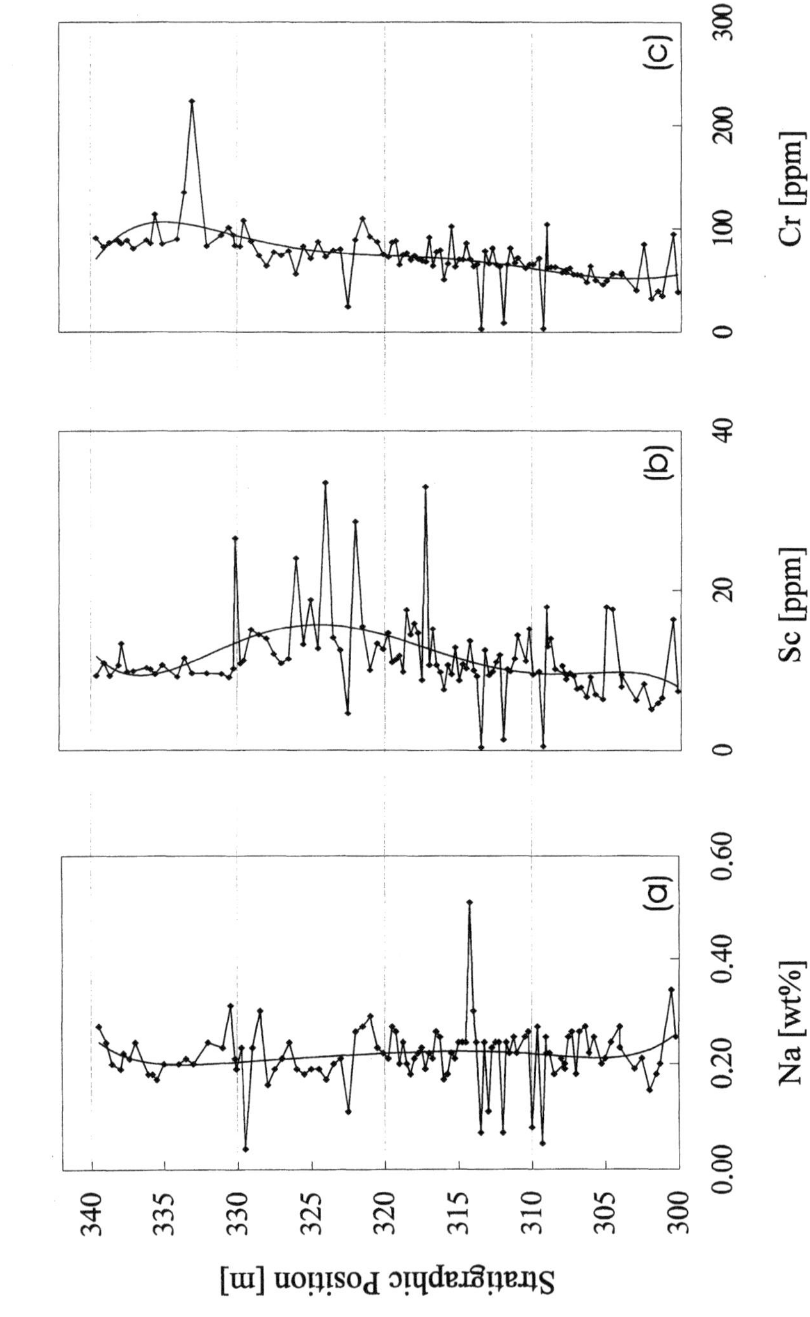

Fig. 7. Stratigraphic level versus elemental abundances for **(a)** Na, **(b)** Sc, **(c)** Cr. At 309.3 m, 312 m, and 331.5 m almost all elements show significant depletions.

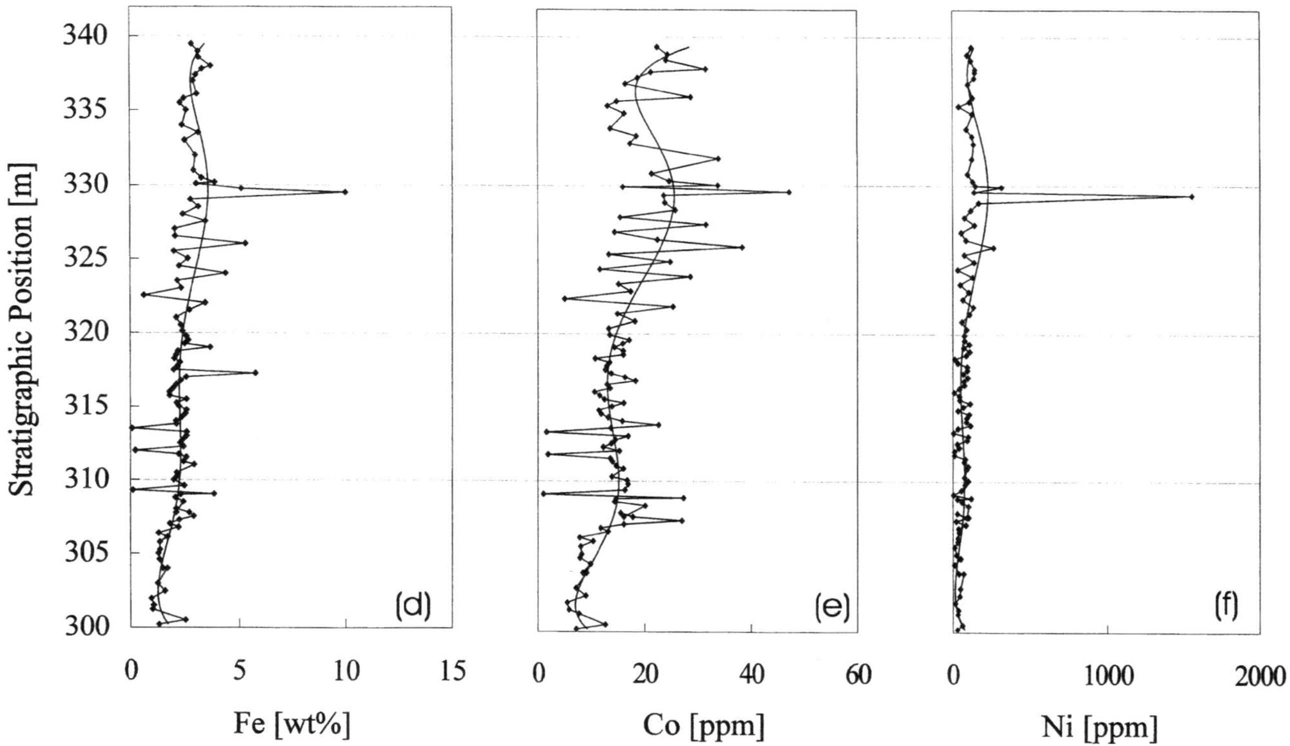

Fig. 7. Continued. Stratigraphic level versus elemental abundances for **(d)** Fe, **(e)** Co, **(f)** Ni. At 309.3 m, 312 m, and 331.5 m almost all elements show significant depletions.

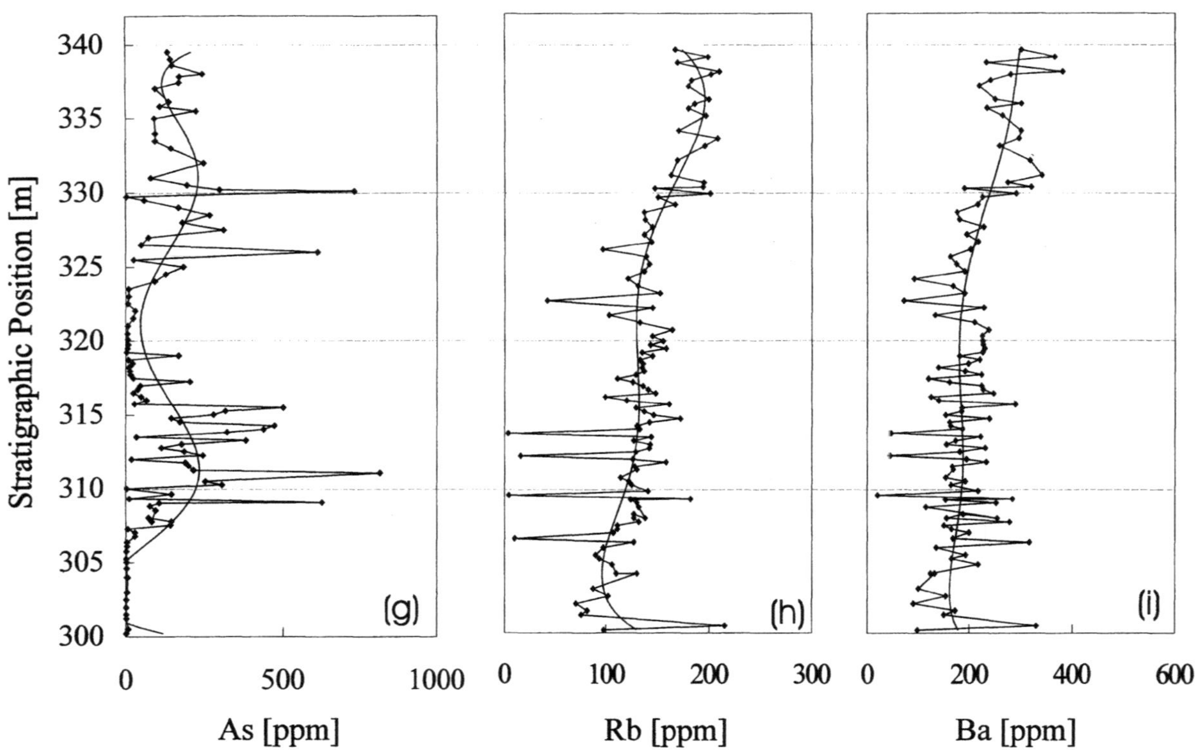

Fig. 7. Continued. Stratigraphic level versus elemental abundances for **(g)** As, **(h)** Rb, **(i)** Ba. At 309.3 m, 312 m, and 331.5 m almost all elements show significant depletions.

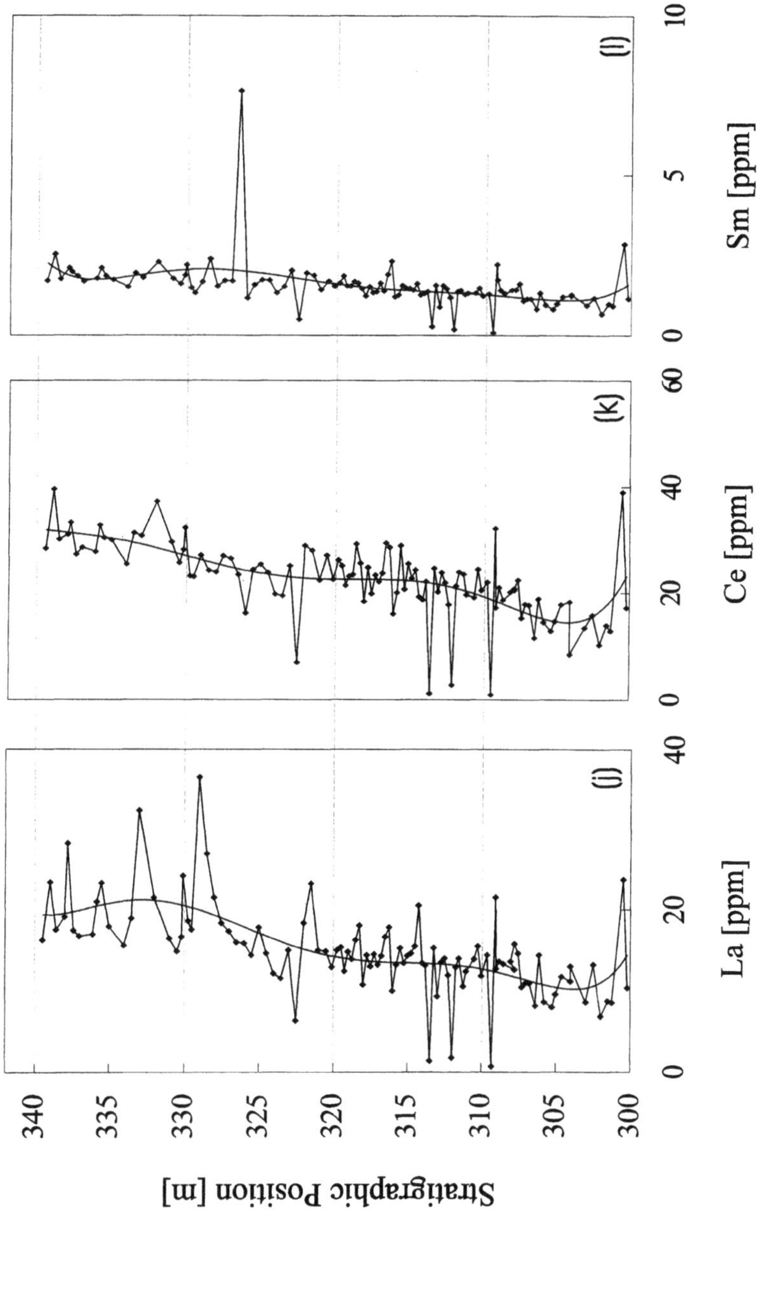

Fig. 7. Continued. Stratigraphic level versus elemental abundances for **(j)** La, **(k)** Ce, **(l)** Sm. At 309.3 m, 312 m, and 331.5 m almost all elements show significant depletions.

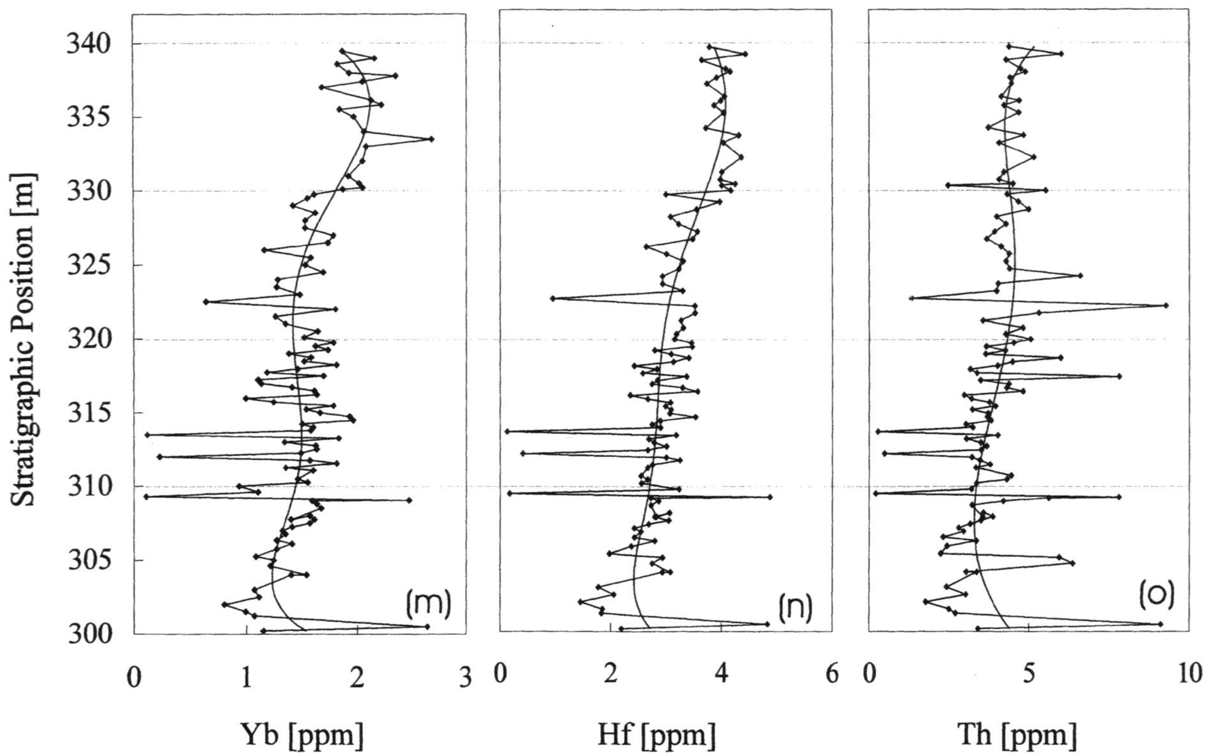

Fig. 7. Continued. Stratigraphic level versus elemental abundances for **(m)** Yb, **(n)** Hf **(o)** Th. At 309.3 m, 312 m, and 331.5 m almost all elements show significant depletions.

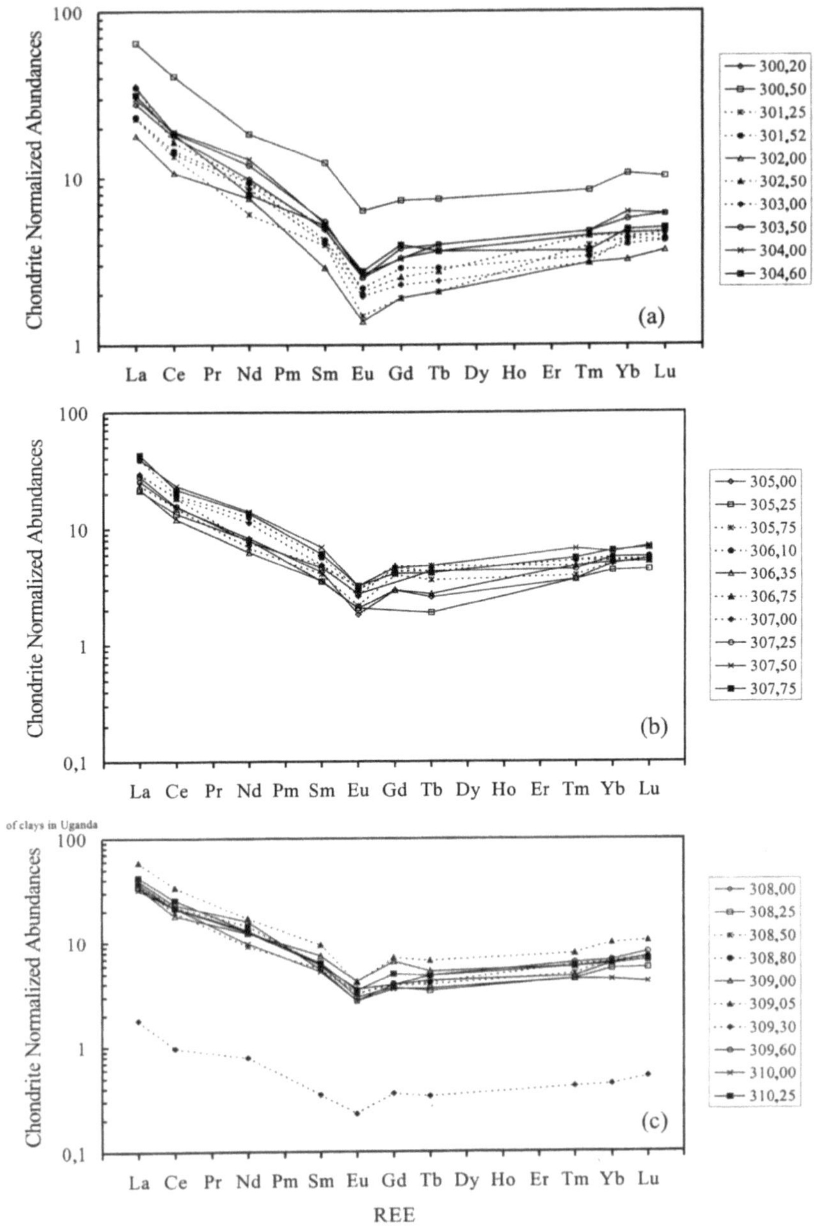

Fig. 8. Chondrite-normalized rare earth element (REE) abundances for samples from different stratigraphic sections: **(a)** from 300.20 m to 304.60 m, **(b)** from 305.00 m to 307.75 m, **(c)** from 308.00 m to 310.25 m.

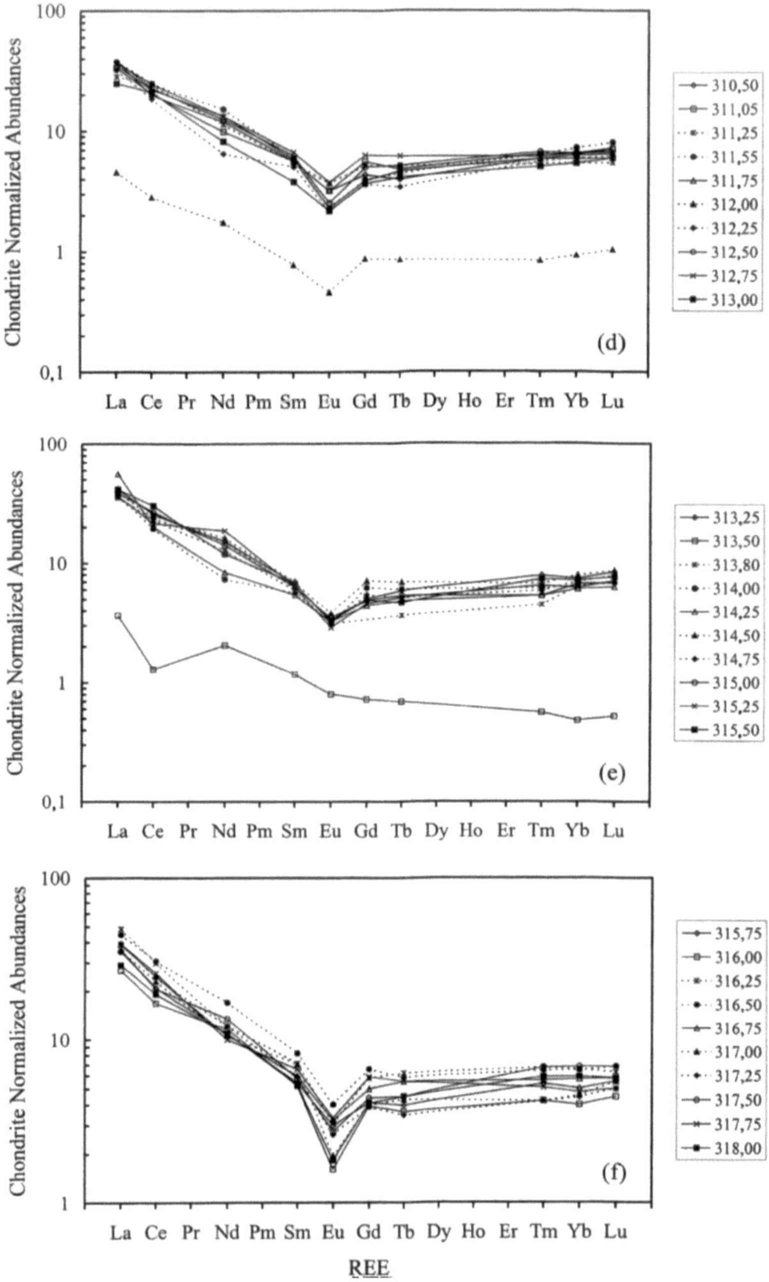

Fig. 8. Continued Chondrite-normalized rare earth element (REE) abundances for samples from different stratigraphic sections: **(d)** from 310.50 m to 313.00 m **(e)** from 313.25 m to 315.50 m, **(f)** from 315.75 m to 318.00 m.

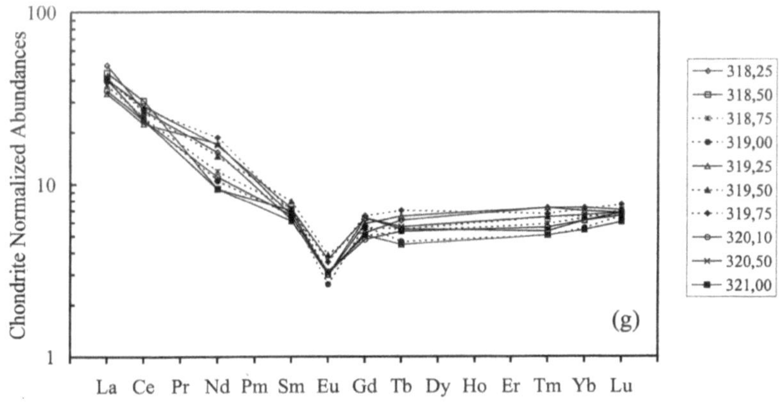

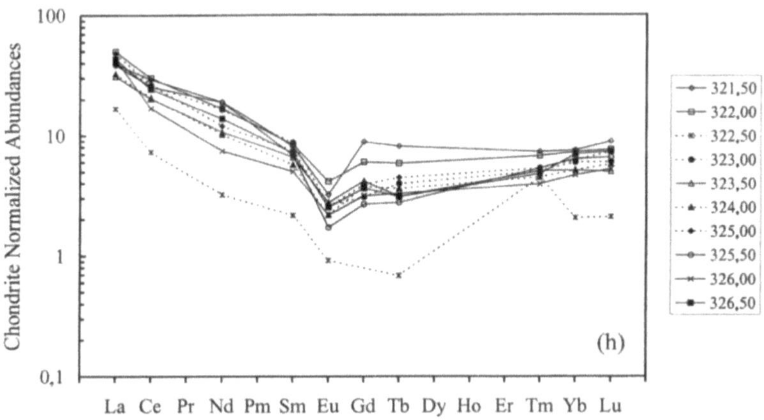

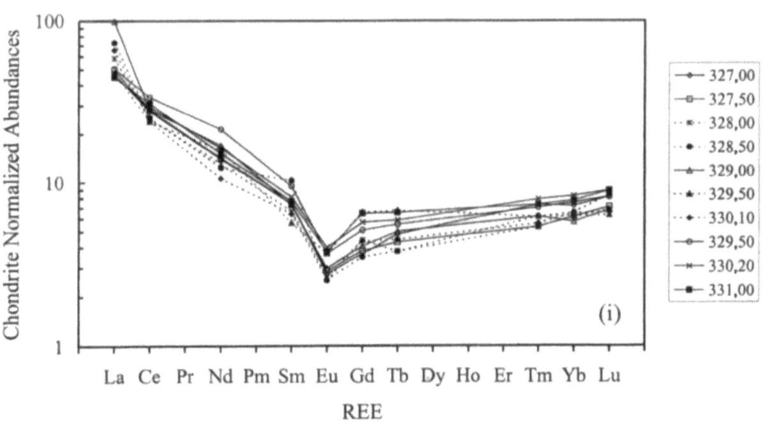

Fig. 8. Continued Chondrite-normalized rare earth element (REE) abundances for samples from different stratigraphic sections: **(g)** from 318.25 m to 321.00 m, **(h)** from 321.50 m to 326.00 m, **(i)** 326.50 m to 330.20 m.

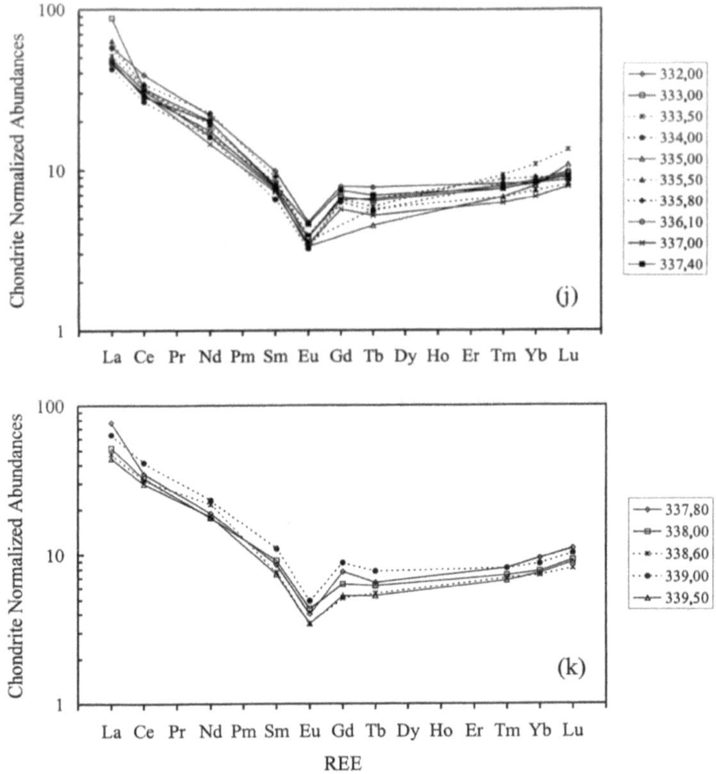

Fig. 8. Continued Chondrite-normalized rare earth element (REE) abundances for samples from different stratigraphic sections: **(j)** from 330.50 m to 336.1 m and **(k)** from 337.0 m to 339.5 m.

At 309.3, 312.0, and at 313.5 m, almost all elements show a significant depletion, in contrast to the generally increasing trend; only Na remains relatively constant. Additional significant enrichments are noticeable at 329.75 m for Co, at 329.5 m for Fe and Ni and at 333 m for Cr and La. La shows another enrichment at 329 m, where the J/K boundary is defined by the first occurrence of *N. steimmannii steimmannii*. According to Cresta et al. (1989), the NJK D zone is defined by the first occurrence of *N. steimmannii minor* at 326 m, where the As content has a maximum.

Almost over the whole sampled sequence, the Ir concentrations (measured at 316.5 keV and 468.1 keV) are too low to gain analytically reliable data, and only three measurements seem to be reliably above the detection limit. One of those is at 333.5 m (close to the J/K definition based upon *Calpionella elliptica*, 334.1 m). No anomalies in the Fe, Co, or Ni abundances are evident at this level, but Cr shows a high value. In the final quarter of the sampled section, no more gradual increase in the trace element contents is noticeable, and the superimposed variations become less significant.

8 Summary and Discussion

So far, only little research has been done on the behavior of $\delta^{13}C$, $\delta^{18}O$, and trace element abundances across J/K boundary sections, with most work having been aimed at paleotemperature (Ditchfield 1997) or rock-forming and recrystallization processes (Brenneke 1997), rather than at the characterization of detailed stratigraphic profiles. Thus, only few data sets are available for comparison with our measurements.

Weissert et al. (1998) found a slight shift in $\delta^{13}C$, and Adatte et al. (1996) found significant changes of both $\delta^{13}C$ and $\delta^{18}O$ at the base of magnetochron CM18r. This correlates to 318 m at the Bosso river section, at which level our data indicate only moderate disturbances. However, our results show more significant disturbances for both stable isotope ratios at 333.5 m, which correlates with a significant drop in the $\delta^{13}C$ value also recorded by Weissert and Channell (1989) and Weissert et al. (1998). This is close to the biostratigraphic definition of the J/K boundary based upon Calpionellids (Cresta et al. 1998). In this region the general shape of the isotope ratio curves changes more conspicuously than at the magnetostratigraphy-based J/K boundary definition at the base of magnetochron CM18 (Ogg and Lowrie 1986). Furthermore, this change correlates with an Ir spike, but the limited precision of the Ir data, the fact that the anomaly is restricted to only one sample, and the observation that Au values are very high (maybe as a result of inclusion of trace amounts of sulfide minerals) precludes any unambiguous conclusion.

According to Rampino and Haggerty (1996), a biotic crisis in general should result in a sudden shift in the isotope ratio record. This claim refers to the general trend of the whole data set, rather than to single outlying values, as the latter could result from diagenesis (McLaren and Goodfellow 1990). In our data set, this might be the case for $\delta^{18}O$ at 312 m and for both C and O values at 306.35 m and 315.75 m.

In the Bosso section, no paleontological evidence for biotic crises at the J/K boundary has been identified so far (cf. Montanari and Koeberl 2000), but, in contrast to the paleontological record, the $\delta^{13}C$ and $\delta^{18}O$ values represent general conditions in the biosphere, and not just single species. The drop of the isotope ratios at 333.5 m and the concurrent Ir spike might indicate a sudden event. Following arguments by McLaren and Goodfellow (1990) and Rampino and Haggerty (1996), the high level of Ir might indicate extraterrestrial origin, the sudden decrease in $\delta^{18}O$ points to a brief global warming period, and a sudden decrease in $\delta^{13}C$ reflects losses of biomass and productivity resulting in a change to organic-rich sediments in the sedimentary record. The subsequent recovery of the biosphere is noticeable by gradual increase of the isotope ratios and probable introduction of new taxa, of which *Calpionellid elliptica* might be one.

The present data, with isotopic shifts at about 333 m, a possible corresponding siderophile element anomaly, and another siderophile element anomaly at about 329 m, hint at an extraterrestrial signal, but the effects are not significant enough to allow any unambiguous conclusions. The Bosso River Gorge samples have not yet been analyzed for the presence of other impact-characteristic evidence, such as shocked minerals or Ni-rich spinels, but it appears that such studies might be useful in the future.

Acknowledgements

This research was supported by the Austrian FWF (Fonds zur Förderung der wissenschaftlichen Forschung, project Y58-GEO, to C.K.), the ÖNB Jubiläumsfonds (project no. 7915, to C.K.), a research scholarship from the University of Vienna (to G.K.), and the Austrian-Italian Scientific and Technical exchange program, project No. 14 (to C.K. and A.M.). Partial support came from ÖNB project 7915. We are grateful to R. Rocchia and an anonymous colleague for critical and helpful reviews of this paper. This is Wits Impact Cratering Research Group Contribution No. 18.

References

Adatte T, Stinnesbeck W, Remane J, Hubberten H (1996) Paleoceanographic changes in the Jurassic-Cretaceous boundary in the western Tethys, northeastern Mexico. Cretaceous Res 17: 671-689

Alvarez LW, Alvarez W, Asaro S, Michel HV (1980) Extraterrestrial cause for the Cretaceous-Tertiary extinction. Science 208: 1095-1108

Ball JD, Crowley SF, Steele DF (1996) Carbon and oxygen isotope ratio analysis of small carbonate samples by conventional phosphoric acid digestion: Sample preparation and calibration. Rapid Communications in Mass Spectrometry 10: 987-995

Benton MJ (1985) Mass extinction among non-marine tetrapods. Nature 316: 811-814

Benton MJ (1995) Diversification and extinction in the history of life. Science 268: 52-58

Brand WA (1996) High precision isotope ratio monitoring techniques in mass spectrometry. Journal of Mass Spectrometry 31: 225-235

Bralower TJ, Monechi S, Thierstein HR (1989) Calcareous nannofossil zonation of the Jurassic-Cretaceous boundary interval and correlation with the geomagnetic polarity time scale. Mar Micropaleontol 14: 153-235

Brennecke JC (1997) A comparison of the stable oxygen and carbon isotope composition of the Early Cretaceous and Late Jurassic carbonates from DSDP sites 105 and 367. Initial Reports Deep Sea Drilling Project 41: 937-955

Cecca F, Pallini G, Erba E, Premoli-Silva I, Coccioni R (1993) Haiterivian-Barremian chronostratigraphy based on ammonites, nannofossils, planktonic foraminifera and magnetic chrons from the Mediterranean domain. Cretaceous Res 15: 457-467

Channell JET, Grandesso P (1987) A revised correlation of Mesozoic polarity chrons and calpionellid zones. Earth and Planetary Science Letters 85: 222-240

Coplen TB, Kendall C, Hopple J (1983) Comparison of stable isotope reference samples. Nature 302: 236-238

Corfield RM, Cartlidge JE, Premoli-Silva I, Housley RA (1991) Oxygen and carbon isotope stratigraphy of the Palaeogene and Cretaceous limestones in the Bottaccione Gorge and Contessa Highway sections, Umbria, Italy. Terra Nova 3: 414-422

Corner B, Reimold WU, Brandt D, Koeberl C (1997) Morokweng impact structure, South Africa: geophysical imaging and shock petrographic studies. Earth and Planetary Science Letters 146: 351-364

Cresta S, Monechi S, Parisi G (1989) Stratigraphia del mesozoico e cenozoico nell'area Umbro Marchigiana. Memori descrittive della carta geologica d'Italia 39: 125-129

Ditchfield PW (1997) High northern paleolatitude Jurassic-Cretaceous paleotemperature variation: new data from Kong Karls Land, Svalbard. Palaeogeography, Palaeoclimatology, Palaeoecology 130: 163-175

Friedman I, O'Neil J, Cebula G (1982) Two new carbonate stable-isotope standards. Geostandards Newsletter 6: 11-12

Gonfiantini R, Stichler W, Rozanski K (1995) Standards and intercomparison materials distributed by the International Atomic Agency for stable isotope measurements. IAEA TECDOC 825: 13-29

Gradstein FM, Agterberg FP, Ogg JG, Hardenbol J, van Veen P, Thierry J, Huang Z (1994) A Mesozoic time scale. J Geophys Res 99: 24051–24074

Hart RJ, Andreoli MAG, Tredoux M, Moser D, Ashwal LD, Eide EA, Webb SJ, Brandt D (1997) Late Jurassic age for the Morokweng impact structure, southern Africa. Earth and Planetary Science Letters 147: 25-35

Hynes A (1990) Two-stage rifting of Pangea by two different mechanisms. Geology 18: 323-326

Housa V, Krs M, Krsová M, Man O, Pruner P, Venhodová D (1999) High-resolution magnetostratigraphy and micropalaeontology across the J/K boundary strata at Brodno near Zilana, western Slovakia: summary of results. Cretaceous Res 20: 699-717

Kent DV, Gradstein FM (1985) A Cretaceous and Jurassic geochronology. Geol Soc Am Bull 96: 1419-1427

Koeberl C (1993) Instrumental neutron activation analysis of geochemical and cosmochemical samples: a fast and reliable method for small sample analysis: J Radioanal Nucl Chem 168: 47-60

Koeberl C, Armstrong RA, Reimold WU (1997) Morokweng, South Africa, A large impact structure of Jurassic-Cretaceous boundary age. Geology 25: 731-734

Lowrie W, Channell JET (1983) Magnetostratigraphy of the Jurassic-Cretaceous boundary in Maiolica limestone (Umbria, Italy). Geology 12: 44-47

McLaren DJ, Goodfellow WD (1990) Geological and biological consequences of giant impacts. Ann Rev Earth Planet Sci 18: 123-171

Montanari A, Koeberl C (2000) Impact Stratigraphy – The Italian record. Lecture Notes in Earth Sciences, vol. 93. Springer Verlag, Heidelberg, 364 pp

Odin GS (1994) Geological time scale. Comptes Rendus de l'Académie des Sciences: 59-71

Ogg JG, Lowrie W (1986) Magnetostratigraphy of the Jurassic-Cretaceous. Geology 14: 547-550

Ogg JG, Hasenyager RW, Wimbledon WA, Channell JET, Bralower BJ (1991) Magnetostratigraphy of the Jurassic-Cretaceous boundary interval, Tethyan and English faunal realms. Cretaceous Res 12: 455-482

Podlaha OG, Mutterlose J, Veizer J (1998) Preservation of $\delta^{18}O$ and $\delta^{13}C$ in belemnite rostra from the Jurassic/Early Cretaceous successions. American Journal of Science 298: 324-347

Rampino MR, Haggerty BM (1996) Impact crises and mass extinctions: A working hypothesis. In: Ryder G, Fastovsky D, Gartner S (eds) New Developments Regarding the KT Event and Other Catastrophes in Earth History. Geological Society of America, Special Paper 307, pp 11–30

Raup DM, Sepkoski JJ (1986) Periodic extinction of families and genera. Science 231: 833-836

Reimold WU, Armstrong RA, Koeberl C (2000) New results from the deep borehole at Morokweng, North West Province, South Africa: Constraints on the size of the J/K boundary age impact structure. Lunar Planet Sci 31, Abs. #1074 (CD-ROM)

Remane J (1991) The Jurassic-Cretaceous boundary: Problems of definition and procedure. Cretaceous Research 12: 447–453

Renne PR, Ernesto M, Pacca IG, Coe RS, Glen JM, Prévat M, Perrin M (1992) The age of Paraná flood volcanism, rifting of Gondwanaland and the Jurassic-Cretaceous boundary. Science 258: 975-979

Ryder G (1996) The unique significance and origin of the Cretaceous-Tertiary boundary: Historical context and burdens of proof. In: Ryder G, Fastovsky D, Gartner S (eds) New Developments Regarding the KT Event and Other Catastrophes in Earth History. Geological Society of America, Special Paper 307, pp 31–38

Ryder G, Fastovsky D, Gartner S (eds) (1996) The Cretaceous-Tertiary Event and other Catastrophes in Earth History. Geological Society of America, Special Paper 307, 576 pp

Scotese CR (1991) Jurassic and Cretaceous plate tectonic reconstructions. Palaeogeography, Palaeoclimatology, Palaeoecology 87: 493-501

Scotese CR (1997) http://www.scotese.com/late1.htm

Sepkoski JJ (1992) A Compendium of fossil marine animal families, 2nd edition. Milwaukee Public Museum Contributions in Biology and Geology 83, 156 pp

Taylor SR, McLennan SM (1985) The continental crust: It's composition and evolution. Blackwell Scientific Publications, Oxford UK: 312 pp

Vergara M, Levi B, Nyström JO, Cancino A (1995) Jurassic and Early Cretaceous island arc volcanism, extension and subsidence in the coast range of central Chile. Geol Soc Am Bull 107: 1427-1440

Weissert H, Channell JET (1989) Tethyan carbonate carbon isotope stratigraphy across the Jurassic-Cretaceous boundary: An indicator of decelerated global carbon cycling? Paleoceanography 4: 483-494

Weissert H, Lini A, Föllmi KB, Kuhn O (1998) Correlation of Early Cretaceous carbon isotope stratigraphy and platform drowning events: A possible link? Palaeogeography, Palaeoclimatology, Palaeoecology 137: 189-203

Phytoplankton Blooms in the Jurassic-Cretaceous Boundary Beds of the Barents Sea Possibly Induced by the Mjølnir Impact

Morten Smelror[1], Henning Dypvik[2], and Atle Mørk[3]

[1]Geological Survey of Norway, N-7491 Trondheim, Norway. (morten.smelror@ngu.no)
[2]Department of Geology, University of Oslo, P.O.Box 1047 Blindern, N-0316 Oslo, Norway. (henning.dypvik@geologi.uio.no)
[3]SINTEF Petroleum Research, N-7465 Trondheim, Norway. (atle.mork@iku.sintef.no)

Abstract. The Mjølnir Crater was formed in the latest Jurassic by a meteorite impact in the moderately deep (300-400 m) palaeo-Barents Sea. Palynological studies of the ejecta-bearing strata of corehole 7430/10-U-01, located 30 km north-east of the Mjølnir Crater, have revealed a high abundance of marine Prasinophyceae algae (mainly *Leiosphaeridia*) and a minor abundance peak of freshwater algae (juvenile *Botryococcus*). The highest concentrations of algae occur in the ejecta unit recognised by high iridium concentrations, occurrences of shocked quartz and traces of possible altered glass.

The marine Mjølnir impact probably caused major shock waves and tsunamies, resulting in mixing of water-masses and sudden introduction of nutrients into the water column. Increased concentrations of nutrients (ammonia, nitrates, phosphates, iron) most likely enhanced reproduction of the algae and caused blooms of stocks of prasinophyte phytoplankton. After tsunami withdrawal from the surrounding land areas, freshwater may have flushed the shelf. This would explain the conspicuous concentration of juvenile freshwater algae in the post-impact beds.

Low algal diversity and the total dominance of morphologically uniform leiospheres is typical for eutrophic water conditions where a few species could quickly utilise the newly released nutrients (mainly N and P) in the water-column. Similar algal blooms can also be traced in contemporaneous deposits on Svalbard and in the offshore Troms III area. Thus, the algae bloom can be recognised as a regional marker horizon in the Jurassic-Cretaceous boundary sediments of the western Barents Sea.

1 Introduction

The Mjølnir bolide (1-3 km in diameter) hit the fairly shallow, epicontinental palaeo-Barents Sea at a time close to the Jurassic-Cretaceous boundary (Gudlaugsson 1993; Dypvik et al. 1996; Tsikalas et al. 1998a). The present-day crater, which is centered at 73°40' N, 29°40' E in the Barents Sea (Fig. 1), was first described by Gudlaugsson (1993). Based on seismic interpretations, in combination with gravimetry and magnetic data, he suggested that the circular Mjølnir structure had been formed by a comet or meteorite impact. Detailed studies on the mineralogy and geochemistry of a shallow core (7430/10-U-01) drilled 30 km north-east of the Mjølnir structure disclosed positive evidence for an impact origin, i.e., grains of shocked quartz and Ir-enrichments (Dypvik et al. 1996). Recent geophysical analyses by Tsikalas et al. (1998a, b, c) and

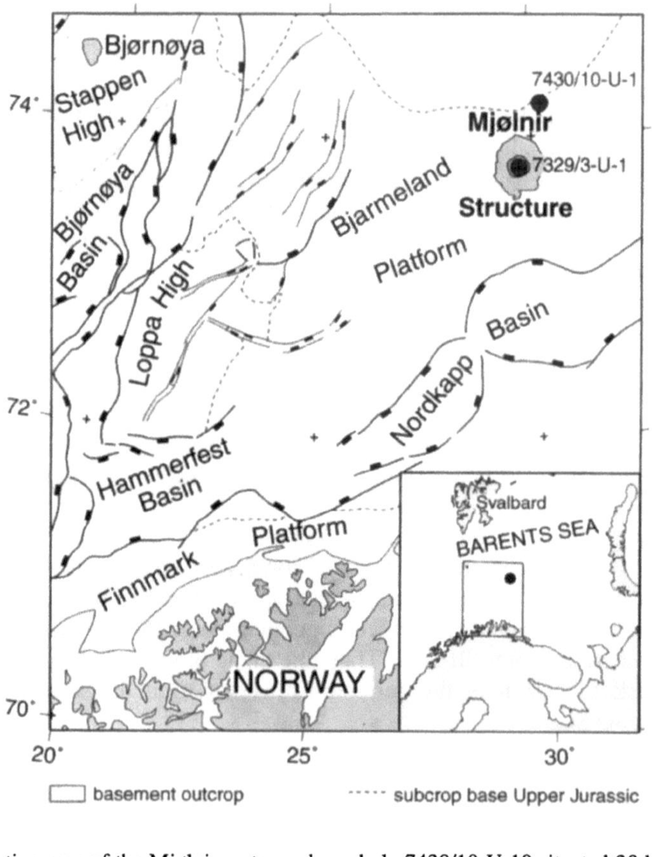

Fig. 1. Location map of the Mjølnir crater and corehole 7430/10-U-10 situated 30 km north-east of the crater (from Tsikalas 1998a).

mineralogical /geochemical analyses by Dypvik and Ferrell (1998) and Dypvik and Attrep (1999) have confirmed this interpretation and improved our understanding of the Mjølnir impact.

The crater is 40 km in diameter (Fig. 2), and has affected an at least 3.6-km-thick Mesozoic shelf section. The crater structure displays a distinct radial zonation, including a 12-km-wide outer zone, with a marginal fault zone and an elevated ring. A 4-km-wide annular depression surrounds the uplifted central high, which has a diameter of 8 km. The crater has been subjected to extensive secondary, post-impact deformation (Tsikalas 1998b). Tsikalas et al. (1998c) estimated the energy released to be in the range of $2.4–53x10^{20}$ J, corresponding to an earthquake of magnitude 8.3. Based on seismic studies they inferred a disturbed volume of 850–1400 km^3 from the impact, with an estimated ejecta volume of 140–180 km^3.

A reconnaissance study of the marine microflora of the Upper Jurassic-Lower Cretaceous strata in core 7430/10-U-01 revealed a conspicuous abundance peak of prasinophycean algae in the ejecta-bearing strata, and of the freshwater algal input at specific levels (Smelror et al. 1999). In the present paper we document the observed algae blooms in core 7430/10-U-01, and present a hypothesis for impact-induced environmental changes that probably caused the observed phytoplankton blooms.

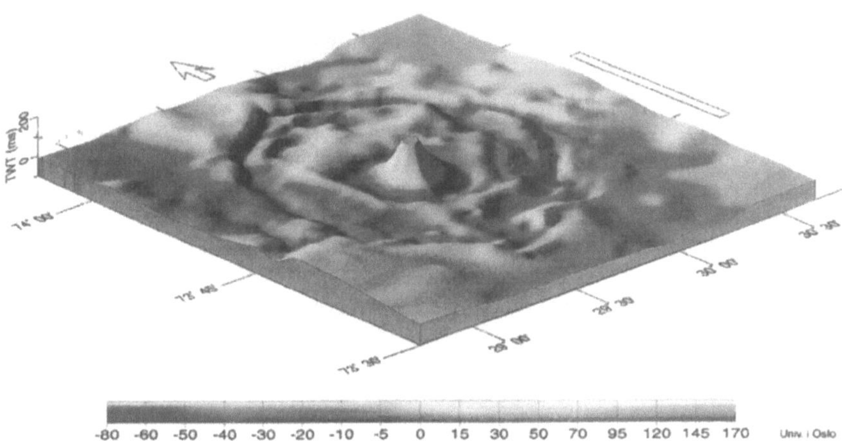

Fig. 2. Perspective diagram of residual two-way traveltime to the lower Barremian seismic reflector across the Mjølnir crater (from Dypvik et al. 1996).

2
Material and Methods

Samples for palynological analyses were taken from different levels within the interval 56.7–42.65 m of core 7430/10-U-01 (Fig. 3). The samples were processed using standard palynological preparation methods, including treatment with HCl and HF. The palynoflora and kerogen was examined by both transmitted light and fluorescence microscopy. Semi-quantitative data were obtained from counts of 150-200 terrestrial and marine palynomorphs from each sample.

Descriptions of core 7430/10-U-01 have been published by Dypvik et al. (1996) and Dypvik and Attrep (1999). According to their descriptions, the 67.6-m-long core consists of 57.1 m of dominantly dark shale of Jurassic-Cretaceous age, below 10.5 m of Quaternary sediments. The lowermost Kimmeridgian-Berriasian part of the core, from level 67.6 m to 41 m, is dominated by dark grey shale assigned to the Hekkingen Formation. The level 67.6-56.6 m is composed of Kimmeridgian dark grey laminated shale, which are overlain by Volgian silty to sandy shale (56.5–50 m), with several minor fining-upward units (2-5 cm thick) follow on top. The upper Volgian to lower Berriasian strata above can be divided into a lower unit (50-47.6 m) of dominantly grey, partly laminated shale, and an upper unit (47.6-41 m) of sandier shale, which is moderately bioturbated and in part glauconitic. The base of the upper unit is marked by a mudflake conglomerate 19 cm thick (47.6-47.4 m). This conglomerate contains grains of shocked quartz and enrichments of Ir. The main Ir peak is, however, located at level 46.85 m (Dypvik et al. 1996; Dypvik and Attrep 1999). Another erosion surface is present at 45 m. Above this level more carbonate-rich sediments appear.

These fine-grained, siliciclastic sediments of the Hekkingen Formation are overlain by a 8.9-m-thick unit of condensed carbonates of Valanginian to early Barremian age, i.e., the Klippfisk Formation of Smelror et al. (1998).

A general sedimentological log of core 7430/10-U-01 is shown in Fig. 3. The age of the strata studied has been determined by the ammonites, bivalves, dinoflagellates, foraminifera and radiolaria recovered. A time versus depth plot of the key biostratigraphic events recognised in the core is shown in Fig. 4.

Fig. 3. (Opposite page) General sedimentological log of core 7430/10-U-01, iridium (maximum anomaly marked in red) and total organic carbon (TOC) measurements, and relative abundance of marine prasinophytes (mainly *Leiosphaeridia*), dinoflagellate cysts, freshwater algae (*Botryococcus*) and terrestrial palynomorphs (pollen and spores) in the Volgian to lower Berriasian strata.

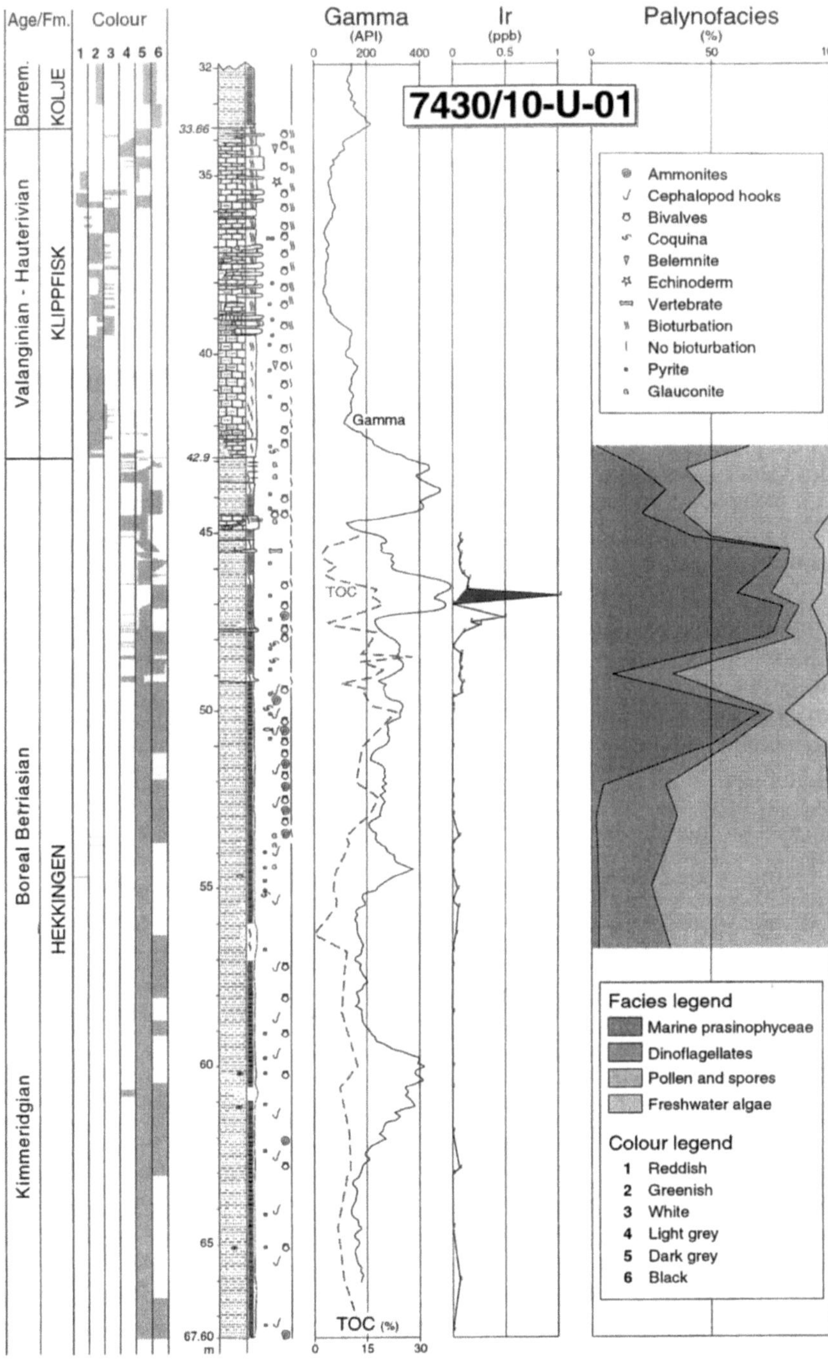

Age/Fm.
Colour
1 2 3 4 5 6
Barrem.
KOLJE
Valanginian - Hauterivian
KLIPPFISK
Boreal Berriasian
Kimmeridgian
HEKKINGEN
Gamma
(API)
Ir
(ppb)
Palynofacies
(%)
7430/10-U-01
Ammonites
Cephalopod hooks
Bivalves
Coquina
Belemnite
Echinoderm
Vertebrate
Bioturbation
No bioturbation
Pyrite
Glauconite
Gamma
TOC
TOC (%)
Facies legend
Marine prasinophyceae
Dinoflagellates
Pollen and spores
Freshwater algae
Colour legend
1 Reddish
2 Greenish
3 White
4 Light grey
5 Dark grey
6 Black
33.66
42.9
67.60
m

3 Microfloras of the Volgian and lower Berriasian Strata (Including the Ejecta-Bearing Unit)

The analysed samples from well 7430/10-U-01 contain well-preserved, and moderately rich assemblages of both marine and terrestrial palynomorphs. The palynofacies of the Kimmeridgian to Berriasian fine-grained sediments comprise dominantly amorphous organic matter, frequently speckled with very small particles of woody and coaly phytoclasts. Pyrite occurs abundantly. This palynofacies type is characteristic for the Upper Jurassic, dark grey shale found along the Norwegian Shelf, on the western Barents Shelf, on Svalbard, East Greenland and in the North Sea area.

A distribution chart of dinoflagellate cysts in core 7430/10-U-01 has been published previously by Smelror et al. (1998), and will not be discussed here. The relative abundance of marine microplankton (dinoflagellate cysts, prasinophytes), freshwater algae (*Botryococcus*) and terrestrial palynomorphs (pollen and spores) in the Volgian to lower Berriasian strata of core 7430/10-U-01 is shown in Fig. 3.

The relative abundance of pollen and spores compared to phytoplankton is mainly a reflection of proximity to terrestrial sources and the productivity of fossilising organic-walled microplankton (Tyson 1993). Pollen and spores are normally deposited in larger amounts in deltaic and near-shore sites. Further offshore, the declining absolute input of pollen and spores and the increased proportions of microplankton, result in a decrease in the relative abundance of terrestrial palynomorphs. Abundances of dinoflagellate cysts normally also vary considerably from nearshore to offshore areas, but often reach a maximum value in continental slope sediments. Stratified basins, in which black organic rich muds are deposited, may often contain high relative abundances of pollen and spores because of low production of fossilising marine palynomorphs. In the Upper Jurassic-Lower Cretaceous of core 7430/10-U-01, terrestrial palynomorphs generally comprise more than 30% of the overall palynoflora, with the highest relative proportions found in the pre-impact strata at 52.14 m and below.

The present investigation shows a distinct increase in the relative abundance of prasinophycean algae at two core levels from 51.0-50.1 m and 48.0-45.15 m (Fig. 3). The lowermost peak is located at the first occurrence of altered glass and ejecta material (Dypvik and Attrep 1999). The upper algal peak corresponds to the core-interval that contains shocked quartz and Ir-peaks (Dypvik and Attrep 1999). Both algal enrichments are nearly monospecific in composition and dominated by an unidentified species of the genus *Leiosphaeridia* (Fig. 5). The Prasinophyceae are characterised by having two distinct phases in their life history, i.e., a motile and a non-motile stage, and cyst formation is well known in many recent genera (Guy-Ohlson 1996). The life cycle of *Halosphaera*, which is believed to be a modern relative of the fossil genus *Leiosphaeridia*, has been studied in detail by Parke and Hartog-Adams (1965) and Boalch and Mommaerts (1969). The latter authors in addition describe a monospecific bloom of *Halosphaera parkeae* occurring in the western part of the English Channel. Fossil non-motile cysts of the genus *Leiosphaeridia* are preserved in the sediments studied in core 7430/10-U-01.

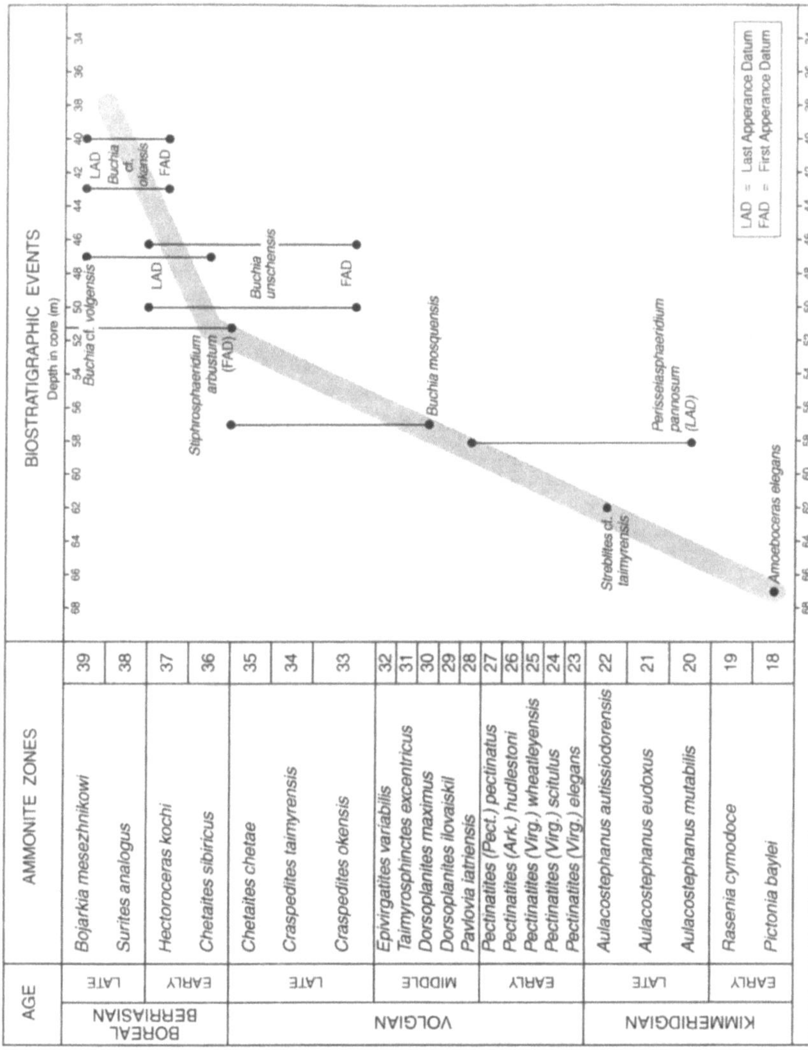

Fig. 4. Time versus depth plot of biostratigraphic events in core 7430/10-U-01, and age determination of the cored strata. Ammonite zones (18-39) are according to the Boreal zonation presented in Nagy and Basov (1998).

Studies of the life cycle and growth of modern prasinophytes have shown that the entire process from the time of cyst formation to the release of new motile cells requires only up to six hours (Guy-Ohlson 1996). Such short life cycles may have been a major comparative advantage for rapid stock growth and maximum utilisation of the large amounts of nutrients that were probably released by the Mjølnir impact.

Another conspicuous palynological observation is the presence of the freshwater alga *Botryococcus* at core-level 52.14-45.15 m, and the marked abundance peak of specimens at 51.0 m (Fig. 3). Modern *Botryococcus* is widely dispersed in temperate and tropical regions, and is known to tolerate seasonally cold climates (Batten and Grenfell 1996). This alga lives and reproduces exclusively in fresh water (lakes, bogs and ponds etc.), but it is also reported to appear in considerable abundance in environments of variable salinity (i.e., brackish habitats). The presence of *Botryococcus* in marine sediments may be a result of fluvial or tidal transport into coastal regions. The oily composition of the *Botryococcus* algal colonies makes them highly buoyant. They can be transported over large distances and become a minor component of even open shelf sediments (Tyson 1993). The presence of juvenile *Botryococcus* in the ejecta-bearing strata was probably caused by significant water mixing and freshwater influx to this location on the central shelf area. The occurrence of mostly juvenile specimens indicates that the input of fresh water was of limited duration, and that the restoration of more normal saline condition was lethal to the stocks of freshwater algae.

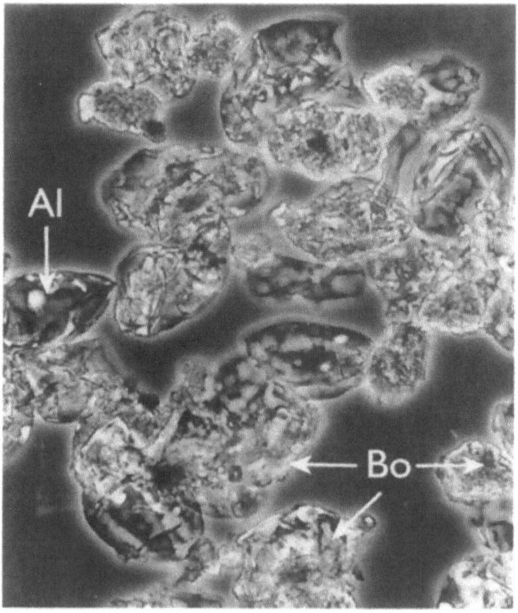

Fig. 5. Prasinophycean algal bloom from 51.0 m in core 7430/10-U-01. Al: *Leiosphaeridia* , Bo: *Botryococcus*.

4 Environmental Consequences of the Mjølnir Impact

During the Late Jurassic to earliest Cretaceous the western palaeo-Barents Sea area was a Boreal shallow shelf sea, forming an apex of the "Kimmerian Ocean" (Kimmeridge Clay sea of Miller 1990), which also covered the present North Sea and Norwegian Sea/East Greenland areas. In this shallow sea clays and silts accumulated under dysoxic and anoxic depositional conditions. The dark claystones and shales are rich in organic matter and have source rock potential for hydrocarbons (Leith et al. 1993). Several models have been proposed to explain the deposition of these organic-rich shales on the Barents Shelf and Svalbard (Dypvik et al. 1991; Leith et al. 1993; Mørk and Bjorøy 1994; Smelror et al. 2001). Most models include restricted bottom-water circulation and poorly oxygenated sea-bottom conditions, caused by stratified water-masses (i.e., a pycnocline), sea-floor topography and/or unique palaeogeographic, palaeoclimatic, and palaeoceanographic settings.

When the Mjølnir meteorite hit the fairly shallow palaeo-Barents Sea huge tsunamis were generated and traveled across the shelf. Calculations by Tsikalas et al. (1998b) suggest that the tsunami reached a height of 60 m at the location of core 7430/10-U-01. According to Dypvik and Attrep (1999), however, even higher waves could have reached this location, which was only 30 km away from the site of the impact. The tsunami, soil slides, and density currents created significant disturbances in the water masses, including a temporary pycnocline breakdown by mixing oxygen-rich surface water with oxygen-depleted bottom water. At shallow shelf and coastal locations the tsunami deeply eroded the sea bed. This erosion and the turnover of the water-masses, reworked large amounts of clastic material, organic matter and nutrients. Simultaneously, ejecta from the impact and fireball material were deposited and mixed with the normal marine background deposits and sediments reworked by tsunami.

The first traces of altered impact glass and ejecta have been recovered from level 51 m in core 7430/10-U-01. Here Dypvik and Attrep (1999) also found an abrupt change in the indicator elements of anoxic conditions (Cr, Ni, Cu, Zn), which could have been associated with a high influx of possible ejecta-related matter. The Mjølnir impact apparently terminated the long-lasting Late Jurassic period with restricted bottom-water circulation and deposition of laminated clays. This environmental change in oceanic bottom-water conditions is also reflected in the increased bioturbation of post-impact setting sediments, and by the overall change in palynofacies. The latter is seen as a change from palynofacies dominated by amorphous organic matter to palynofacies dominated by palynomorphs, cuticles and woody/coaly phytoclasts. This change may reflect an oceanic change from dysoxic and partly anoxic depositional conditions to more ventilated and oxygenated bottom conditions.

Tyson (1993) argued that generally high relative abundances of Prasinophyceae could be explained by the decline of *in situ* phytoplankton (i.e., dinoflagellate cyst) production in permanently stratified basins. According to him the ratio of prasinophycean algae to dinoflagellate cysts, therefore, represents an index of hydrographic stability as it increases into more typically stratified basin facies. Following Tyson's (1993) model the distinct increase of prasinophycean algae at level 51.0-50.1 m and 48.0-45.15 m in the present core could indicate more stable dysoxic to anoxic bottom conditions. However, according to the geochemical information on the succession, the situation appears to be the opposite of this, the more ventilated conditions being associated with the impact indicators, i.e., at the stratigraphic levels in which abundance peaks of the prasinophycean algae occur (Fig. 3).

Tsunamies related to major marine impacts could be a result of the impact itself, as well as by later phases of crater collapse (Melosh 1989). Tsikalas et al. (1998b) argued that the Mjølnir structure suffered extensive post-impact collapse and deformation owing to the moderately deep shelf location. They suggested that the collapse of the impact-induced cavity, and the subsequent surge of seawater into the crater, transported large amounts of ejecta and "crater wall" material back into it. Dypvik and Attrep (1999) suggested that waves generated by the collapse of the crater could explain the presence of grains of shocked quartz in the mud-flake conglomerate at level 47.6 m.

Recent phytoplankton growth and distribution are to a large extent controlled by the availability of nitrogen and phosphorus, and common constituents as CO_2, Na^+, K^+, Mg^{2+}, Ca^{2+} and SO_4^{2-}, which usually are available in sufficient amounts. Elements, such as Fe, Zn, Mn and Cu, are also essential for phytoplankton growth (Brand et al. 1983). Increased availability of nutrients by upwelling may result in improved living conditions for some species and related phytoplankton blooms (Raymont 1980). Under such conditions the species composition of the stocks might change and lead to the dominance of few species. The blooms of prasinophytes in the ejecta bearing strata of core 7430/10-U-01 suggest that the Mjølnir impact caused changes in the oceanic conditions with re-mobilisation of essential nutrients.

The occurrences of fresh water algae (*Botryococcus*) also require special consideration. Their sudden incoming and abundance peaks can be explained by tsunami withdrawal processes from the surrounding land areas. Coastal freshwater basins were flushed in the low relief landscape and the sediments reworked into the shelf areas. Such processes created temporary brackish water conditions on parts of the shelf. Increased precipitation related to the post-impact effects could have resulted in additional fresh water run-off and influx of fresh- and brackish water species over large shelf areas.

5 Regional Extent of the Prasinophyceae Algae Bloom Related to the Mjølnir Impact

Algal blooms comparable to those found in the Jurassic-Cretaceous boundary beds of corehole 7430/10-U-01 have also been traced in contemporaneous deposits in the Janusfjellet section on Svalbard (Dypvik et al. 2000) and in the offshore Troms

III area (pers. obs.). This algal bloom can, therefore, be recognised as a distinct marker horizon in Jurassic-Cretaceous boundary strata over major parts the western Barents Shelf.

Preliminary analysis of the very first post-impact beds (i.e., lowermost Berriasian) from the recent core recently drilled within the Mjølnir crater (7329/03-U-01) have proven a similar bloom of prasinophytes (*Leiosphaeridia*) (Smelror et al. 1999). In the Mjølnir core samples the abundance of prasinophyceaen algae is extremely high. These preliminary observations support the observations and interpretations of the association of phytoplankton blooms and geochemical/mineralogical anomalies (Dypvik and Attrep 1999; Dypvik et al. 2000) of core 7430/10-U-01.

However, it is still premature to relate the algal blooms described with full confidence (and solely) to the Mjølnir impact events. Other factors such as the global relative sea-level drop at the Jurassic-Cretaceous boundary, changes in ocean circulation patterns, plate tectonic activity and climatic changes may also have been important in controlling the phytoplankton distributions. The effects of the Mjølnir impact could then have been superimposed on the other major regional and global effects. These questions are being addressed in ongoing studies aimed at providing more detailed information on the relationships between the prasinophycean algal blooms, the eutrophication of the palaeo-Barents Sea at the Jurassic-Cretaceous boundary, and the Mjølnir meteorite impact.

6 Conclusions

The Mjølnir crater of the central Barents Sea was formed close to the Jurassic/Cretaceous boundary by a meteorite impact into the moderately deep shelf sea (300-400 m water depth). Palynological studies of the ejecta-bearing strata of corehole 7430/10-U-01, 30 km north-east of the Mjlønir crater, have revealed high abundance of marine Prasinophyceae algae (leiosphaeres) and a minor enrichment peak of freshwater algae (i.e., juvenile *Botryococcus*).

Low diversity of algal species and the total dominance of uniform leiospheres, are characteristic of eutrophic water conditions, where a few species might quickly utilise the enrichment of nutrients in the water-column. The leiosphere blooms discovered in samples from core 7430/10-U-01 most likely developed as a result of a sudden increase in supply of nutrients to the water column caused by both the meteorite impact and the subsequent crater collapse. Sedimentological, palynological and geochemical data indicate that the Mjølnir impact caused mixing of previously stratified water-masses, and a breakdown of the shelf sea pycnocline. This resulted in an overall change from dysoxic/anoxic to more ventilated bottom water conditions.

A short-lived, but significant, influx and bloom of juvenile freshwater algae in the central parts of the basin happened in response to a sudden and heavy influx of fresh water. This freshwater influx was probably caused by tsunami-triggered coastal flushing/erosion during water withdrawal of tsunami flood-water from adjacent land areas of low relief. It could also have resulted from increased precipitation caused by the local climatic changes following by the impact.

Algal blooms comparable to those encountered in the Jurassic-Cretaceous boundary beds of the 7430/10-U-01 core and in the earliest Berriasian deposits of the Mjølnir core (7329/03-U-01), have also been traced in contemporaneous deposits on Svalbard and in wells in the offshore Troms III area. These can apparently be recognised as a distinct stratigraphic marker horizon the Jurassic-Cretaceous boundary strata over major parts the western Barents Shelf.

References

Batten DJ, Grenfell HR (1996) Chapter 7D. *Botryococcus*. In Jansonius, J. and McGregor, D.C. (eds.), Palynology: principles and applications. American Association of Stratigraphic Palynologists Foundation, Vol 1, 205-214

Boalch GT, Mommaerts JP (1969) A new punctate species of *Halosphaera*. Journal of the Marine Biological Association of the United Kingdom 49: 129-139

Dypvik H, Attrep M Jr. (1999) Geochemical signals of the late Jurassic, marine Mjølnir impact. Meteoritics and Planetary Science 34: 393-406

Dypvik H, Ferrell RE Jr. (1998) Clay mineral alteration associated with a meteorite impact in the marine environment (Barents Sea). Clay Minerals 33: 51-64

Dypvik H, Nagy J, Eikeland TA, Backer-Owe K, Johansen H (1991) Depositional conditions of the Bathonian to Hauterivian Janusfjellet Subgroup, Spitsbergen. Sedimentary Geology 72: 55-78

Dypvik H, Gudlaugsson ST, Tsikalas F, Attrep M Jr., Ferrell RE Jr, Krinsley DH, Mørk A, Faleide JI, Nagy J (1996) Mjølnir structure: An impact crater in the Barents Sea. Geology 24: 779-782

Dypvik H, Kyte FT, Smelror M (2000) Iridium peaks and algal blooms – The Mjølnir impact. Lunar and Planetary Science 31: abstract #1538 (CD-ROM)

Gudlaugsson ST (1993) Large impact crater in the Barents Sea. Geology 21: 291-294

Guy-Ohlson D (1996) Chapter 7B. Prasinophycean algae. In: Jansonius J, McGregor DC (eds) Palynology: principles and applications. American Association of Stratigraphic Palynologists Foundation, Vol 1, 181-189

Leith TL, Weiss HM, Mørk A, Århus N, Elvebakk G, Embry AF, Brooks PW, Stewart KR, Pchlina TM, Bro EG, Verba ML, Danyushevskaya A, Borisov AV (1993) Mesozoic hydrocarbon source-rocks of the Arctic region. In: Vorren TO; Bergsager E, Dahl-Stamnes OA, Holter E, Johansen B, Lie E, Lund TB (eds); Arctic Geology and Petroleum Potential. NPF Special Publication 2, 1-25. Elsevier, Amsterdam

Melosh J (1989) Impact Cratering - A Geological Process. Oxford Monographs in Geology and Geophysics, 11, Oxford University Press, New York, USA, 245 pp

Miller RG (1990) A paleoceanographic approach to the Kimmeridge Clay Formation. In: Huc AY (Ed) Deposition of organic facies. AAPG Studies in Geology 30: 13-26

Mørk A, Bjorøy M (1984) Mesozoic source-rocks on Svalbard. In: Spencer AM, Johnsen SO, Moerk A, Nysaether E, Songstad P, Spinnangr A, (eds) Petroleum Geology of the North European Margin, 371-382. Graham and Trotman, London

Nagy J, Basob VA (1998) Revised foraminiferal taxa and biostratigraphy of Bathonian to Ryazanian deposits in Spitsbergen. Micropaleontology 44: 217-255

Parke M, den Hartog-Adams I (1965) Three new species of *Halosphaera*. Journal of the Marine Biological Association of the United Kingdom 45: 537-557

Raymont, JEG (1980) Plankton and productivity in the oceans. 2nd edition, Vol. 1 Phytoplankton, Pergamon Press, 489 pp

Smelror M, Dypvik H, Mørk A (1999) Phytoplankton "blooms" related to the Late Jurassic Mjølnir Meteorite Impact (Barents Sea): Preliminary results. In: Buffetaut E, Le Loeuff J (eds) Workshop on Geological and Biological Evidence for Global Catastrophes. Esperaza/Quillan (Aude, France), 26-30 Sept. 1999, Programme, Abstracts and Field Guide, p 71

Smelror M, Mørk A, Monteil E, Rutledge D, Leereveld H (1998) The Klippfisk Formation - a new lithostratigraphic unit of Lower Cretaceous platform carbonates on the Western Barents Shelf. Polar Research 17: 181-202

Smelror M, Mørk MBE, Mørk A, Løseth H, Weiss HM (2001) Middle Jurassic-Lower Cretaceous transgressive-regressive sequences and facies distribution of northern Nordland and Troms (Norway). In: Martinsen OJ, Dreyer T (eds) Sedimentary environments offshore Norway-Palaeozoic to Recent. NPF Special Publication 10, Elsevier Science B.V., Amsterdam, pp 211-232

Tsikalas F, Gudlaugsson ST, Eldholm O, Faleide JI (1998a) Integrated geophysical analysis supporting the impact origin of the Mjølnir structure, Barents Sea. Tectonophysics 289: 257-280

Tsikalas F, Gudlaugsson ST, Faleide JI (1998b) Collapse, infilling, and postimpact deformation at the Mjølnir impact structure, Barents Sea. Geological Society of America, Bulletin 110: 537-552

Tsikalas F, Gudlaugsson ST, Faleide JI (1998c) The anatomy of a buried complex impact structure: The Mjølnir Structure, Barents Sea. Journal of Geophysical Research 103: 30469-30483

Tyson RV (1993) Palynofacies analysis. In: Jenkins DG (ed) Applied Micropaleontology. Kluwer Academic Publishers, The Netherlands, p 153-191.

A Geographic Database Approach to the KT Boundary

Wolfgang Kiessling[1] and Philippe Claeys[2]

Institut für Mineralogie, Museum für Naturkunde, Invalidenstr. 43,
D-10115 Berlin, Germany.
[1]Present address: Department of Geophysical Sciences, University of Chicago, 5734 S Ellis Avenue, Chicago IL 60637, USA. (kiessl@geosci.uchicago.edu)
[2]Present address: Department of Geology, Vrije Universiteit Brussel, Pleinlaan 2, B-1050 Brussels, Belgium. (phclaeys@vub.ac.be)

Abstract. A comprehensive database on the Cretaceous-Tertiary (KT) boundary is presented. The database (KTbase) is designed to evaluate spatial patterns in data that are important to constrain the causes and mechanisms of the K/T event. KTbase contains paleontological, sedimentological, mineralogical, and geochemical data on currently 350 KT sites. Although still incomplete, KTbase can already be used to summarize global aspects of the ecological catastrophe coincident with the KT boundary, to suggest some regional patterns related to the impact of an extraterrestrial body in the Gulf of Mexico and to document the distribution of ejecta debris from the Chicxulub crater.

1 Introduction

Progress towards solving the Cretaceous-Tertiary (KT) boundary mass extinction problem demands a multidisciplinary approach and cooperation between geologists, paleontologists, geochemists, and mineralogists. A major biotic event at or near the boundary both on land and in the oceans had long been recognized by paleontologists (Phillips 1860). The most significant advance in the searches for the possible causes of the event came out of geochemical analyses of the KT clay layer and the discovery of the now famous Ir positive anomaly (Alvarez et al. 1980). The suggestion of a bolide impact as the major cause of the KT mass extinction event triggered intense research by all Earth-science disciplines. After much scepticism in the early 80s, the discovery of the >200 km in diameter Chicxulub crater in Yucatan (Mexico) convinced most of the scientific community that a large extraterrestrial object impacted the Earth precisely at the time of the KT boundary (Ryder 1996).

However, although a tremendous amount of data was collected, the current discussion on the KT boundary is still polarized by a major controversy regarding the exact role played by the Chicxulub bolide impact in the mass extinction scenario. The opinions range from a complete refusal of the impact as a major trigger of extinction (Glasby and Kunzendorf 1996) to the interpretation of Chicxulub being the sole ultimate cause of the Late Cretaceous faunal change (e.g., Smit 1990, Sheehan et al. 2000). A close examination of published literature reveals that besides the given incompleteness of the fossil record, these interpretations are often based on a limited utilization of available data, or selective citing (MacLeod et al. 1997; Hudson 1997). Often the limited utilization of data is probably not intentional, but due to the fact that publications are so numerous, so widely scattered, and that they cover such a broad range of topics, that one can hardly be familiar with all aspects of the question. Another problem is that many authors focus on a single, perhaps "favorite" group of organisms and then try to expand their conclusions to the whole Late Cretaceous biosphere. The KT mass extinction question can only be answered by considering the global picture of the biotic demise, including its biological and physical processes.

Computer databases offer the opportunity to organize, overview and evaluate large amounts of data easily and more objectively than other data collection means. Therefore, in the framework of an ambitious program on mass extinction and evolutionary transformations undertaken at the Museum of Natural History in Berlin, we are currently compiling a multidisciplinary relational database on the KT boundary (KTbase) based on the vast existing literature. Owing to the various aspects involved in the discussion surrounding the KT boundary, we are collecting and evaluating together paleontological, sedimentological, mineralogical, and geochemical data. The combination of the database with graphic presentations on paleogeographic maps using GIS (Geographic Information System) software provides a better understanding of spatial patterns in the KT event.

Our goals in this paper are: i) to summarize the worldwide distribution of impact debris, ii) to present a balanced review of extinction patterns observed in various invertebrate marine fossil groups, the only organisms where so far our data base is complete enough, iii) to demonstrate the potential of KTbase for unravelling the mechanisms of the impact-extinction scenario, and iv) to point out some particular research topics worthy of further studies. Although the review on extinction patterns is similar to that pioneered by MacLeod et al. (1997), our approach is drastically different. Rather than trying to neglect its role, we consider as a working hypothesis to be tested that the KT impact at Chicxulub is the cause of the mass extinction event. We also address the question from a multidisciplinary point of view and base our study on an updated bibliography. In this paper, when the 'impact hypothesis' is mentioned, it should be referred to as the hypothesis of an impact causing the mass extinction of organisms observed at the KT boundary. The occurrence of the Chicxulub impact itself must now be seen as an established fact rather than a hypothesis.

2 The Database

The development of a comprehensive database starts with a literature survey and the elaboration of an appropriate database structure. So far, more than 2000 references on the KT boundary have been collected and evaluated. A locality-based entry format for the main table was defined to use this database in connection with a geographic information system (GIS) analysis. Paleontological tables linked to the main table are taxon-based. Species level taxonomy and quantitative data were applied whenever possible.

Currently 350 KT boundary sites are catalogued in KTbase. Only localities with at least the late Maastrichtian and/or the Danian preserved were considered as potential KT boundary sites and included in KTbase. A minimum distance of 20 km between different boundary sections to be counted separately was defined to obtain an even treatment of data.

The completeness of KT sections is often debated. Some authors advocate that most KT sections contain hiatuses, mostly in the basal Danian (MacLeod and Keller 1991). One criterion used by many paleontologists for completeness is the presence of the basal Paleocene planktonic foraminifer Zone P0. However, the short duration of this zone renders it difficult to detect in sediments with low sedimentation rates (Olsson and Liu 1993). Even more importantly, P0 may not be a globally significant biozone (Norris et al. 1999). The exact location of the boundary is rather equivocal and varies significantly depending on the paleontological dating method used (compare, e.g., Keller et al. 1994, and Pospichal 1996a for the Mimbral section in Mexico, the different opinions in Keller et al. 1997, or the detection of Danian coccospheres in Maastrichtian sediments, Mai 1999). We decided to rely on the first occurrence of impact products (impact glass/spherules for proximate sites, iridium anomaly for distal sites) to define precisely the KT boundary, but also to evaluate stratigraphic completeness. The record in a section of an event as short as the deposition of ejecta debris is probably one of the best indications of stratigraphic completeness available. In addition, a synchronous deposition of ejecta is more likely than the synchronous global first occurrence of a newly evolved species. Based on this convention, more than 25% of all localities in KTbase are stratigraphically complete, and can thus be considered as true KT boundaries. Another 20% of localities record the latest Maastrichtian and earliest Danian plankton zones and are considered as fairly complete although impact ejecta were not recorded (Fig. 1). For 55% of the localities in KTbase, the completeness is either unknown or a significant sedimentation break is evident. Although the incomplete sites cannot be used to test the abruptness of extinction in relation to the impact, they can still be useful for the analysis of global biotic changes, especially in terms of geographic patterns.

Sites from all paleoenvironments are considered (Fig. 2). The great majority of KT boundary sites is located in marine environments (84%); many of them are in deeper water with only microfossils described (30% of all localities). This fact addresses a major shortcoming in all discussions about the KT boundary mass

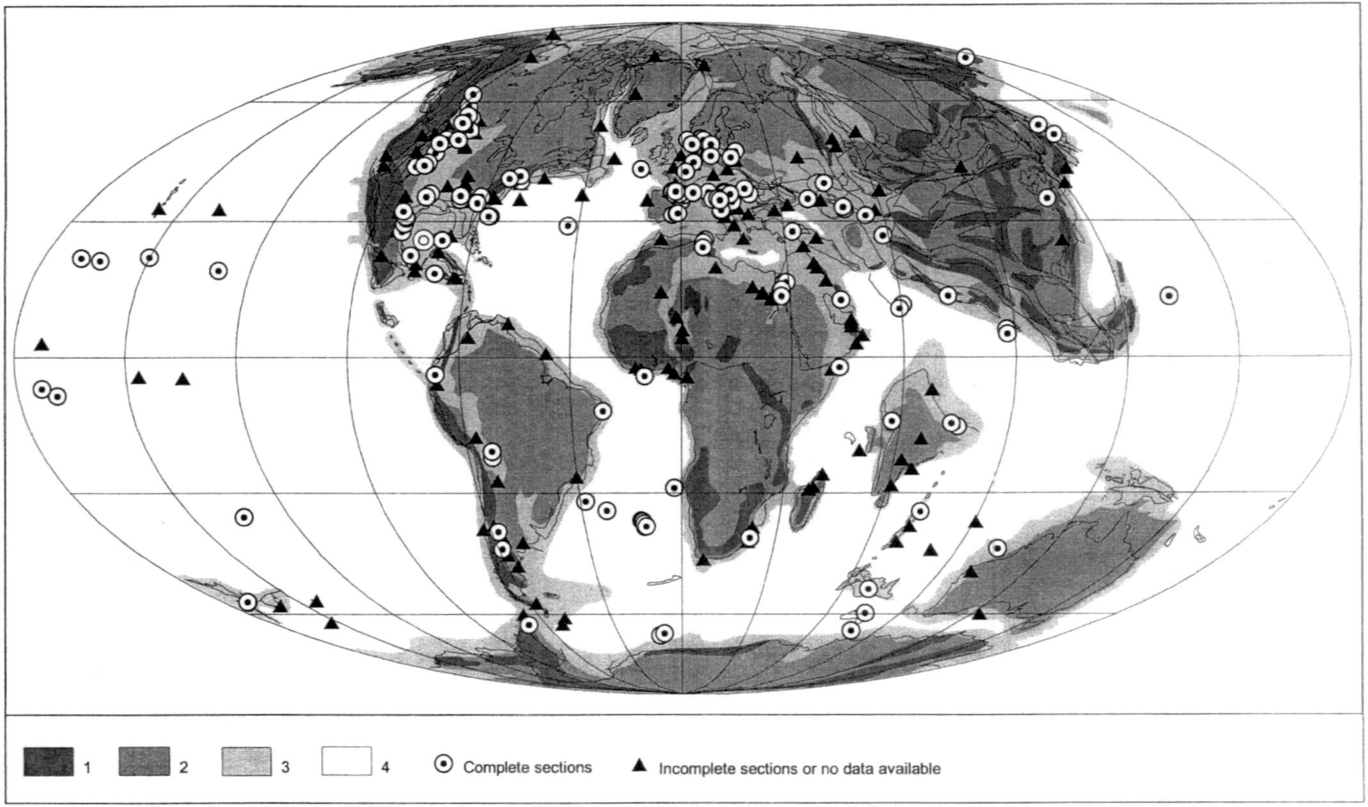

Fig. 1. Stratigraphic completeness of Cretaceous-Tertiary boundary sections plotted on a paleogeographic map based on the 65 Ma plate tectonic reconstruction (Golonka, pers. comm. 1998). 1- Mountains; 2- Lowland; 3- Shelf sea; 4- Ocean. Complete refers to sections with planktonic foraminiferal zones P0 and/or P1a present (marine) or impact tracers preserved (terrestrial and marine). Note that complete sections are highlighted and symbols may cover indications of incomplete sections. The Chicxulub impact site is marked by two concentric circles.

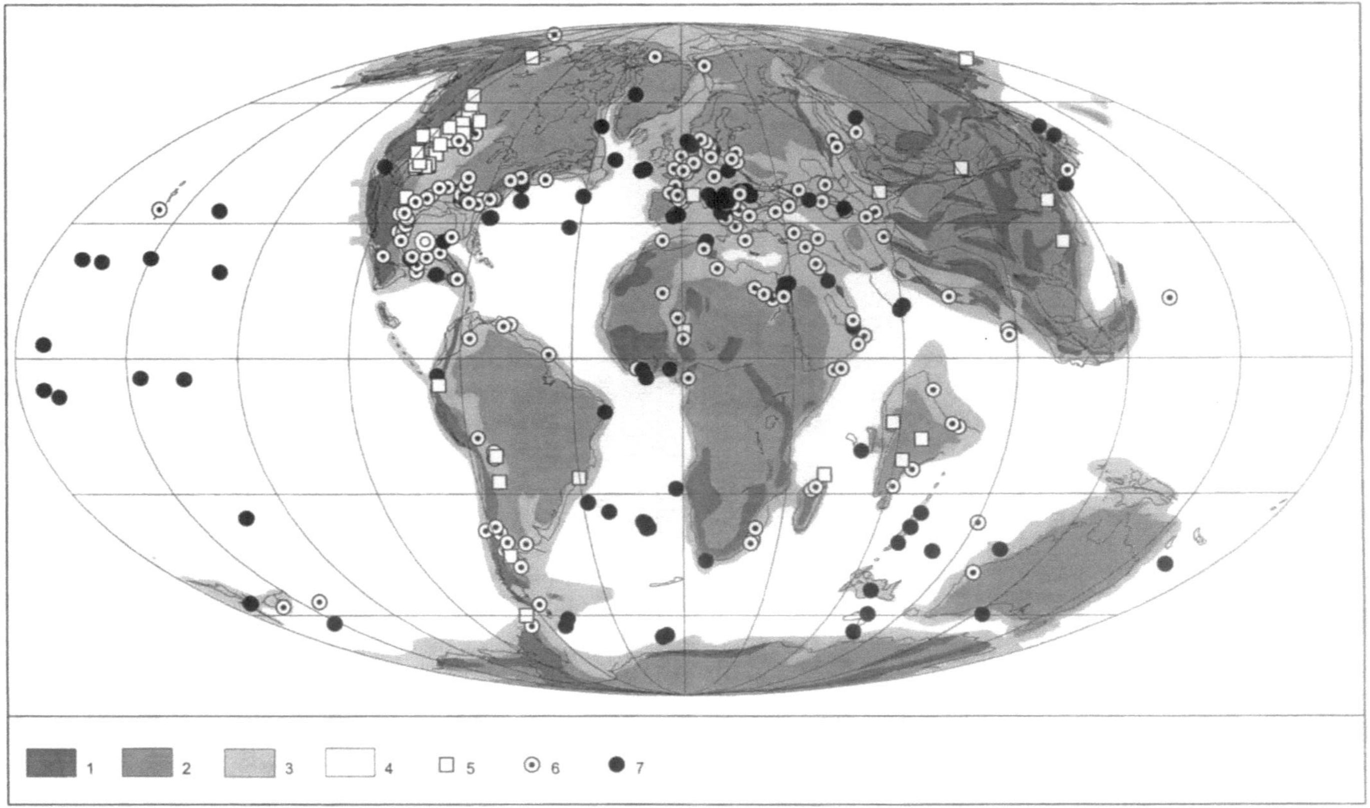

Fig. 2. Paleoenvironmental setting of Cretaceous-Tertiary boundary sections. 1- Mountains; 2- Lowland; 3- Shelf sea; 4- Ocean; 5- Terrestrial section; 6- Neritic to outer shelf section; 7- Bathyal to abyssal section

extinction: Abundant KT boundary localities are known worldwide, but paleontological information is often limited to a few groups of organisms. Fifty-six percent of all marine KT sites have at least moderate or better information on the stratigraphical distribution of calcareous plankton (planktonic foraminifera and nannoplankton). Only 44% of marine sections contain any indication of metazoan invertebrates and as little as 12% of all localities have information on the vertebrate record (including fishes). The picture is getting worse, if the quality of the fossil record is considered. While most of the microfossil data can be regarded as excellent or very good for unraveling the extinction patterns, good data on higher fossil groups are scarce. For example, 10% of all localities have data on the vertical distribution of lophophorates (bryozoans and brachiopods), but only one section (= 0.4%) in northern Denmark (Stevns Klint and Nye Kløv, respectively) exhibits a high quality record that can be used to evaluate the rapidity and intensity of extinction in this fossil group (Surlyk and Johansen 1984, Hakansson and Thomsen 1999). Therefore, it is presumptuous to claim that the actual extinction pattern of lophophorates is known. This is especially true for the spatial pattern in extinction, one of the crucial elements to precisely document an impact-extinction link.

The database is designed to detect global patterns in extinction and ecological changes and for this purpose is less biased than local or regional studies, or investigations limited to only one group of organisms. The GIS approach also aids to reduce the statistical noise that is commonly involved in species level extinction analyses. KTbase currently contains species level data for calcareous nannoplankton, radiolarians, corals, algae, ammonites, bivalves, gastropods, echinoids, bryozoans, and mammals. So far 4000 species are recorded from two or three time slices in the late Maastrichtian and the Danian.

3 Historical Geology

3.1 The Maastrichtian World

Many geoscientists interpret the Maastrichtian as a time of major changes in oceanographic and climatic parameters. What remains to be clearly demonstrated is that they are more substantial than environmental changes taking place in other Cretaceous stages of similar duration.

In general, the Maastrichtian is geologically much related to the Campanian as well as the Danian. The often cited regression at or just prior to the KT boundary (Schmitz et al. 1992; Archibald 1996a, b; Keller and Stinnesbeck 1996) is very minor as compared to the regression that is evident at the Selandian-Thanetian boundary (Haq et al. 1988; Hardenbol et al. 1998). Additionally, the initial data for regression come from the KT boundary section at Braggs, Alabama (Donovan et al. 1988; Baum and Vail 1988; Habib et al. 1992). In this section the sediments, interpreted as a transgressive sequence tract resting on a sequence boundary may instead reflect tsunami deposits related to the nearby Chicxulub impact and thus have no relation to eustatic sea level. Actually Sloss (1963, 1988), when compiling the megasequences on the North American craton, included the Danian in his Zuni Megasequence (Zuni III Subsequence). The complex pattern of sea-level fluctuations for the late Maastrichtian evoked by some authors is based on unconstrained and contradictory data (Speijer and van der Zwaan 1996). Local regression (e.g., Keller et al. 1998) is opposed by local transgressions (Pardo et al. 1999) even in tectonically stable cratonic areas.

The plate tectonic configuration for the Maastrichtian-Danian is well constrained (Fig. 1). The Arctic Ocean was almost completely isolated from the world's ocean (Marincovich 1993). The Atlantic was wide already and during the Maastrichtian, the tectonic sills separating the South from the North Atlantic disappeared (Zachos and Arthur 1999), but deep-water exchange remained limited even in the late Maastrichtian (Kucera and Malmgren 1997). The Drake Passage between South America and the Antarctic was still closed and the separation of Australia and Antarctica just began. Although North and South America were widely separated, deep-water exchange may have been hampered by tectonic ridges related to subduction along the eastern margin of the Caribbean plate (Pindell 1994). Compared to earlier Cretaceous stages, the longitudinal interior seaways were reduced In North America, the western interior seaway became shallower and progressively terrestrial, due to the onset of Laramide Orogeny (Frank and Arthur 1999).

Carbonate platforms were widespread in the northern Tethys. Their real extent was considerably reduced in the late Maastrichtian owing to regional regression or tectonic drowning (Philip et al. 1995). There is no evidence for ocean anoxic events in the Maastrichtian and in comparison to the mid-Cretaceous, oxygen-depleted environments are rare. The oceanic $^{87}Sr/^{86}Sr$ ratio rises in the early Maastrichtian but is more or less constant in the late Maastrichtian (see compilation of Barrera and Savin 1999). The $\delta^{13}C$ signatures of benthic and planktonic foraminifera exhibit a similar pattern, with increasing values in the early Maastrichtian but stabilizing in the late Maastrichtian (Barrera and Savin 1999). The $\delta^{13}C$ gradient between shallow-dwelling planktonic foraminifera and deep water benthic foraminifera declines through the Maastrichtian at many DSDP Sites such as Sites 525, 528 (South Atlantic); Site 690 (Antarctic); and Site 750 (Kerguelen Plateau). The early Maastrichtian gradient of around 2‰ drops to 1‰ in the latest Maastrichtian suggesting a gradual decline in bioproductivity in the southern oceans. However, no such reduction is detected in tropical latitudes or in the northern oceans, demonstrating that no global productivity reduction occurred in the latest Maastrichtian (Barrera and Savin 1999). The $\delta^{13}C$ gradient decline of the late Maastrichtian must then be viewed as local phenomena related to paleoceanographic changes limited to the southern oceans.

The partial pressure of CO_2 (pCO_2) declined through the Cretaceous. Berner (1994) modelled a pCO_2 of about 1.8 relative to the Recent for the Maastrichtian. The major pCO_2 decline is noted in early Late Cretaceous and pCO_2 did not change considerably in the latest Cretaceous stages. In major ocean basins (Pacific,

Indian, South Atlantic), the CCD became shallower to some 2000 m in the early Maastrichtian (Van Andel 1975), whereas it dropped by nearly 3000 m in the North Atlantic (Thierstein and Okada 1979; Kaminski et al. 1999).

A significant cooling trend in the Campanian-Maastrichtian oceans was recognized by Barrera and Savin (1999). As cooling was more pronounced in high latitudes, the latitudinal thermal gradient increased significantly in the Maastrichtian (Barrera and Savin 1999) compared to earlier Cretaceous times. Mean temperatures were possibly lower in the early than in the late Maastrichtian. Miller et al. (1999) even suggested that glaciation may have occurred in the early Maastrichtian. Probably in relation to the onset of Deccan Trap volcanism, surface and intermediate waters warmed by 2-4° C some 0.5 Ma before the KT boundary (Li and Keller 1998; Kucera and Malmgren 1998; Barrera and Savin 1999). In parts of North America and in India, some evidence was presented for increasing aridity in the latest Maastrichtian (Buck and Mack 1995; Khadkikar 1999).

Besides the Indian Deccan traps and perhaps the South Atlantic Walvis Ridge (Courtillot 1990; Courtillot 1999), there is no evidence for enhanced volcanism in the Maastrichtian. Sea floor spreading rates were decreasing during the Upper Cretaceous (Gaffin 1987) and volcanism related to convergent tectonics was not markedly different from other Late Cretaceous stages (Golonka, pers. comm 1999).

Maastrichtian deep water was possibly generated in the northern North Atlantic and the North Pacific (Frank and Arthur 1999). Barrera et al. (1997) suggested strong undulations in the thermohaline circulation regime of the world's oceans during the Maastrichtian and Frank and Arthur (1999) proposed that a major reorganization of the oceanic circulation patterns occurred at the early/late Maastrichtian boundary (see also Saltzman and Barron 1982). All these suggestions are well constrained and may be an explanation for most of the background extinctions observed within the Maastrichtian. The near-extinction of inoceramids during the mid-Maastrichtian can be explained by enhanced production of cold, oxygen-rich deep waters in polar latitudes preferentially affecting this low-oxygen adapted bivalve group (MacLeod et al. 1996). Deep-water benthic foraminifera also show pronounced changes at the early/late Maastrichtian boundary (Thomas 1990).

In spite of the mid-Maastrichtian oceanic events, there is no hard evidence supporting unusual stress for either the marine or the terrestrial biosphere in the Maastrichtian. Opposing the view of Glasby and Kunzendorf (1996), there is neither evidence for a significant long-term drop in sea-level or a frequent development of anoxia in the Maastrichtian, nor is the gradual decline in pCO_2 likely to cause severe stress to the biosphere.

3.2 The Danian World

Much less literature is available on the Danian as compared to the Maastrichtian or Thanetian. The Danian world in general appears much like the Maastrichtian, except for the first 500 ky when the global biosphere recovered from the KT event. Formerly, the Danian was often included in the Cretaceous system owing to faunal similarities with the Maastrichtian and differences from the Selandian (see e.g., discussions in Rosenkrantz 1960; Voigt 1960; Yanshin 1960). No significant sea-level changes are noted in the Danian, but a major regression occurs at the Selandian (formerly included in the Danian)-Thanetian boundary (Haq et al. 1988; Hardenbol et al. 1998). Climate was similar to the Maastrichtian for most of the Paleocene until an abrupt warming occurred at the Paleocene-Eocene transition (Bains et al. 1999). This trend is demonstrated by oxygen-isotope analysis of both deep water (Miller et al. 1987), and shallow water foraminifera from various paleolatitudes (Frakes et al. 1994) and leaf margin and size analysis (Wolfe and Upchurch 1987). Frakes et al. (1994) data compilation suggests that although tropical surface waters were slightly cooler in the Danian than in the Maastrichtian, high latitude (60°) waters were warmer, and hence the temperature gradient was considerably reduced in the Danian. The pCO_2 was about the same or slightly higher than in the Maastrichtian (Berner 1994).

The chemistry of the oceans was not significantly different from the Maastrichtian. Calcite cements prevail possibly indicating a low Mg/Ca ratio in seawater (Hardie 1996). The $^{87}Sr/^{86}Sr$ spike reported from the basal Danian (e.g., MacDougall 1988) is still disputed (McArthur et al. 1998, Vonhof and Smit 1997, MacLeod et al. 2001). The Danian strontium isotopic ratio is similar to or slightly higher than in the late Maastrichtian. Apart from the short-term decline in the basal Danian most likely linked to the KT event, the $\delta^{13}C$ isotope record exhibits a positive trend from the latest Cretaceous to the Paleocene (Veizer et al. 1999).

4
Distribution Patterns of Impact Ejecta Debris

The Chicxulub impact event has produced a broad range of ejecta material, which can be easily located in almost any complete Upper Cretaceous section of the stratigraphic record, thanks to its close association with the KT boundary faunal changes. After years of debates, most authors agree today that the ejecta debris and geochemical signals, coincide exactly with the mass extinction horizon and are located precisely at the KT boundary forming at distal sites from Chicxulub, the famous KT boundary clay layer. When completed, KTbase will contribute significantly to our understanding of impact ejecta distribution on Earth and the other planets, and might permit to link impact debris found at other levels in the sedimentary record with a specific crater or vice versa.

The KT boundary clay is marked by a strong enrichment in Ir. Several studies have demonstrated that the KT boundary clay is enriched in all the platinum group elements (PGE) including Pt, Pd, Os, Ru (Kyte et al. 1980; Turekian 1982; Evans et al. 1993). KTbase contains essentially Ir data as most authors have reported the geochemical anomaly in terms of Ir concentration because this element is easier to analyze by neutron activation methods than the other PGE. KT confirms that the Ir anomaly is indeed global (Fig. 3). It is detected in all known parts of the Late

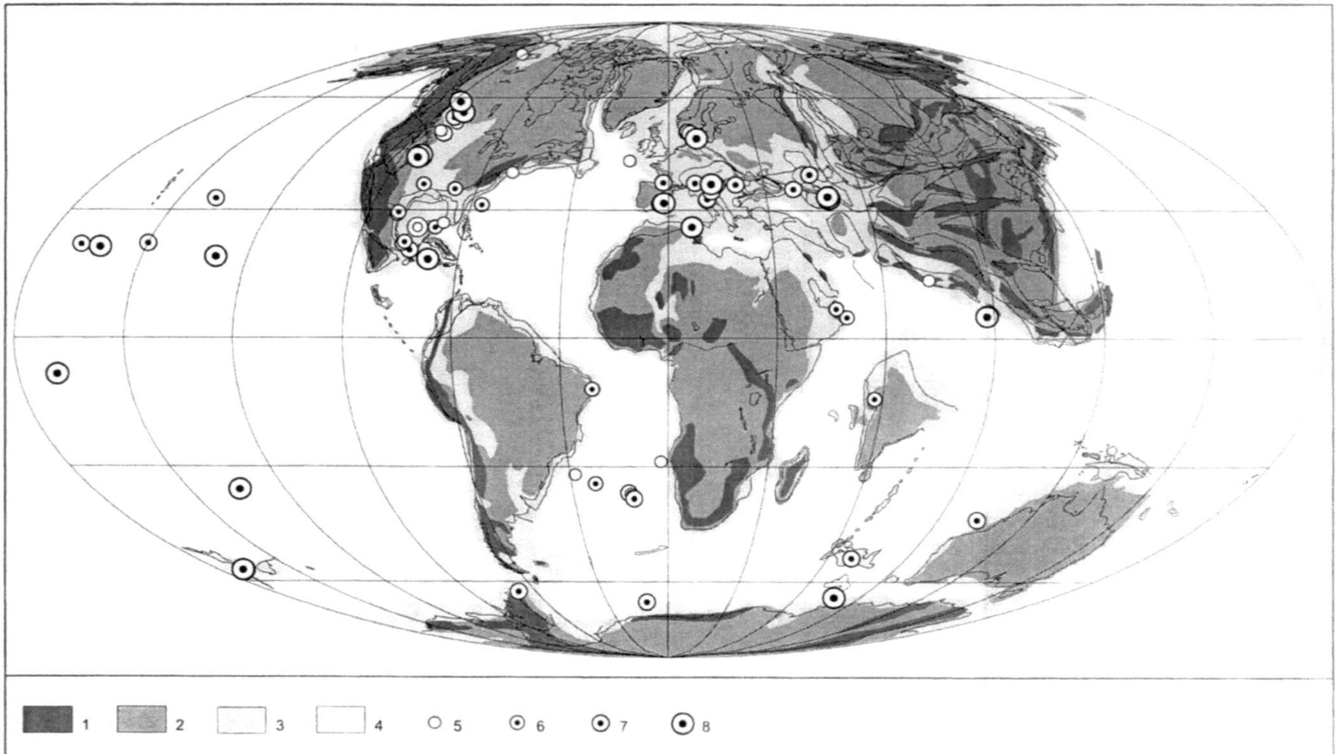

Fig. 3. Maximum iridium concentrations measured in K/T boundary sites. 1- Mountains; 2- Lowland; 3- Shelf sea; 4- Ocean; 5- Iridium anomaly indicated but no precise data available; 6- less than 1.5 ppb; 7 - 1.5 - 10 ppb; 8- More than 10 ppb. Note absence of a distinct pattern in iridium concentrations in relation to Chicxulub and global distribution of iridium. Concentrations around the crater are lower owing to high dilution.

Cretaceous world, including the high latitudes. So far, it was measured in eighty-four KT boundary sites, covering a broad range of depositional environments from deep marine to continental settings (Fig. 3). The concentration ranges from 0.5 to > 50 ppb and does not seem vary with distance from the impact site, even when normalized for sedimentation rates. The Ir from the impacting bolide was transported to the upper atmosphere by the fireball raising from the crater. The distribution of Ir indicates that, after the impact the whole Earth was engulfed in a nearly homogenous cloud of vapor and dust particles.

The high-pressure shock wave generated by a meteorite impact produces in quartz several sets of very fine glassy silicate lamellae known as planar deformation features. Shocked quartz are found in abundance in the KT boundary sediments in the US Western Interior, in Pacific ocean deep sea cores (Bohor et al. 1984; Bohor et al. 1987; Izett, 1990; Kyte et al., 1996), at sites in the Western Atlantic (Klaver et al., 1987; Norris et al., 1999) and Brazil (Albertão and Martins, 1996). The KT sites in the western interior of North America contain abundant larger than 0.5 mm in size shocked quartz grains (Bohor et al. 1984; Pilmore and Flores 1987; Izett 1990; Bohor 1990). Shocked quartz are also abundant but rarely reach size over 100 microns in all the seven drill holes on the Pacific Plate in which the KT boundary has been found (Bostwick and Kyte 1996; Kyte et al. 1996). West of the crater KT shocked quartz grains are found as far as 10,000 km from Chicxulub crater, at ODP Site 596 in the southwestern Pacific. Although located at comparable distance from the crater, the European and North African KT boundaries have yielded fewer, and generally very small sized, shocked quartz grains (Bohor and Izett 1986; Montanari 1991). This skewed distribution was first pointed out by Alvarez et al. (1995) and is confirmed by this study (Fig. 4). Alvarez et al. (1995) attributed this asymmetric distribution of the shocked quartz to the rotation of the Earth which affect differently the ballistic trajectory and orbit of the eastbound and westbound particles (see Fig. 1 in Alvarez et al. 1995 and Alvarez 1996). Schultz and D'Hondt (1996) viewed the shocked quartz distribution as the result of an oblique impact originating from the southwest.

The shocked minerals show a clearly different distribution pattern than the fireball Ir and cosmic spinels. Most of the bolide Ir remains in the upper part of the rapidly raising and expanding fireball until it is spread homogeneously into the upper atmosphere and distributed globally. The shocked grains originated from the deeper Yucatan basement (> 3 km) and most likely were transported on steep, high-velocity ballistic trajectories.

Ni-rich magnesioferrite cosmic spinels characterize most KT sites worldwide (Robin et al. 1991, 1992). Robin et al. (1993) have pointed out geographic differences in the spinel chemistry. Spinels from the Pacific ocean are enriched in Al and Fe^{3+} compared to their counterparts in Europe and North America. The origin of the chemical segregation of the spinel is unclear (see arguments in Robin et al. 1994, Kyte and Bohor 1995).

Two types of spherules have been identified at the KT boundary. Smit et al. (1992) have provided an extensive description of the two types. The first spherules are altered microkrystites, composed initially of clinopyroxene and rimmed with Ni-rich spinels. They are always well rounded, usually less than 400 μm in size

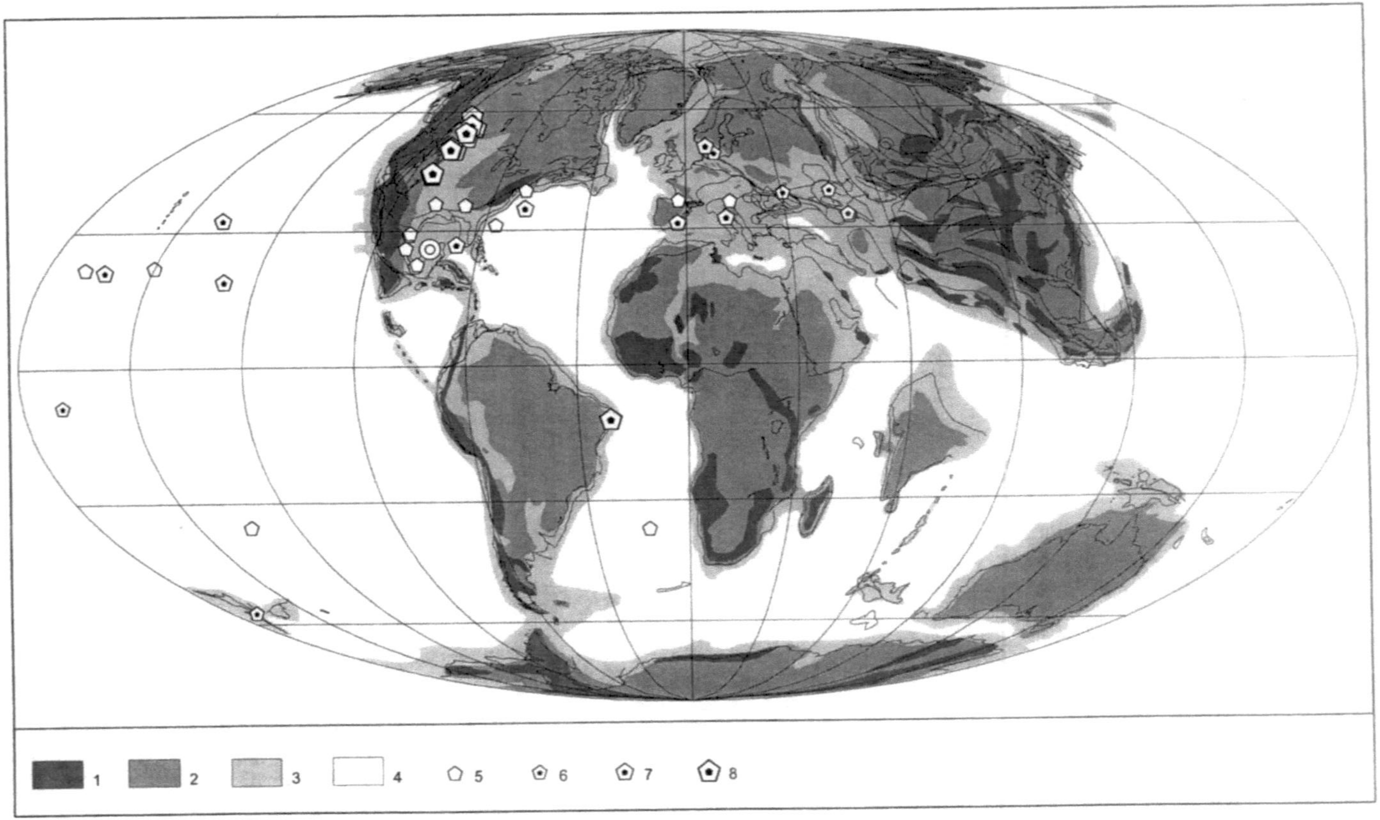

Fig. 4. Shocked quartz distribution and sizes in K/T boundary sites. 1- Mountains; 2- Lowland; 3- Shelf sea; 4- Ocean; 5- Shocked quartz present, but no maximum size indicated; 6- Maximum diameter less than 150 μ m; 7- 150-300 μ m maximum diameter; 8- More than 300 μ m maximum diameter. Note abundance of large shocked quartz crystals in North America and in the North Pacific. No shocked quartz was reported from Africa, India, eastern Asia and Australia.

and formed of internal crystallites. They occur worldwide, so far there is no clearly documented variation in abundance pattern or size with distance from Chicxulub.

The second type of spherules is composed of smectite, chlorite, and/or montmorillonite and is derived form the alteration of impact glass originating from the Chicxulub crater. They are larger up to 2 –3 cm in size, and have rounded but also often elongated splash-form shapes reminiscent of tektites morphologies. These spherules are limited to within a 4000 km radius from Chicxulub, at KT boundary sites in Haiti, Southern Mexico, Northeastern Mexico, Texas and Alabama, the US Western Interior and at DSDP sites in the Northwestern Atlantic. At sites in Haiti and Mexico, these spherules can contain a preserved impact glass core (Izett 1990; Sigurdsson et al. 1991; Smit et al. 1992; Claeys et al. 1993).

Microscopic diamonds were painstakingly extracted out of the KT boundary clay in the US Western Interior and in NE Mexico (Gilmour et al. 1992; Hough et al. 1997). So far not enough information exists on the exact geographic distribution of impact diamond at the KT boundary for KTbase to draw conclusions concerning their distribution pattern, transport and global occurrence. The soot believed to originate from global KT boundary wildfire seems to be widespread (Wolbach et al. 1988). However, fire appears to an important component of latest Cretaceous terrestrial ecosystems (Scott et al. 2000) and many charcoal fragments in the boundary layer show evidence of biodegradation prior to being coalified (Jones and Lim 2000). More details about the distribution of the Chicxulub ejecta will be the subject of another article when more information is entered in KTbase.

5 Marine Extinction Patterns

The most recent update on extinction rates for higher marine animal taxa was given by Sepkoski (1996). According to his data there is a 47% ± 4.1% loss in genera, a 39% ± 2.2% loss in multiple-interval genera, and a 16% ± 1.5% decline in families at the KT boundary. This makes the KT ranking at the fifth place of the Phanerozoic "Big Five" mass extinction events. However, 76 ± 5% extinction in global marine species was cited by Jablonski (1991) based on rarefaction from generic data. This indicates that on a species level, the KTB ranks equal to the Triassic/Jurassic mass extinction but still well below the mass extinctions in the latest Ordovician, and at the Frasnian-Famennian and Permian/Triassic boundaries.

In the following chapters, we present an evaluation of the magnitude of extinction for major marine invertebrate clades based on a comprehensive literature survey. We also try to judge the quality of data on which the extinction rates are based. We especially emphasize the paleogeographic distribution of the information and occasionally the paleogeographic aspects of extinction patterns to provide a global and comprehensive view of the biotic changes.

5.1 Phytoplankton

5.1.1 Calcareous Nannoplankton

The KT boundary event eliminated 88% of calcareous nannoplankton species, 81% of the genera and 48% of the families (calculated from KTbase). These extinction values are resting on firm grounds owing to limited taxonomic inconsistencies and the global distribution of good KT section with nannoplankton data (Fig. 5). The nannofossil extinctions are widely accepted as rapid and no significant difference in extinction rate between high and low latitudes is observed (Pospichal 1996b, Gardin and Monechi 1998) and no significant turnover prior to the KT boundary is evident (Gartner 1996). Most scientists analyzing KT nannoplankton changes agree that the extinction is catastrophic and fits the impact scenario (Jiang and Gartner 1996; Pospichal 1996a, b; Henriksson 1996; but see Perch-Nielsen et al. 1982). Cretaceous species found in Danian sediments are usually interpreted as reworked (Pospichal 1994, Henriksson 1996; Gardin and Monechi 1998), but survivorship cannot be disregarded.

Information on the taxonomic composition and relative abundance of nannoplankton floras is stored in KTbase using three time slices: (1) CC 26 nannofossil zone (latest Maastrichtian); (2) NP 1 (earliest Danian); (3) NP 2-4 (early to middle Danian). By choosing these time slices we are able to evaluate the pre-extinction healthy ecosystem (CC 26), the devastated post-impact ecosystem (NP 1) and the recovery ecosystem (NP 2-4). The information in each time slice was averaged from quantitative data in the literature. Species level nannoplankton data for 54 localities are currently listed in KTbase. For 43 localities, detailed species counts are available but only 32 sections provided data on all three time slices. Extinction was uniformly intense around the globe, no distinct pattern is evident. Globally, a total of 229 species are recorded in CC 26 and nearly as much in NP 1 (207). The great majority (79%) of species found in NP 1 are Cretaceous species. Only very rare species in the latest Maastrichtian have never been reported in the earliest Danian. In NP 2-4 zones 212 species are globally reported, 62% of which are still Cretaceous species. All Cretaceous species, however, decrease rapidly in abundance at all localities, except where reworking is evidently substantial (Huber 1996).

Opportunistic survivors and newly evolved Danian taxa successively bloomed following the event. The fact that blooming occurred at different times and different places, may indicate unstable and variable oceanographic conditions in the post KT oceans (Fig. 6). The recovery of nannoplankton as defined by the relative abundance of incoming (= newly evolved) species shows a complicated pattern. The Prinsiaceae, a typical Tertiary family, form an important Paleocene stratigraphic marker (Wise, pers. comm.. 1999). The relative abundance of species in this family exhibits a distinct geographic-temporal pattern (Fig. 7). Prinsiceaens are well diversified and abundant already in NP1 in the western Mediterranean area, in India and in the southern Ocean. However, in the Gulf of Mexico region and, especially, the Pacific the family appears to occur significantly later and to be less abundant in the early Danian. If the evolution and success of the Prinsiaceae are considered markers of recovery, we must acknowledge a delayed recovery of calcareous nannoplankton in these regions.

5.1.2 Other Phytoplankton

Dinoflagellates enter the fossil record in the form of either organic-walled cysts or so-called calcispheres. The organic-walled cysts are known to have a patchy fossil record that is mostly limited to organic-rich deposits. Therefore, reliable extinction patterns can rarely be determined The most detailed work was done in Seymour Island, Antarctica (Askin and Jacobson 1996), Alabama and Georgia, USA (Habib et al. 1992, 1996; Moshkovitz and Habib 1993), Austrian Alps (Kuhn and Kirsch 1992) and El Kef, Tunisia (Brinkhuis and Zachariasse 1988). No significant extinction has been noted in any of these sections. Maastrichtian and/or Danian organic-walled dinoflagellate cysts are known from a variety of other localities, but are mostly not useful for the KT discussion. Variations in diversity and taxonomic composition are usually interpreted to follow regional sea-level fluctuations rather than being related to any global event. What is remarkable about dinoflagellates is that they record the short term cooling after the KT event as well as the subsequent warming in subtropical latitudes (Brinkhuis et al. 1998).

The calcitic cysts of dinoflagellates are commonly treated with calcareous nannoplankton owing to similar preparation methods applied to both groups. Mesozoic and Cenozoic calcispheres, however, are clearly dinoflagellates. Here, we mention only studies that explicitly refer to calcispheres. Other references are discussed in the nannoplankton section. Detailed studies on the vertical distribution of calcispheres through the KT boundary are available from the Maastricht area in the Netherlands (Willems 1996), Northern Germany and Denmark (Kienel 1994) and from the Weddell Sea (Fütterer 1990). While Willems (1996) recorded no significant changes in floral composition, Fütterer (1990) reported a complete turnover in the calcispheres and even regarded the often cited survivor *Thoracosphaera operculata* as a newly evolved species. Kienel (1994) came to the conclusion that most calcispheres are survivors, a few disappeared before or after the KT boundary and many new species evolved in the Danian. The most notable feature in calcispheres is the occasional bloom of *Lentodinella danica* and *Obliquipithonella (=Thoracosphaera) operculata* and related forms directly following the KT boundary event (Fig. 6).

Very little is known about diatoms in general and planktonic diatoms in particular at the KT boundary. The same is true for silicoflagellates. Only one KT section in Antarctica yielded some data (Harwood 1988) suggesting that diatoms were little affected by the KT event in terms of extinction rate. However, an ecological signal is evident by the increase of resting spores at the KT boundary in Antarctica (Harwood 1988) and a relative increase of diatoms with respect to radiolarians (Hollis et al. 1995) in New Zealand.

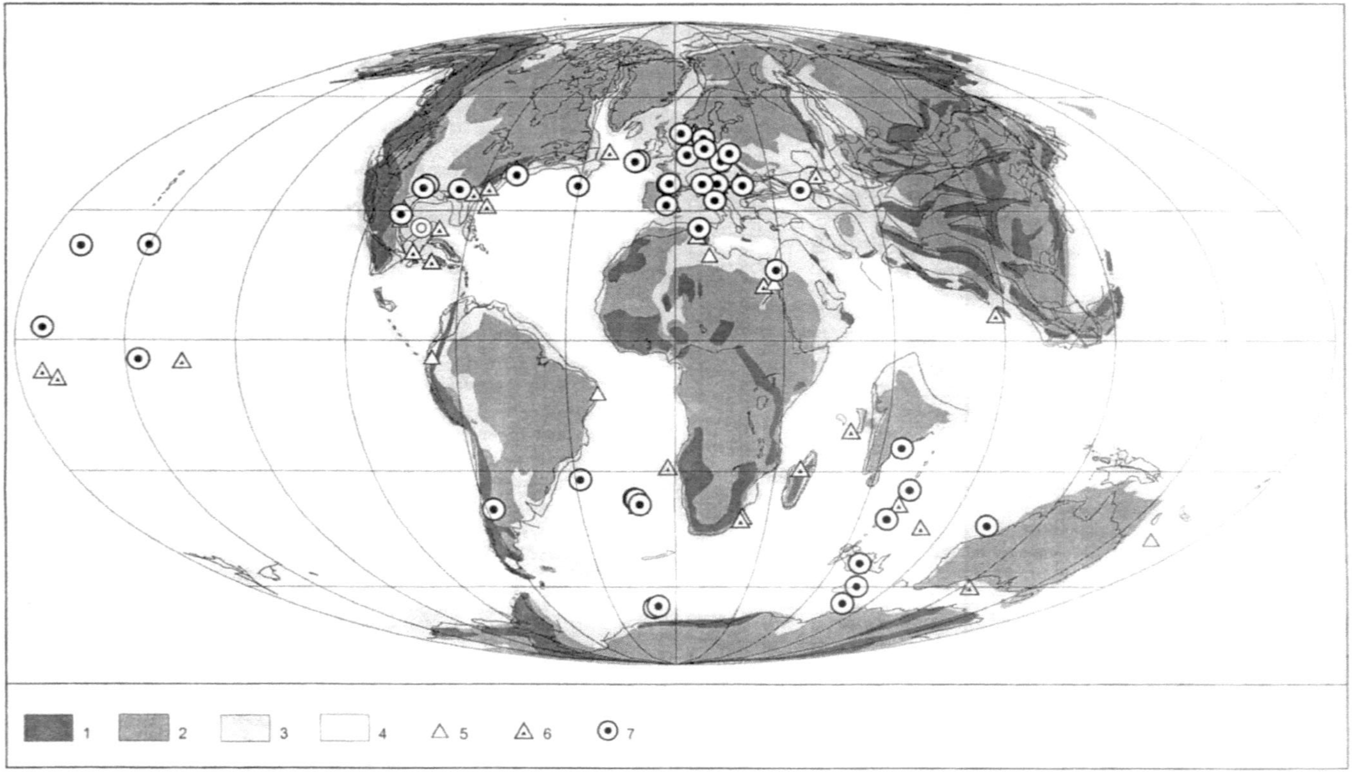

Fig. 5. Geographic distribution of nannoplankton data across the K/T boundary. 1- Mountains; 2- Lowland; 3- Shelf sea; 4- Ocean; 5- Poor record or very limited data available; 6- Moderate data with qualitative distribution data across the boundary or on one side of the boundary; 7- Good data with detailed quantitative data from several horizons across the boundary.

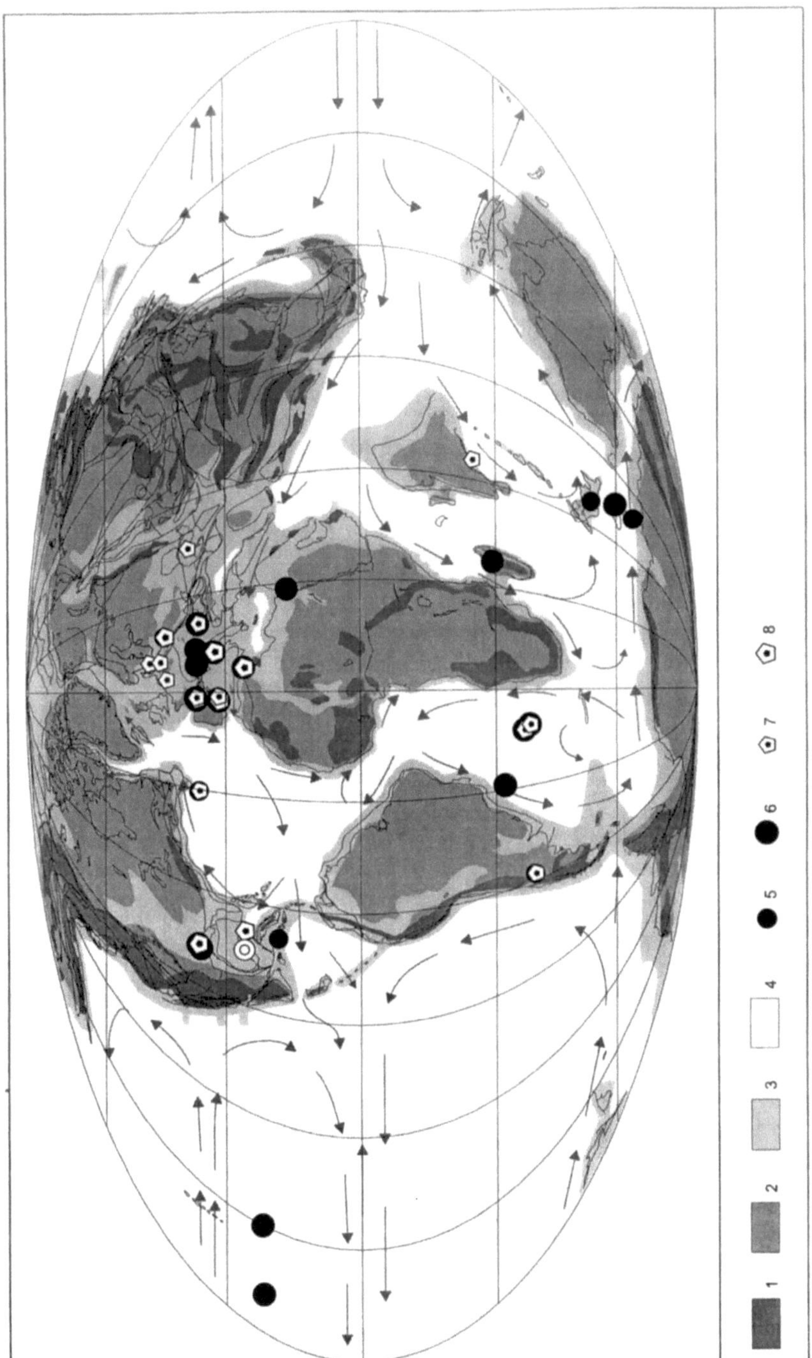

Fig. 6. Spatial distribution of calcisphere (*Thoracosphaera* = *Obliquipithonella*) and *Braarudosphaera* blooms in nannoplankton zone NP1. Blooms occur at different times and different places indicating a complex and unstable oceanic system. 1- Mountains; 2- Lowland; 3- Shelf sea; 4- Ocean; 5- Moderate calcisphere blooms; 6- Strong calcisphere blooms; 7- Moderate *Braarudosphaera* blooms; 8- Strong *Braarudosphaera* bloom.

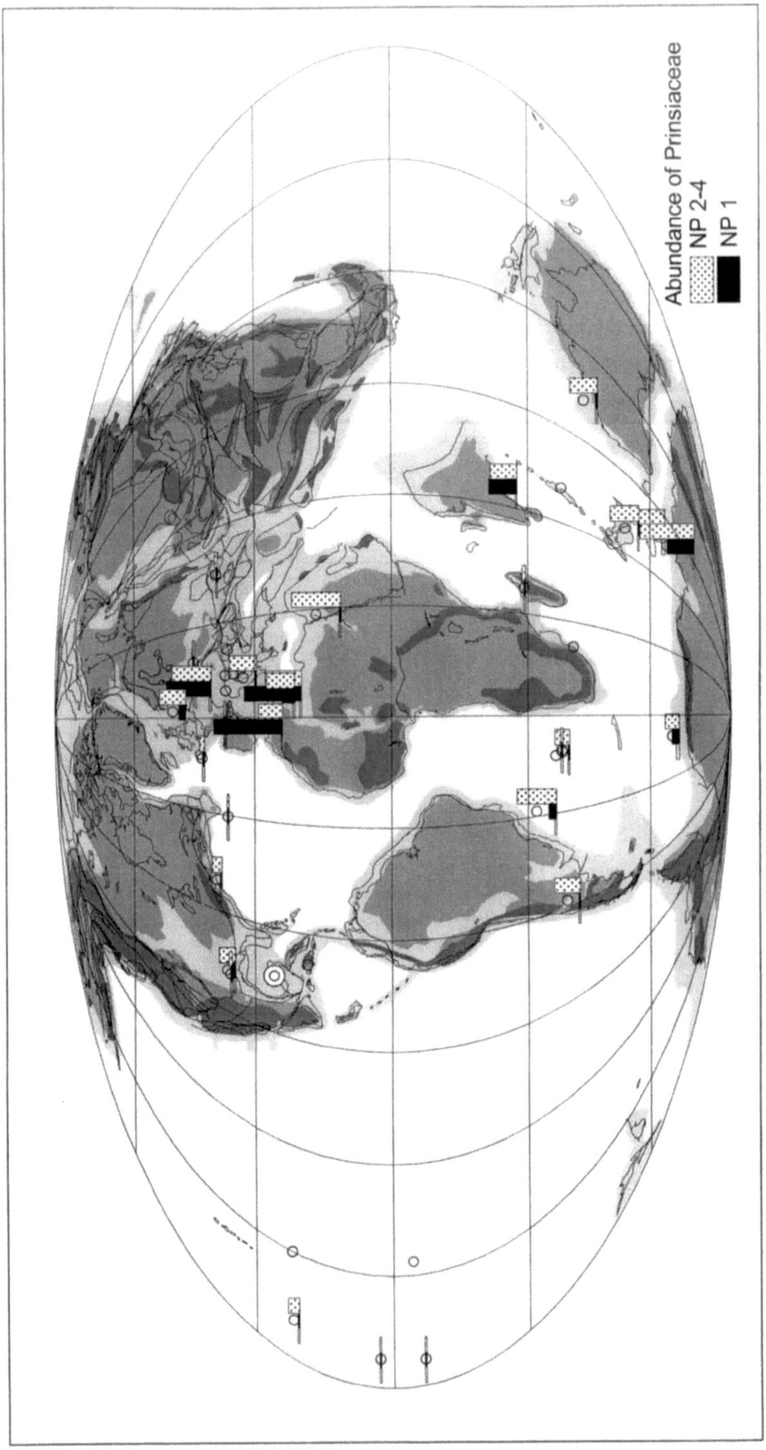

Fig. 7. Origination pattern of the Tertiary nannoplankton (coccolithophorid) family Prinsiaceae. Height of bars indicates abundance of this family in the indicated nannoplankton zones. Note obviously delayed recovery in the Gulf of Mexico area and the Pacific.

5.2 Zooplankton

5.2.1 *Planktonic Foraminifera*

Extinction patterns from planktonic foraminifera are probably the best documented of all organisms at the KT boundary. High resolution data exist for sections covering virtually all marine environments and major regions of the Earth. Planktonic foraminifera are said to suffer the most dramatic species extinctions among marine organisms across the KT boundary (Canudo et al. 1991).

In spite or perhaps because of the good data there is, however, no consensus as to how the extinction pattern should be interpreted. Major discrepancies concern the rapidity of extinction and the percentage of species going extinct exactly at or shortly after the KT boundary. Published extinction ratios vary between 39% (Pardo et al. 1999), 70% (Molina et al. 1998) and more than 90% (D'Hondt et al. 1996).

For a non-specialist, it is very difficult to follow the discussions and concerns around the extinction patterns of planktonic foraminifera. Several impact-critical reports cite multiple extinction waves, prior, at, and subsequent to the KT boundary (stepwise extinction, Keller 1988a, 1989; MacLeod 1996; MacLeod et al. 1997; Abramovich et al. 1998), whereas classical studies and impact-supporters (Bramlette 1965; Herm et al. 1981; Luterbacher and Premoli Silva 1964; Smit 1982; Olsson and Liu 1993; Smit 1990) endorse a massive extinction exactly at the boundary. The results of the so-called El Kef blind test (Ginsburg 1997) are disappointing in that they could not clarify the problems and proved the still heterogeneous taxonomic concepts. Unfortunately, this experiment does not fully qualify as an unbiased blind test (Aubry 1999) and there is no clear discussion of the results presented so far in literature.

The comparison of planktonic foraminifera pattern with those of calcareous nannoplankton, exhibits no major differences: (1) Few extinctions exist prior to the KT boundary and the late Maastrichtian faunas were very diverse and highly structured both geographically and ecologically; (2) a major extinction event is located more or less exactly at the boundary (depending on stratigraphic resolution and reworking of fossils), although many of the vanishing species can be found in the earliest Danian in rapidly decreasing abundance; (3) some species tend to reach high abundance in the earliest Danian; and (4) recovery is mostly characterized by newly evolving species and was fairly rapid.

The ongoing extinctions in early Danian planktonic foraminifera are often cited as evidence against an impact-triggered extinction (Keller 1989; MacLeod et al. 1997). However, post-impact extinctions can well be related to the impact itself, although no direct physical effects of the impact may have persisted longer than a few decades (Pope et al. 1994). However, the induced perturbation may still affect the climate and especially the oceans long after the direct physical effects of the impact have disappeared. This is particularly clear in view of the recent data showing that the mode of oceanic circulation is fragile and subject to rapid changes (von Blankenburg 1999). Moreover, the loss of keystone species could have had long-lasting effects on the biosphere. Recovery in planktonic foraminifera took approximately the same time as the recovery in the calcareous nannoplankton but obviously was dependant on paleolatitude. The incoming Paleogene fauna dominated assemblages in Zone P1a in low and middle latitudes but not earlier than P1b to P1c in high latitudes (MacLeod and Keller 1994).

5.2.2 Radiolarians

The evaluation of radiolarian extinction, though based on only one publication (Foreman 1968), often indicated a major decline in radiolarian populations (Thierstein 1982). However, detailed work on sections in New Zealand (Hollis et al. 1995; Hollis 1997) and Ecuador (Keller et al. 1997) have shown that virtually no radiolarians went extinct at the KT boundary. Nevertheless, all of these studies indicate an ecological change in the radiolarian faunas. This is exemplified by increasing radiolarian abundance and a spumellarian bloom in the early Danian suggested to indicate increasing primary productivity across the boundary (Hollis 1996, 1997).

The knowledge of vertical radiolarian distribution patterns across the KT boundary must be regarded as insufficient. All sections with good radiolarian data are located in the Pacific region (Fig. 8). Only few radiolarian-bearing sections are known from other regions and none of them records a complete record across the boundary. In addition to the references above and the comprehensive references provided by MacLeod et al. (1997) only one paper is noteworthy. Bubik et al. (1999) and Bak (1999) described a diverse radiolarian fauna in latest Maastrichtian Flysch (*Micula prinsii* zone) of the Czech Carpathians. This assemblage is remarkable since it is the only record of Tethyan late Maastrichtian radiolarians. Faunal similarity with the intermediate-high latitude faunas is limited Many species reported by Bak (1999) are currently unknown from Paleocene strata. If it turns out that those actually vanished at the KT boundary, the global species extinction rate would rise considerably (28% generic extinction). Current data suggest that radiolarian extinction was weak or delayed in high-nutrient settings (upwelling, high latitudes), but higher in oligotrophic environments.

5.2.3 Plankton Summary

Although the only phytoplankton group experiencing significant extinctions at the K/T boundary is the calcareous nannoplankton (especially coccolithophorids), nearly all plankton groups considered here exhibit pronounced and rapid changes in their assemblages coincident with the KT boundary. These changes are exemplified by very high extinction rates (calcareous nannoplankton, planktonic foraminifera) or ecological changes (diatoms, radiolarians, dinoflagellates).

Owing to the small size of the micro- and nannoplankton considered here, many secondary biases on their fossil record are possible. The most widely cited bias is perhaps reworking due to oceanic currents or bioturbation. Many occurrences of Cretaceous calcareous nannoplankton species may actually be due to reworking, as reworking is commonly observed in the nannofloras (Pospichal 1994, 1996b). However, for planktonic foraminifera. MacLeod (1996) provided convincing evidence that some of the Cretaceous species actually survived into the earliest Danian, which does not contradict an abrupt extinction event. What is remarkable is that the amount of those survivors decreases markedly upward in almost all studies.

5.3 Benthic Foraminifera

Benthic foraminifera are commonly cited as having gone through the KT bottleneck without significant extinctions (Beckmann 1960; Miller 1982). However, in terms of extinction rate, there is a great deal of selectivity apparently related to the habitat and ecology of benthic foraminifera (Lipps and Hickman 1982; Kaiho 1992; Coccioni and Galeotti 1994, 1998). Larger foraminifera thriving in Maastrichtian shallow carbonate platforms are strongly affected by the KT boundary extinction (Tambareau et al. 1997; Hottinger 1997). Benthic foraminifera living in the upper part of the intermediate water (upper bathyal) have been reported to exhibit very low extinction rates (10 – 25%: Kaiho 1992; Speijer and van der Zwaan 1996), although environmental collapse may result in the transient disappearance of more than 50% of species (Keller 1988b; Speijer and van der Zwaan 1996). Calcareous benthic foraminifera living in deep water depths experienced low total extinctions (Dailey 1983; Thomas 1990; Nomura 1991; Widmark and Malmgren 1992). Agglutinated foraminifers exhibit a significant decrease in diversity and abundance (Kuhnt et al. 1998). Even for deep-water benthic assemblages environmental collapse related to the KT event has been reported (Coccioni and Galeotti 1994). Although the agglutinated foraminifer decline started below the KT boundary, the subsequent recovery pattern appears to be linked to recovery in the phytoplankton (Kuhnt and Kaminski 1993).

Although data on extinction patterns for benthic foraminifera are globally available (Fig. 9), information on larger foraminifera at the KT boundary are scarce. Late Maastrichtian assemblages are known from the Southern Alps (Premoli Silva and Luterbacher 1966), the Northern Alps (Tragelehn 1996), the Pyrenees (Tambareau et al. 1997), the Carpathians (Köhler and Borza 1984), the Dinarides (Drobne and Barattolo 1995), Iraq (Al-Ameri and Lawa 1986), Guatemala (Fourcade et al. 1997), Turkey (Sirel 1996) and Somalia (Cerchi et al. 1993). All these studies reported fairly diverse orbitoid assemblages in the late Maastrichtian. Larger foraminifera were abundant and diverse in late Maastrichtian carbonate platforms and there is no evidence for a significant diversity decline prior to the KT boundary. Only one Maastrichtian genus (*Laffitteina*) survived the KT boundary and the full recovery of larger foraminifera did not take place before the Eocene (Hottinger 1997). MacLeod et al. (1997) state that no obvious geochemical or isotopic anomaly is associated with the extinction of any Maastrichtian larger foraminifer taxon. Our recent observations show the disappearance of a rich larger foraminiferal fauna in platform carbonates of Southern France (Larcan section) directly associated with the iridium anomaly (Rocchia, pers. comm. 1999).

5.4 Ostracods

Ostracods are described from a number of KT sections. However, little detailed information on extinction patterns is available. Few extinctions were reported from the Arctic (Brouwers and De Deckker 1993), but intense environmental change and high endemism obscures the pattern. In low and mid latitude faunas a progressive pattern of faunal replacement during the Maastrichtian is often favored (Maddocks 1985; Whatley 1990) and extinction intensity at the KT boundary itself does not appear to be strongly enhanced (Braccini and Peypouquet 1995). Rare citations indicating abrupt changes and significant extinctions at the KT boundary (Liebau, 1982, Bertels-Psotka 1995), should be seriously considered. Ostracod diversity and density is rising in the late Maastrichtian at some ODP sites (Majoran and Widmark 1998; Majoran 1999) and stable Maastrichtian diversities were reported from Denmark (Jorgensen 1979). Hence, if future studies verify significant extinctions, those should have taken place directly at the KT boundary.

5.5 Molluscs

5.5.1 Cephalopods

Ammonites are long known to be victims of the KT event. Former studies suggested that ammonites became extinct well before the KT boundary (Wiedmann 1973 1988, Ward 1983). Conservative approaches state that ammonites were well in decline prior to the KT boundary as exemplified by a diversity decline and increasingly patchy distribution in the during the Late Cretaceous (Wiedmann 1988, 1996; MacLeod et al. 1997). Most of the current discussion concerns the rapidity of extinction and the number of species reaching up to the KT boundary layer. Marshall and Ward (1996) have statistically shown that many ammonite species previously thought to disappear before the KT boundary are likely to have reached the boundary. Although Marshall and Ward (1996) could not disregard the possibility of a pre-KT extinction event for ammonites, their calculated stratigraphic confidence intervals are well in accordance with a major extinction event exactly at the KT boundary. Although ammonites are continuously reported to disappear slightly before the KT boundary, the recent detection of an ammonite enriched in iridium now indisputably assigns the last living ammonite to the KT boundary itself (Rocchia et al. 1999).

Good data on ammonites are not reported from many sections, but these sections are widely distributed and can be regarded as recording the global picture when summarized (Fig. 10). Late Maastrichtian ammonite faunas are known from North America (Kennedy et al. 1997; Kennedy et al. 1998), South America (Stinnesbeck 1996); Antarctica (Zinsmeister et al. 1989), Europe (Kennedy et al. 1986; Kennedy 1986; Wiedmann 1988b; Birkelund 1993; Ivanov 1993, Machalski 1996), India (Kennedy and Henderson 1992), Far East Russia (Zakharov 1996), Australia (Henderson and Mcnamara 1985), and Japan (Shigeta 1989). KTbase reveals that 94 species (36 genera) were present in the late Maastrichtian (including species not formally described), and at least 21 species lived in the latest Maastrichtian.

MacLeod et al. (1997) gave the erroneous impression that belemnites disappeared from the Tethyan Realm well before the KT event. Although in decline belemnites repeatedly invaded the Tethyan Realm even in the Maastrichtian (Christensen 1997a, b). The KT event affected only three genera (including *Fusiteuthis*). To put this figure into context one must be aware that only one genus of belemnites survived the Campanian/Maastrichtian boundary (Christensen 1997a), and hence the quoted belemnite decline was clearly in the Campanian.

Nautilids appear to pass the boundary without extinctions. The taxa known from Maastrichtian and Danian strata are all survivors (e.g., *Cimomia, Eutrephoceras, Hercoglossa*). K-strategy (low reproduction rates, but brood care) may have been responsible for the preferential survival of nautilids as opposed to r-strategy in ammonites (Kennedy 1989).

5.5.2
Bivalves

Bivalves were very diverse in the Maastrichtian and suffered a significant extinction at the KT boundary. Jablonski and Raup (1995) reported a 70-80% extinction rate on the species level and 50% diversity decline on the genus level. Excluding rudists, KTbase currently indicates a 17% extinction rate on a family level, 49% on a genus level and 92% on the species level. Inoceramids and rudists are the most prominent victims among the bivalves. However, recent studies indicate that the inoceramids experienced considerable extinctions in the mid-late Maastrichtian, well before the KT event. With the exception of the problematic inoceramid *Tenuipteria*, inoceramids became totally extinct in Europe some 3.5 Ma before the boundary (Chauris et al. 1998) and exhibit a major extinction during the mid-Maastrichtian on a global scale (MacLeod et al. 1996). Inoceramids disappear even earlier in Antarctica (Crame and Arthur 1997). Similar patterns were reported for rudists, the principal reef builders in the Late Cretaceous. In the Caribbean, a short-term middle Maastrichtian rudist extinction is reported that left only a few lineages behind and no rudists are evident in the latest Maastrichtian (Johnson and Kauffman 1996). One must not forget, however, that the absolute peak in rudist diversity was in the Maastrichtian (Jones 1986; Kauffman, pers. comm. 1999) and rudists were still common in the late Maastrichtian Tethys (Hanna 1995; Plenicar et al. 1995; Drobne et al. 1997;

Vecsei and Moussavian 1997). Even in the Caribbean region, late Maastrichtian carbonate platforms with diverse rudists are known directly beneath the impact-related deposits (Fourcade et al. 1997, 1999). This indicates that although a mid-Maastrichtian extinction may have affected the rudists regionally, they made it well up to the KT boundary, but could not survive the KT mass extinction.

Although bivalves are common in many shallow marine KT boundary sections, detailed studies indicating their vertical distribution are rare (Fig. 11). The best data exist in Antarctica (Zinsmeister et al. 1989), the Gulf Coast region (Toulmin 1977, Koch 1996), and Denmark (Heinberg 1999).

5.5.3
Gastropods

Sohl (1987) presented one of the most detailed reviews on Cretaceous gastropods and extinctions at the KT boundary. Three gastropod families become extinct at the KT boundary: the nerinoids, cassiopidids, and the acteonellids (Sohl 1987). Only the latter two groups are still globally distributed in the late Maastrichtian and may, therefore, be affected by the KT event. However, on a sub-family level intense extinctions are noted at the KT boundary, especially within the Tethys and in taxonomic units that originate in the early Mesozoic or Early Cretaceous (Sohl 1987). A gastropod group heavily affected by the KT event is the Aporrhaidae, a family characterized by strongly modified apertural margins. Roy (1994, 1996) indicated a 76% generic extinction in this mesogastropod family at the KT boundary, whereas the related Strombidae where not affected As already pointed out by MacLeod et al. (1997) a significant radiation in carnivorous gastropods evident in the Late Cretaceous is not interrupted by the KT event (Taylor et al. 1980).

KTbase indicates high extinction rates on a genus level (58%), but this may be due to taxonomic misconceptions. The family level extinction rate of 15% agrees well with the average extinction rate for marine organisms (Sepkoski 1996).

5.6.
Lophophorates

5.6.1
Bryozoans

Cretaceous and Tertiary bryozoan faunas are dominated by two orders (Cheilostomata and Cyclostomata). Bryozoans are very diverse and abundant in KT sections of Denmark, but were rarely recorded from other sites. Owing to the complex taxonomy of Maastrichtian/Danian bryozoans only very recently a detailed study on extinction patterns has been published. Håkansson and Thomsen (1999) listed 195 cheilostome bryozoan species from the Nye Klov section in northern Denmark. Their tabulation suggests a species level extinction rate of 79% exactly at the KT boundary. Detailed taxonomic data on cyclostome bryozoans are still missing. Former reports suggested low extinction rates (ca. 25%, Hakansson and Thomsen 1979) but a recent compilation of generic diversity and abundance suggests that cyclostome diversity increased throughout the Cretaceous and declined by more than 50% at the KT boundary (McKinney et al. 1998). The relative abundance of cyclostomes is enhanced in the early Danian as compared to the cheilostomes suggesting a delayed recovery in the latter group (McKinney et al. 1998). The great emphasis on the analysis of bryozoans from Denmark is likely to bias actual extinction patterns owing to abundant endemic elements, at least in the Danian (Voigt 1999).

Other KT areas where bryozoans have been analyzed to a certain degree are Limhamn, Sweden (Cheetham 1971); Maastricht area, Netherlands (e.g., Voigt 1987, 1995), Majunga Basin, Madagascar (Brood 1976), Paris Basin and Pyrenees, France (Braga and Bignot 1986; Peybernes et al. 1998), and Alabama (Toulmin 1979). However, most studies did only describe few species or even only indicated some morphological features of bryozoans. Thus, the extinction pattern is only known from one locality in high paleolatitudes and for one bryozoan order.

5.6.2 Brachiopods

As for bryozoans, only one locality has been documented in full detail on the KT extinction pattern of brachiopods. This one section (Nye Klov, Denmark) has been repeatedly reported to evidence a sudden and significant extinction in all Cretaceous brachiopod groups (Surlyk and Johansen 1984; Johansen 1987, 1988; 1989a, b). The extinction ratio was high on the species and genus levels (70% and 40%, respectively), but only one family disappeared. Brachiopods have been recorded and described from a few other KT localities (e.g., Reiskind 1973; Kennedy and Klinger 1975; Simon 1998), but none produced any additional data on the extinction patterns.

5.7 Echinoderms

5.7.1 Echinoids

The KT faunal change in echinoids is the best explored among the Echinodermata. A recent survey summarized the global database on echinoid extinction at the KT boundary (Jeffery and Smith 1998; Smith and Jeffery 1998; Jeffery 2001). Based on a revised taxonomy, the authors found a 36% extinction on the genus level. Although this value is well below previous estimates (Roman 1984), the extinction is still severe. On a species level, 60% of echinoids are thought to become extinct at the KT boundary (Jeffery 1997a). Extinction was strongest in the holectypoids (irregular echinoids) and least pronounced in the cidaroids (regular echinoids). Although regular echinoids are mostly epifaunal and irregular echinoids are predominantly infaunal, no overall selectivity for life habit is evident. Other selectivity patterns of sea urchin extinctions suggested by Smith and Jeffery (1998) and Jeffery (2001) are discussed below.

Jeffery (1997b) suggested that a major ecological turnover occurred in echinoids prior to the KT boundary and Smith and Jeffery (1998) pointed to a "slow squeeze" rather than a sudden extinction event for echinoids. However, diverse and healthy echinoid assemblages thrived in the Tethys during Maastrichtian times (Smith et al. 1995) and appear to reach up to the latest Maastrichtian (Jeffery 1997a). Only few detailed studies on the vertical distribution of echinoids through the KT boundary are known to the authors (Jeffery 1997a; Smith et al. 1999). The currently available extinction pattern is, therefore, poorly constrained

5.7.2 Other Echinoderms

Faunal change of crinoids, holothurians, asteroids and ophiuroids is poorly known and mostly applicable to European sections. The classical work of Rasmussen (1979) indicates little change at the KT boundary of Denmark but a major evolutionary transformation at the Danian-Selandian boundary. Based on his work the genus extinction ratios would be insignificant namely 10% for crinoids, 4% for asteroids, and 0% for ophiuroids. On a species level extinction ratios of 56% for crinoids, 82% for asteroids, and 33% for ophiuroids arise from the study of Rasmussen (1979). Based on observations in northeast Belgium and the Netherlands, Jagt (1995) found the highest crinoid diversity in the latest Maastrichtian and recorded a dramatic diversity decline in the basal Danian.

All these statements are questionable as they originate from observations in a single region. A few additional Maastrichtian and Danian faunas were described but none of them are sufficient to evaluate extinction intensity. In a monograph on holothurians, Gilliland (1993) reported no evidence of a diversity drop at the KT boundary but admitted that the fossil record around the boundary was too poor to give conclusive statements.

5.8 Corals

One extensive review on the extinction of Cretaceous scleractinian corals exists (Rosen and Turnsek 1989; updated by Rosen 2000) and forms the base of many discussions in subsequent papers. Rosen and Turnsek (1989) used different databases in their analysis. Their Treatise-derived database (Wells 1956) is not considered useful for resolving KT extinctions, due to the coarse stratigraphic resolution. The own database of Rosen and Turnsek (1989) compiled species-level data on Maastrichtian and Paleocene corals from 21 localities considering 35 references (S-data). Although this database is the best available, it is clearly biased by incomplete geographic coverage, non-phylogenetic taxonomy, and little species overlap between localities of one time slice. In addition, there is almost no locality-overlap, that is, Maastrichtian and Paleocene were compared from different localities.

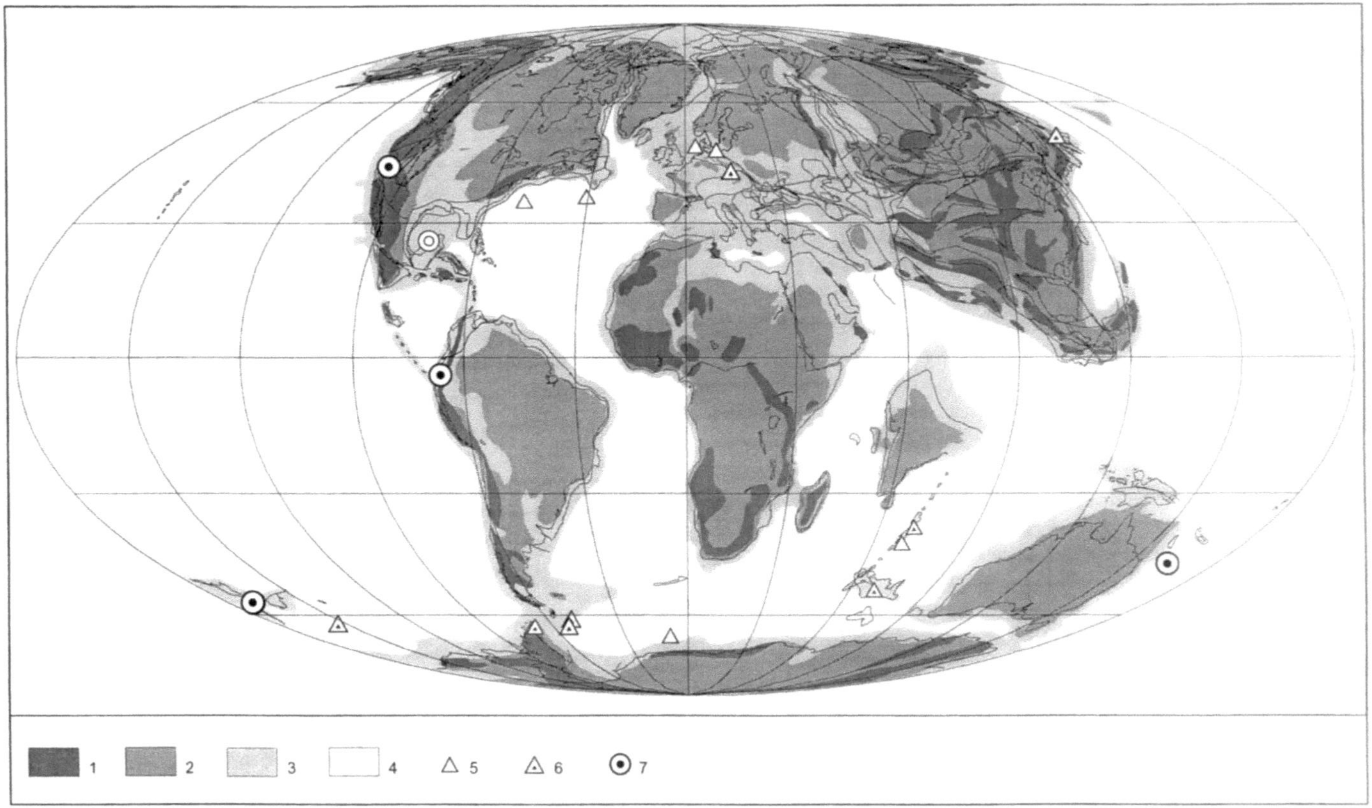

Fig. 8. Geographic distribution of radiolarian data across the K/T boundary. See Fig. 5 for legend. Note that good data are almost exclusively related to cool surface currents (high latitude and eastern boundary currents).

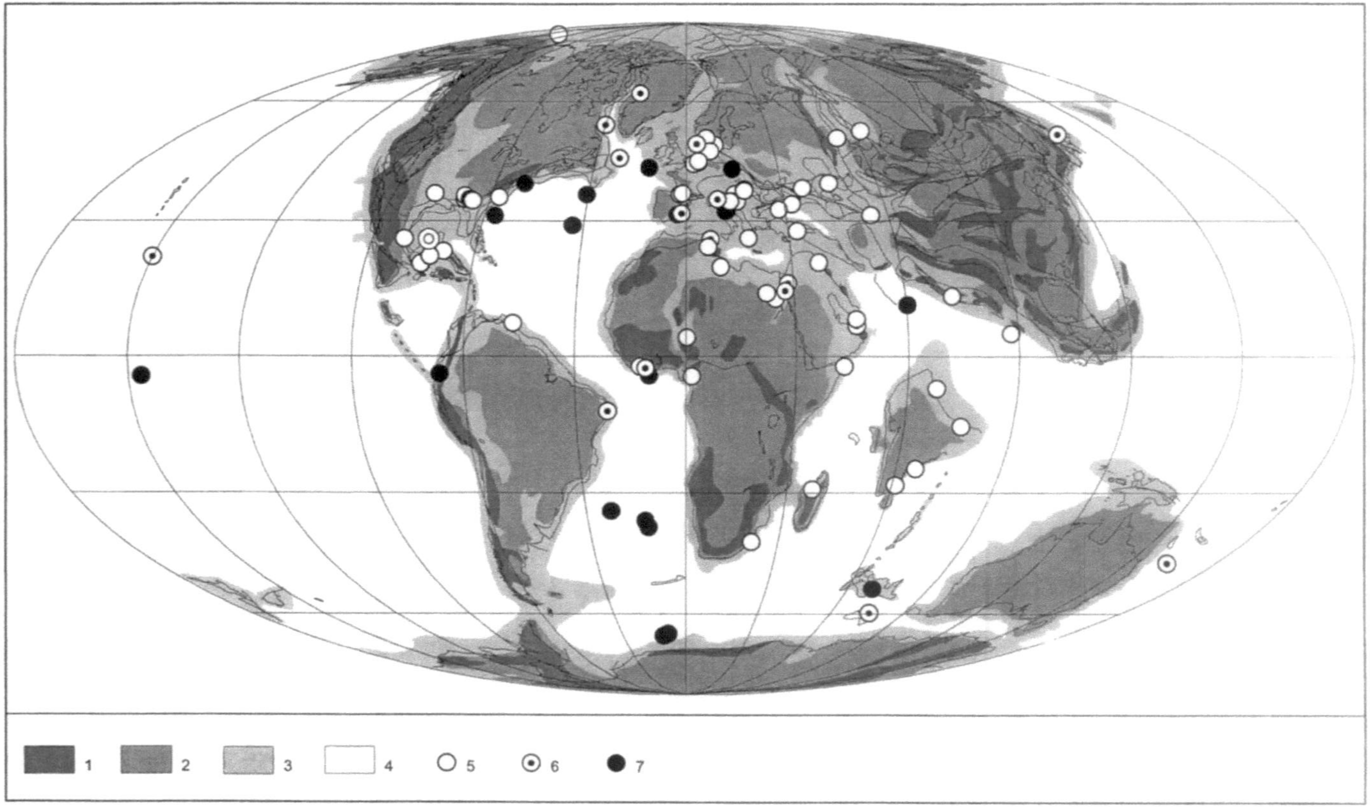

Fig. 9. Data on benthic foraminifers across the K/T boundary. 1- Mountains; 2- Lowland; 3- Shelf sea; 4- Ocean; 5- Shallow water; 6- Intermediate water; 7- Deep water. Note that although data on shallow water faunas are abundant, data quality is mostly poor.

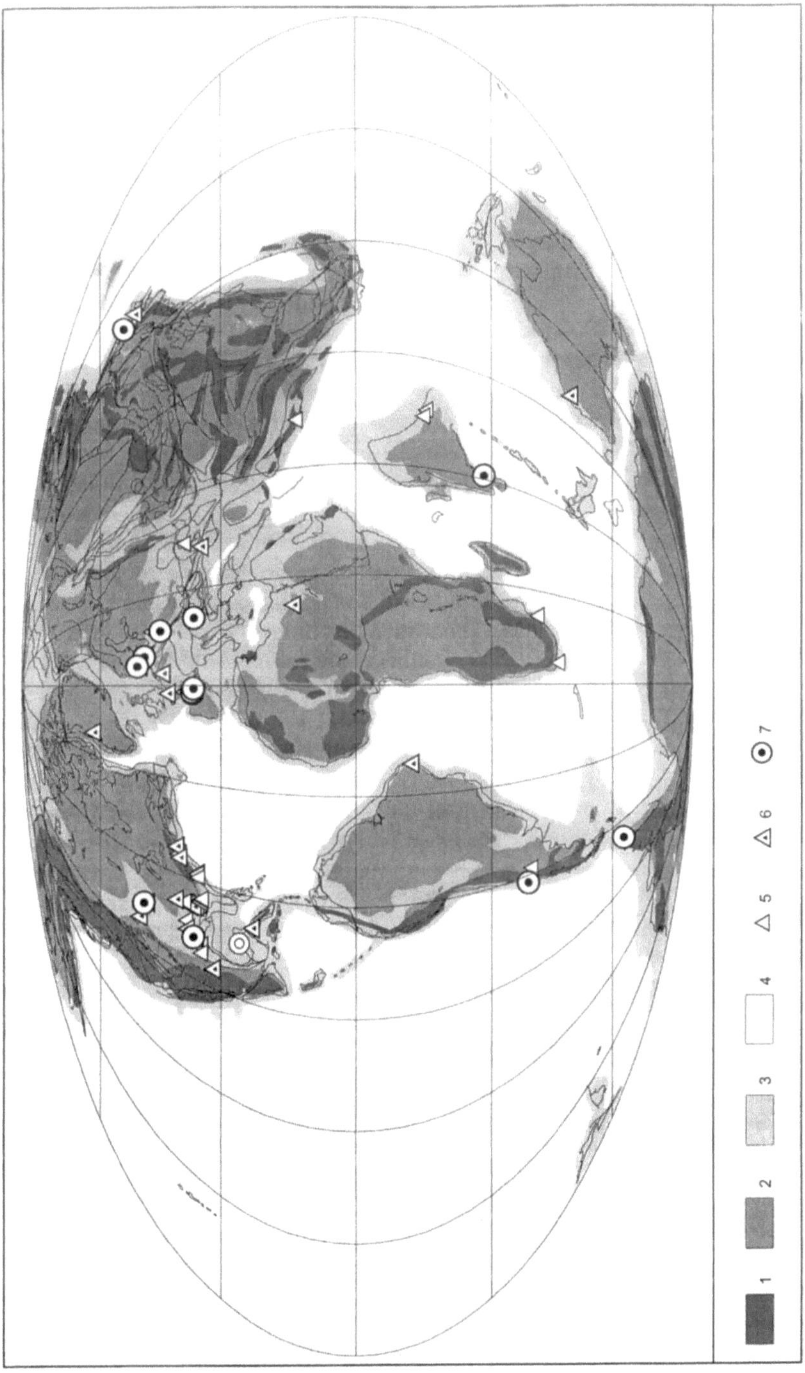

Fig. 10. Geographic distribution of ammonite data in the late Maastrichtian. See Fig. 5 for legend. Quality is evaluated by the taxonomic details presented and by the proximity of reported faunas to the K/T boundary.

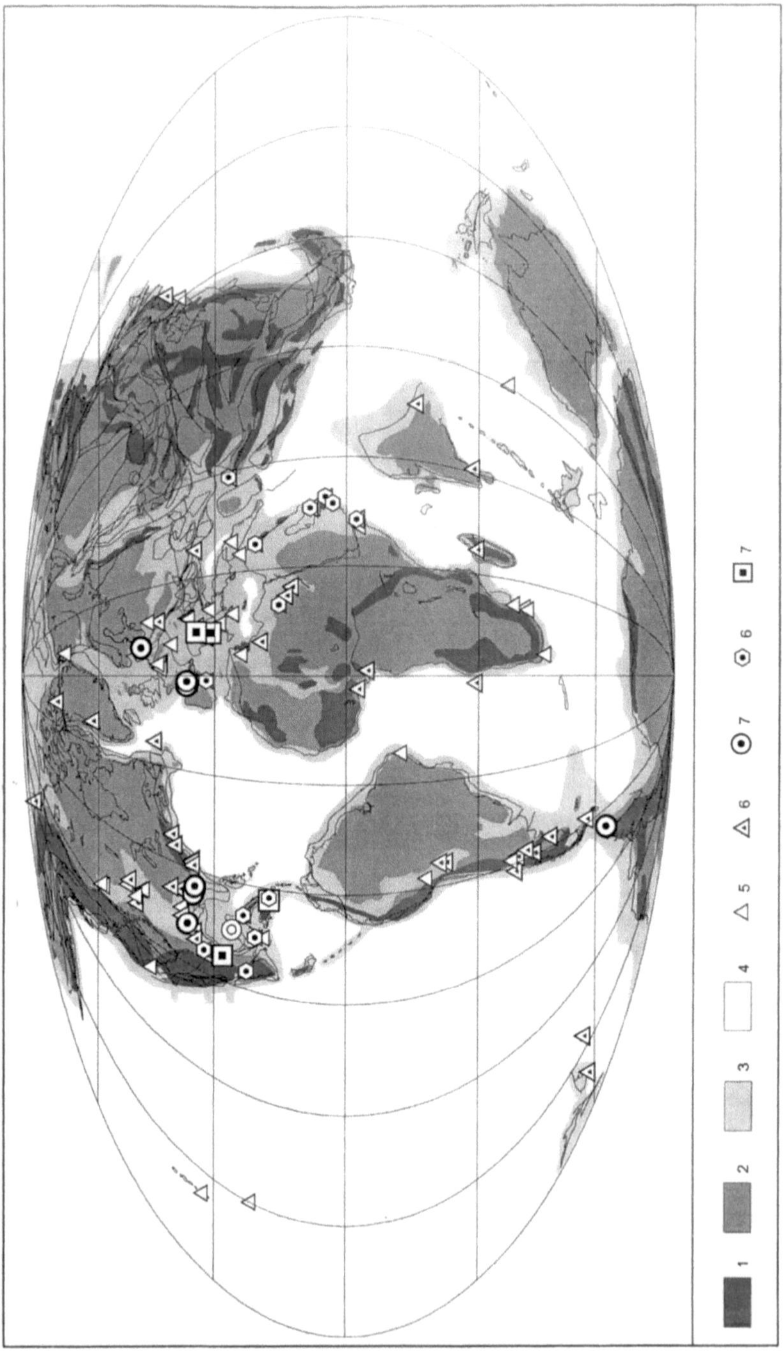

Fig. 11. Geographic distribution of bivalve data across the K/T boundary. 5- few bivalve genera indicated at locality, 6- moderate quality data for non-rudist bivalves; 7- high quality data for non-rudist bivalves; 8- moderate quality data for rudists; 8- high quality data for rudists.

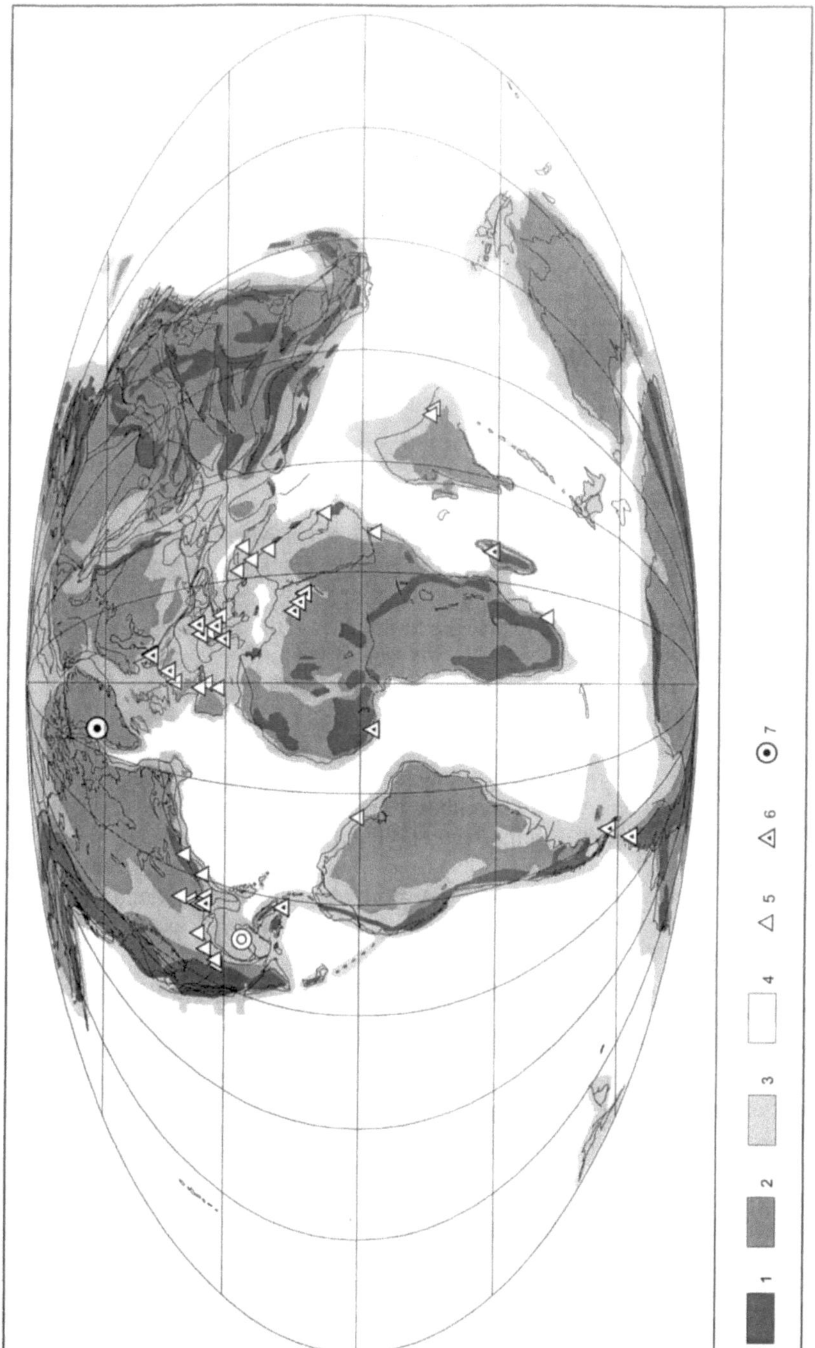

Fig. 12. Geographic distribution of coral data across the K/T boundary. See Fig. 5 for legend.

Therefore, the extinction ratio given by Rosen and Turnsek (1989) should be viewed with caution. An overall generic extinction of 83% and a species extinction of 98% were found in their S-database. These rates are considerably higher than the ones found with KTbase (68% generic extinction, 94% species extinction). Besides the debatable numbers due to the 'chauvinotypy' factor (Rosen 1988), some interesting conclusions were drawn in the analysis: Diversity decline of corals during the Late Cretaceous is insignificant but the Paleocene saw the destruction of the paleolatitudinal coral pattern, a significant diversity loss and a prevalence of non-photosymbiotic corals. Rosen and Turnsek (1989) indicated that percent extinctions were highest in the high diversity coral belt, that was predominated by photosymbiotic corals. Although based on currently weak data (Fig. 12), the preferential extinction of photosymbiotic corals (see also Rosen 2000) would perfectly fit the impact scenario.

5.9 Benthic Calcareous Algae

Although little studied taxonomically, benthic calcareous algae are well known from many shallow-water KT boundary sections, in Maastrichtian as well as in Danian sediments. No mass extinction for benthic calcareous algae was ever cited prior to Aguirre et al. (2000a) who suggested a 65-70% species extinction for both coralline red algae and dasyclad green algae. Although less comprehensive KTbase indicates a much lower extinction rate of 9% on a genus level and 50% on a species level. The discrepancy in the results is difficult to explain, but may be due to the fact that KTbase currently only contains the most common species and is exclusively based on actual occurrence data in boundary sections. Diverse Danian floras are known in many areas, such as France (Deloffre 1980), Lower Austria (Kuehn 1960), Alps (Tragelehn 1996), Adriatic Platform (Drobne et al. 1997), Turkey (Sirel 1996), India (Rao and Pia 1936), Guatemala (Fourcade et al. 1997), Mexico (Deloffre et al. 1989), and Oman (Roger et al. 1998). In most places benthic calcareous algae are even more abundant in the Danian than in the late Maastrichtian, although the basal Danian contains usually few algae (Tragelehn 1996).

Corallinacean algae exhibit a diversity increase in the Cretaceous, especially in the Maastrichtian and continue to diversify in the Paleocene (Steneck 1983; Aguirre et al. 2000b). The same is true for dasycladacean green algae (Tragelehn 1996; Aguirre et al. 2000a). The reason for the insignificant generic extinction in benthic calcareous algae is currently unknown and counterintuitive considering the period of darkness invoked by the impact scenario.

5.10 General Trends in Paleontological Data

There is a relationship between the quality of the fossil record of a particular group and its observed or estimated extinction rates, except the much exaggerated estimations for corals. The more precisely the temporal distribution of a taxon is studied, the more evidence emerges for significant and rapid extinctions or ecological changes exactly at, and subsequent to the KT boundary. One exception are the planktonic foraminifera consistently reported as exhibiting a sharp decline exactly at the boundary by one group of paleontologists and showing few changes but a high background extinction from the latest Maastrichtian to the early Danian by the other group. A new blind test complying with the arguments of Aubry (1999) is clearly required.

6 Ecosystem Patterns

6.1 Pelagic

A sudden and pronounced devastation of the pelagic ecosystem is evident for the KT boundary. The ecosystem collapse is exemplified by a strong decline in $\delta^{13}C$ in pelagic carbonates (D'Hondt et al. 1998) and reduced accumulations of nutrient-type trace elements (Zachos et al. 1989) coincident with the deposition of the boundary clay. The devastation in the pelagic ecosystem can be seen short-lived when only referring to planktonic foraminifera, calcareous nannoplankton and carbon isotopes. High-latitude regions often showing a delayed ecological response and an expanded extinction, also needed longest for recovery. This is especially evident in the radiolarians. In high latitude and eastern boundary settings (regions affected by cool currents along the eastern ocean margins), radiolarians appear to be unaffected by the KT event (see above). Dramatic faunal change end extinction of Cretaceous species is, however, reported for nearly the whole Danian (Hollis 1996, 1997). As an ecosystem can only be regarded as recovered when all fundamental parts of it are, the global pelagic ecosystem needed approximately 3 Ma to recover but was fully established by the mid-Paleocene (approximately Selandian).

Many authors have problems with the long-term "environmental perturbations" seen in the ongoing extinctions throughout the Danian. However, no additional environmental perturbations are required to explain this pattern. The destruction of the pelagic ecosystem by a single event and appended extinctions of ecological keystone species is sufficient to explain ongoing stress and extinction in the pelagic ecosystem.

6.2 Reefs

The collapse of the reef ecosystem at the KT boundary is also widely cited, but it is difficult to determine its exact timing and magnitude. General reviews follow the idea that reefs are most vulnerable during major events and exhibit a significant decline 1 Ma prior to the actual extinction events (Copper 1994). If this holds true for the KT boundary, an abrupt event is unlikely to have caused the reef ecosystem collapse. Actually, some evidence was cited that rudist reefs vanished well before the KT boundary in the late Maastrichtian (Johnson and Kauffman 1996). However, this notion arose from studies in the Caribbean only and may reflect a regional situation. In Iraq, the Aqra Limestone contains buildups of rudists, algae and some corals up to the late Maastrichtian (Al-Ameri and Lawa 1986). Late Maastrichtian rudist associations are also reported from Spain (Pons et al. 1994) and the Dinarides (Polsak 1985; Drobne et al. 1988). Therefore, it seems that the Cretaceous reef ecosystem persisted on a global scale until the late Maastrichtian and perhaps even the latest Maastrichtian and there was no major decline prior to the KT event. It should be noted though that Maastrichtian as well as Campanian reefs were poorly developed Most of the 'reefs' are small biostromes.

The reef ecosystem "collapse" at the KT boundary is not very pronounced A global reef database developed by Kiessling et al. (1999) points to a 50% decline in reef numbers from the Maastrichtian to the Danian. However, this figure is misleading since the stages have different durations. When normalized for the duration of stages, reef growth rates did not differ at all between the Maastrichtian and the Danian. There are reports of continuous reef growth in the Dinarides (Jelaska and Bulic 1975; Polsak 1985) which require further confirmation. Reefs in the widest sense also persisted through the KT boundary at Stevns Klint (Surlyk 1997). Although mound growth was commonly terminated by the KT event, the boundary clay virtually cuts through bryozoan mounds at this locality (Surlyk 1997). This is especially surprising considering the high extinction rate observed in the Danish bryozoan fauna.

Although coral extinction is thought to be severe, algal extinction is not, and encrusting patterns important for reef growth did not change notably (Moussavian 1992). Hence, the major impact on the tropical reef ecosystem appears to be the extinction of rudistid bivalves. Nevertheless, there appears to be a post-KT reef gap in tropical to subtropical latitudes. Recovery in reefs is said to be delayed with respect to other habitats (e.g., Jablonski 1991; Copper 1994). This is no surprise since reefs represent not just an environment, but are complex ecosystems actively built by organisms and require finely tuned biotic interactions. Considering this statement recovery in the tropical reef ecosystem was surprisingly rapid. Tragelehn (1996) documented reef recovery in the Alps. He stated that the first small Danian reefs occur between foraminifer zones P1b and P1c (early Danian) and larger reefs developed in the middle/late Danian.

Danian coral-algal reefs are also well known from the Maiella platform in Italy (Moussavian and Vecsei 1995; Vecsei and Moussavian 1997), France (Deloffre and Snea 1980; Cros and Lucas 1982), Egypt (Schuster 1996), Turkey (Sirel et al. 1986) and the Carpathians (Scheibner 1968). Possible occurrences are also reported from Pakistan (Duncan 1880; Adams 1970). In summary, there was no real devastation of the reef ecosystem and recovery was rapid, except for North America. Reef recovery appears to be correlated to recovery in the pelagic ecosystem in low latitudes, but is independent of it in higher latitudes (Fig. 13).

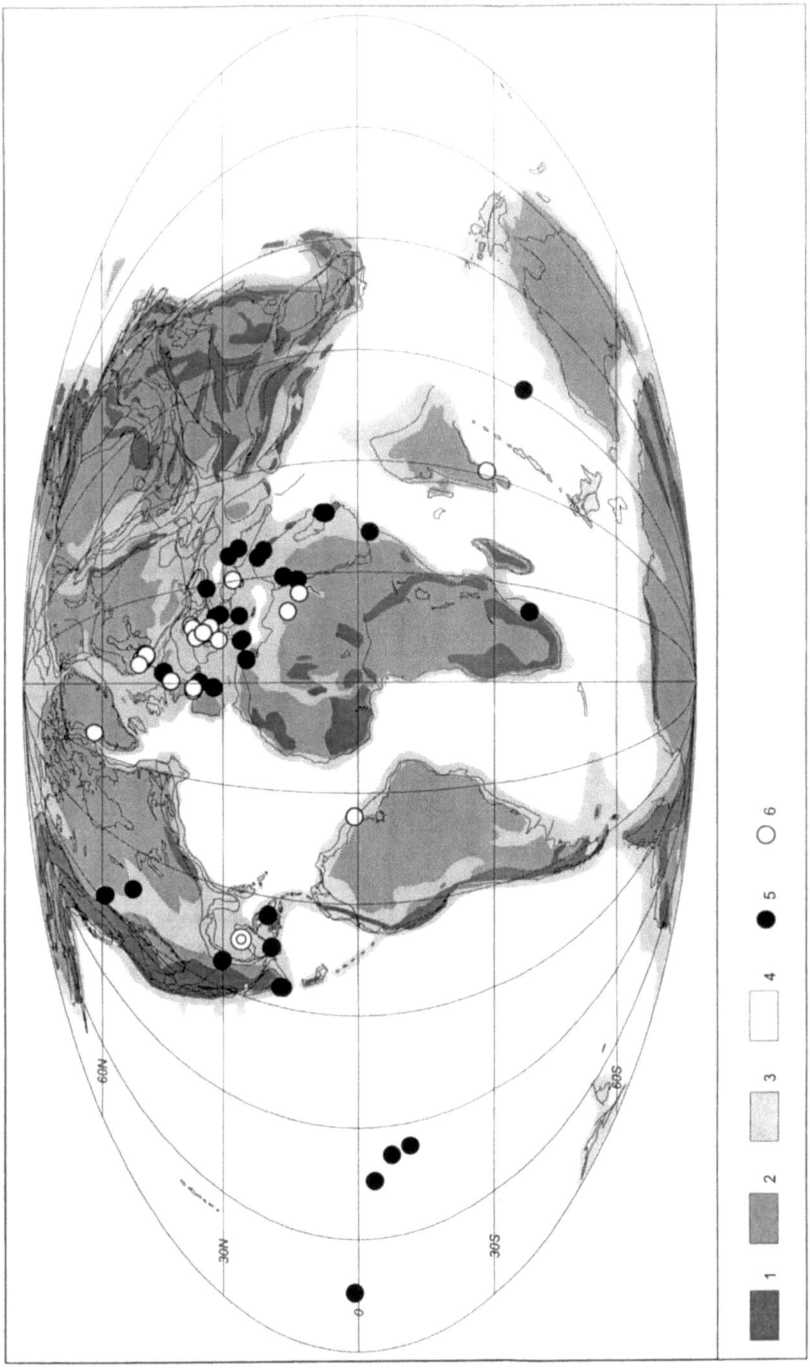

Fig. 13. Reported occurrences of Maastrichtian (black circles) and Danian (white circles) reefs. Note that reefs were absent in the Danian of the Gulf of Mexico region, but occur in the early Danian in Europe.

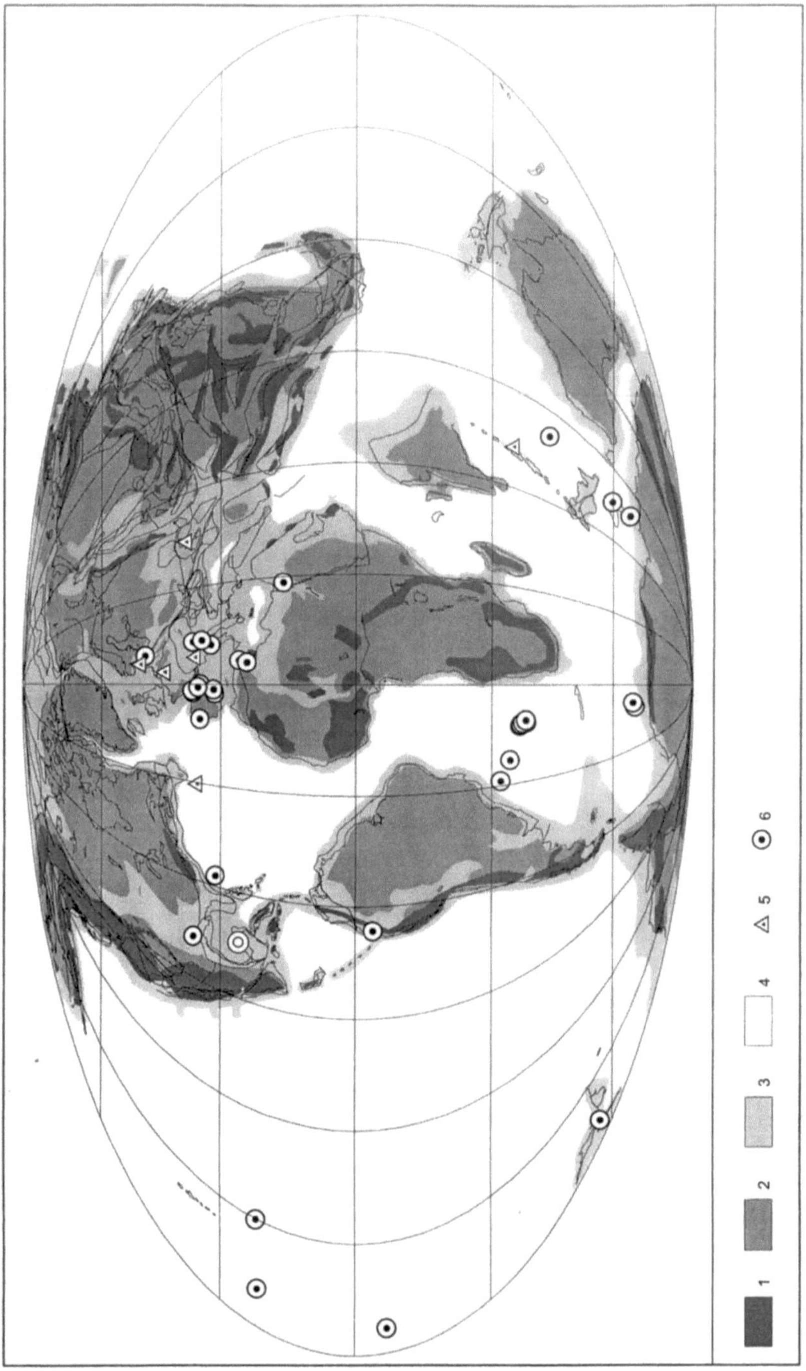

Fig. 14. Evidence for δ^{13}C-decline in pelagic carbonates (data mostly from planktonic foraminifera). The decline (circles) is as pronounced in high latitudes as it is in low latitudes. Several sections without a pronounced drop (triangles) are either incomplete or from shallow water environments.

6.3 Biomass reduction

Probably the most significant paleontological signal associated with the KT event is the strong reduction of standing biomass. The most prominent proxy of biomass reduction is the negative shift in $\delta^{13}C$ of planktonic foraminifera and the decline of the surface to deep-water $\delta^{13}C$ gradient (e.g., Zachos and Arthur 1986; Kaiho et al. 1999). This globally observed $\delta^{13}C$ shift (Fig. 14) points to a 50% decrease in planktonic biomass (D'Hondt et al. 1996).

Although an isotopic anomaly is observed until 3 Ma after the KTB, it is unlikely that biomass reduction was very significant for such an extended time. Rather a smaller fraction of marine primary production reached the early Danian seafloor, probably due to higher rates of picoplankton (mostly bacteria) production (D'Hondt et al. 1998).

A decline in planktonic production is also evident by significantly reduced accumulation rates of pelagic carbonates after the KT event (Zachos et al. 1985; Arthur et al. 1987). The decline could be up to 84% locally (D'Hondt et al. 1996) but KTbase gives an average value of 60% for the drop in global carbonate accumulation rates. A direct measurement of reduced nannofossil accumulation rates at the KT was carried on by Henriksson (1996) who found a 100-fold decrease after the event lasting for some 200 ka. This extreme reduction may again not only be explained by reduced standing biomass (productivity) of the nannoplankton but also be due to the absence of larger planktonic grazers in the ocean. Thus, the production of fecal pellets was also reduced preventing the majority of calcitic nannoplankton elements to reach the seafloor. The global reduction of biomass is not limited to the plankton. Several earliest Danian benthic assemblages are strongly reduced in terms of biomass. The Brazos River section in east Texas records a biomass decline by more than two orders of magnitude (Hansen et al. 1987) and in Denmark a 'dead zone' in the basal Danian (Håkansson and Thomsen 1999) is evident where nothing but some small bivalves, gastropods, and sponge spicules are to found.

6.4 Selectivity of extinctions

6.4.1 Ecological selectivity

Many suggestions have been made concerning the selectivity of extinction and its conformance with the abrupt pattern imposed by the scenario of an asteroid or comet impact event. The major killing agents currently envisaged by the impact scenario are: global darkness for several months, cooling and subsequent greenhouse warming, wildfires, SOx and/or NOx poisoning and acid rain (Alvarez et al. 1980; O'Keefe and Ahrens 1982; Brett 1992; Pope et al. 1994; Toon et al. 1997; Arinobu et al. 1999; Premovic et al. 2000). This is indeed a set of terrifying and complex factors mostly derived from numerical models. Their real effects remain poorly constrained by the paleontological and paleoceanographical existing data.

According to Smit (1999) the effects of the impact released dusting of the atmosphere is a factor severe enough to explain most of the observed effects on the biosphere. Although the relative importance of the killing mechanisms is unknown, some prediction of the extinction selectivity can be made. It would be expected that

1. Terrestrial and marine ecosystems are equally affected;
2. Deep water habitats are less affected than shallow water habitats;
3. Detritus feeders are less affected than suspension feeders, herbivores, and carnivores;
4. Eurytopic species are less affected than stenotopic species;
5. Small organisms with low metabolism are less affected than large organisms;
6. Deep-water infauna should be more affected than epifauna owing to the collapse of the food web and less nutrients reaching the sea-floor, all used-up by the epifauna;
7. Organisms with durable resting stages in their life cycles should do better than forms without.
8. Organisms with siliceous test are most likely to survive acid rain than calcareous ones

No prediction can be made by the impact scenario as to how endemism or global abundance influences survival probability. None of the predicted selectivity patterns are impact-specific, as they can also be caused by other causes. However, the predictions have to be supported by the paleontological data in order to make the impact-extinction link credible.

1) Many papers suggest that terrestrial organisms (plants, vertebrates) were strongly affected by the KT event (Johnson and Hickey 1990; Sheehan et al. 1991); an equally efficient global catastrophe in marine and terrestrial ecosystem appears, therefore, likely. A similarly equal decline in terrestrial and marine environments is only evident at the Permian-Triassic boundary, whereas other mass extinction were apparently more effective in the marine environment.

2) Biota thriving in deeper water appears to be actually less affected by the KT event than shallow-water organisms. This is exemplified by the great differences in extinction between benthic foraminifera from different habitats. The proposed difference in the extinction rate of photosymbiotic (shallow) and non-photosymbiotic (both shallow and deep) corals points to the same conclusion.

3) Several authors suggested that the KT event was a strongly selective for trophic ecology (Sheehan and Hansen 1986; Rhodes and Thayer 1991; Sheehan et al. 1996). Herbivorous organisms (terrestrial as well as marine) were reported to be more strongly affected by the extinction than those feeding on detritus (Arthur et al. 1987; Roy et al. 1990; Hansen et al. 1993). Deposit-feeders are generally cited as less affected than suspension-feeders (Sheehan et al. 1996). Although some deposit-feeding bivalve clades show extinctions as severe as for suspension-feeding clades (Jablonski and Raup 1995; Jablonski 1996), the quantitative comparison of faunas before and after the KT event often shows a shift from faunas dominated by suspension-feeders to faunas dominated by deposit-feeders

(Hansen et al. 1987, but see Heinberg 1999 for the exception at Stevns Klint). Trophic strategy appears to be an important selectivity factor for KT echinoids (Jeffery 2001). One would expect that planktotrophic larvae are more affected than lecitotrophic larvae during the predicted months of global darkness. Yet, Valentine and Jablonski (1986), Heinberg (1999), and Jeffery (2001) reported evidence that no such selectivity exists.

4) The prediction that ecological generalists should do better than specialists is corroborated by benthic foraminifera (Hottinger 1997), echinoids (Markov and Solojev 1998; Smith and Jeffery 1998), corals (Rosen and Turnsek 1989), and brachiopods (Johansen 1988). However, the interpreted pattern in cephalopods contradicts the assumption. Ammonites inhabited a wide range of Maastrichtian environments and had a r-mode reproductive strategy, whereas the more specialized and K-mode selective nautilids survived.

5) Size reduction in post-extinction populations was reported for calcareous nannoplankton (Gardin and Monechi 1998), planktonic foraminifera (e.g., Premoli Silva and Luterbacher 1966; Norris et al. 1999), shallow benthic foraminifera (Hottinger 1997), and echinoids (Smith and Jeffery 1998; Jeffery 2001). In most cases the small forms were already present or are likely to have been present in the late Maastrichtian, but dominate faunas only subsequent to the KT event. This pattern was not confirmed for bivalves and gastropods (Jablonski and Raup 1995; but see Hayami 1997 for an opposing view). Including rudists in the statistical tests would, however, produce a significant reduction in bivalve sizes across the KT boundary.

6) For deep-water benthic foraminifera, both agglutinated and calcareous, there is a pronounced change from epifauna-dominated to infauna-dominated assemblages at the boundary (Thomas 1990; Widmark and Malgren 1992; Kuhnt et al. 1998). This trend is not seen in shallow water communities. For bivalves Heinberg (1999) noted that epifaunal species were even more strongly affected than infaunal forms.

7) Both diatoms and dinoflagellates are little affected by the KT event and commonly have inactive resting stages in their life cycles. Other groups without immobile resting stages (e.g., calcareous nannoplankton) are strongly affected by the KT event. These observations support the prediction above. However, many groups without resting stages, are also little affected (e.g., radiolarians) suggesting that the ability to persist short-term environmental perturbations by metabolically inactive cysts or spores is not crucial.

8) Radiolarians and diatoms – both bearing siliceous tests – were little affected by the KT event, whereas planktonic foraminifera and calcareous nannoplankton show very high extinction rates. Although the acid rain scenario is an unlikely trigger of marine extintions (D'Hondt et al. 1994, 1996), this observation would agree with it.

Contradicting many of the statements above, Jablonski and Raup (1995) found "no selectivity of marine bivalve genera by life position (burrowing versus exposed), body size, bathymetric position on the continental shelf, or relative breadth of bathymetric range". They were also critical of the lower extinction rate in deposit-feeders because it is due to low extinction rates in only two bivalve orders. However, their study relies heavily on Gulf Coast and Atlantic Coastal Plain assemblages, where the most detailed data originated. Extinction patterns observed in these regions may not be representative due to their proximity to the Chicxulub crater.

Another significant selectivity component is commonly seen in the geographic distribution. More widespread taxa are more resistant to extinction than endemic taxa (Erwin 1989; Jablonski 1991, 1996). However, this phenomenon is observed at other mass-extinction boundaries as well, including the ongoing extinction in the Recent. Thus, it cannot be used as a test of the impact hypothesis. Numerous additional selectivity patterns were suggested for the KT event. For echinoids the age of a clade is considered important (Markov and Solojev 1998; Smith and Jeffery 1998) and old and young genera exhibit higher extinction rates than others. Keller et al. (1993) suggested that the effects of the KT boundary event were less severe in high latitudes as they found only one species of planktonic foraminifera extinct exactly at the boundary in Nye Kløv (Denmark). However, all the typical victims of the KT boundary become extinct shortly (200-300 kyr) afterwards.

6.4.2 Spatial selectivity

Are there regions more affected by the KT crisis than others? From various particularities in the extinction scenario, we would expect various spatial inhomogenities in ecosystem devastation and extinction. For instance, stronger immediate effects are expected close to the impact site and environmental consequences should be enhanced in the Northern Hemisphere (Crowley and North 1991).

Schultz and D'Hondt (1996) argue for an oblique impact from the southwest that predicts enhanced ecosystem devastation in North America by immediate effects (shock wave, earthquakes, and tsunamis). If spatial patterns in ecosystem destruction, extinction, and recovery agree with this finding, the role of the impact in extinction could be specified For terrestrial environments, ecosystem devastation actually appears to be significantly higher in North America than elsewhere. Many North American sections contain a fern spike in the lowermost Danian, that is commonly interpreted as a pioneer flora following the complete destruction of vegetation on the continent (Fleming and Nicols 1990). No fern spike was observed in other regions except for Japan (Saito et al. 1986). Only recently, a moss spike has been observed in Europe (Brinkhuis and Schioler 1996), suggesting similarly severe vegetation destruction as in North America.

Thus far, no enhanced extinction rates in North American marine environments as compared to other regions were explicitly reported. However, Smith and Jeffery (1998) stated that although endemism per se does not enhance the extinction risk for Maastrichtian echinoids, endemic echinoids in North America are more prone to extinction than endemic species in other regions. While molluscan extinction rates are similar on all continents, several features of post-KT Gulf Coast molluscan assemblages are unique (Jablonski 1998) and point to an enhanced ecosystem devastation close to the proposed impact site. The Gulf Coast assemblages are characterized by a rapid expansion and decline of "bloom taxa" and have lower proportions of invaders in the early recovery phase as compared to other regions (Jablonski 1998). McClure and Bohonak (1995) found no selectivity in the extinction of bivalves in the Gulf and Coastal Atlantic plain of North America and consequently argue that the mass extinction was so severe that ecological differences in bivalve genera are irrelevant. We argue that this lack of selectivity is due to the vicinity of the analyzed bivalve assemblages to the KT impact. In Denmark, extinction in bivalve faunas is more selective in terms of life habit (Heinberg 1999).

Another interesting aspect concerns the paleogeographic distribution of reefs (Fig. 13). Although few Maastrichtian reefs are known in comparison to earlier Cretaceous stages, they are essentially globally distributed However, Danian recovery of the reef ecosystem was essentially limited to Europe and North Africa, with the exception of coral mounds in Greenland and questionable occurrences offshore Brazil (Carozzi et al. 1981). In the Gulf region, reefs are only known from the late Paleocene (Bryan 1991). Although still speculative, this geographic recovery pattern may be linked to enhanced ecological devastation in the Gulf region due to the proximity of the Chicxulub impact. The observed pattern in the nannoplankton (Fig. 7) points to the same direction but the interpretation that this pattern is related to Chicxulub is even more speculative.

7
Conclusions and Perspectives

The iridium anomaly is globally distributed, and its magnitude seems to be unrelated to distance from the Chicxulub crater. Shocked quartz distribution is more confined and exhibits a marked geographic pattern. Maximum size and abundance of shocked quartz decrease with distance from Chicxulub and is generally enhanced in North America.

Contrasting current opinions in the paleontological community, most marine organisms do not seem to suffer significant extinction before the KT boundary and the evidence for globally stressed ecosystems in the late Maastrichtian is rather unconvincing. This is especially true when compared to other Cretaceous stages, which often showed strong fluctuations in physical parameters (climate and sea level changes, anoxic events, high sea-floor spreading rate) without resulting in major extinctions. Increasing data on macrofossils and refined studies of key sections tend to show that ecological changes and extinctions were concentrated more or less exactly at the KT boundary. The link between the Chicxulub impact and a short-term ecological catastrophe, however, remains most plausible only in the calcareous plankton. So far detailed taxon-quantitative distribution patterns through KT boundary sections are rare and therefore poorly constrained on a global scale. Based on the quality of data available for particular sections and on the global distribution of good sections, we are able to quantify the knowledge of biotic change through the KT boundary. Our understanding of the actual extinction patterns for the fossil groups discussed in this paper can be ordered in a decreasing sequence: Calcareous nannoplankton – planktonic foraminifera – deeper water benthic foraminifera – dinoflagellates – bivalves – echinoids – ammonites – gastropods – radiolarians – brachiopods – bryozoans – larger foraminifera – belemnites – benthic calcareous algae – corals – crinoids – siliceous phytoplankton– asteroids, holothurians, ophiuroids.

Preliminary analysis of geographic patterns observed in extinction and recovery indicates that the ecological devastation in the shallow marine environment appears most pronounced in North America. In this region endemic echinoids were affected strongest, selectivity in mollusc extinction is least significant, recovery in molluscs shows a different pattern than in other regions, and reefs took longer to recover than in Europe. If these observations are confirmed, they will provide a good indication of the direct effects of the Chicxulub impact on proximate environments.

Collection of the vast literature and presentation of the data in a GIS-linked database format allows to organize and combine all the geological, geochemical and paleontological information. When more complete, KTbase can be used as a tool to test several of the prediction made by the impact-extinction scenario, such as selectivity of extinction, or regional patterns of biotic change. The conclusions presented here must then be viewed as preliminary, and we encourage colleagues from all disciplines to contribute to KT base. Ultimately it is our goal to have the database available online.

Important open questions concern the low extinction rate among benthic calcareous algae, the decline of rudists and reefs prior to the KT boundary, and the geographic variation of extinction and recovery. We suggest that future studies should concentrate on these topics rather than on the pattern of still another planktonic foraminifer section. Important hints as to how the impact caused extinction could be gathered by systematically comparing extinction and recovery patterns in North America with other regions. We expect that in North America the diversity decline is highest and least selective in comparison to regions more distant to Chicxulub, especially on the southern hemisphere. To complete our knowledge of global extinction patterns KTbase demands data on early Paleocene radiolarians from the Tethys, on latest Maastrichtian ammonite assemblages outside northern Spain and southern France, on lophophorates from regions outside Denmark, and on echinoderms, corals and algae in all regions. A region lacking any good data in general but crucial for the KT discussion is northern South America. KTbase indicate six known boundary sections in this area, three of which are assumed to be complete. Future studies should concentrate on this area.

References

Abramovich S, Almogi-Labin A, Benjamini C (1998) Decline of the Maastrichtian pelagic ecosystem based on planktic foraminifera assemblage change: Implication for the terminal Cretaceous faunal crisis. Geology 26: 63-66

Adams CG (1970) A reconsideration of the east Indian letter classification of the Tertiary. Bulletin of the British Museum (Natural History) Geology 19: 87-137

Aguirre J, Riding R, Braga JC (2000a) Late Cretaceous incident light reduction: evidence from benthic algae. Lethaia 33: 206-214

Aguirre J, Riding R, Braga JC (2000b) Diversity of coralline red algae: origination and extinction patterns from the Early Cretaceous to the Pleistocene. Paleobiology 26: 651–667

Al-Ameri F, Lawa FA (1986) Palaeoecological model and faunal interaction within Aqra Limestone Formation, north Iraq. J Geol Soc Iraq 19: 7-27

Albertão GA, Martins PPJ (1996) A possible tsunami deposit at the Cretaceous--Tertiary boundary in Pernambuco, northeastern Brazil. Sedimentary Geology 104: 189-201

Alvarez LW, Alvarez W, Asaro F, Michel HV (1980) Extraterrestrial cause for the Cretaceous-Tertiary extinction. Science 208: 1095-1108

Alvarez W (1996) Trajectories of ballistic ejecta from the Chicxulub crater. In: Ryder G, Fastovski D, Gartner S (eds) The Cretaceous-Tertiary Event and Other Catastrophes in Earth History. Geological Society of America, Special Paper 307, pp 141-150

Alvarez W, Claeys P, Kieffer SW (1995) Emplacement of Cretaceous-Tertiary boundary shocked quartz from Chicxulub crater. Science 269: 930-935

Archibald JD (1996a) Testing extinction theories at the Cretaceous-Tertiary boundary using the vertebrate fossil record. In: MacLeod N, Keller G (eds) Cretaceous-Tertiary mass extinctions: Biotic and environmental changes. Norton and Company, New York, pp 373-397

Archibald JD (1996b) Dinosaur extinction and the end of an era: what the fossils say. Columbia University Press, New York, 237 pp

Arinobu T, Ishiwatari R, Kaiho K, Lamolda MA (1999) Spike of pyrosynthetic polycyclic aromatic hydrocarbons associated with an abrupt decrease in $\delta^{13}C$ of a terrestrial biomarker at the Cretaceous-Tertiary boundary at Caravaca, Spain. Geology 27: 723-726

Arthur MA, Zachos JC, Jones DS (1987) Primary productivity and the Cretaceous/Tertiary boundary event in the oceans. Cretaceous Research 8: 43-54

Askin RA, Jacobson SR (1996) Palynological change across the Cretaceous-Tertiary boundary on Seymour Island, Antarctica: Environmental and depositional factors. In: MacLeod N, Keller G (eds) Cretaceous-Tertiary Mass Extinctions. Norton, New York, pp 7-25

Aubry M-P (1999) The Cretaceous-Tertiary boundary: The El Kef blind test. Marine Micropaleontology 36: 65-66

Bains S, Corfield RM, Norris RD (1999) Mechanisms of climate warming at the end of the Paleocene. Science 285: 724-727

Bak M (1999) Uppermost Maastrichtian Radiolaria from the Magura Nappe deposits, Czech Outer Carpathians. Annales Societatis Geologorum Poloniae 69: 137-159

Barrera E, Savin SM, Thomas E, Jones CE (1997) Evidence for thermohaline-circulation reversals controlled by sea-level change in the latest Cretaceous. Geology 25: 715-718

Barrera E, Savin SM (1999) Evolution of late Campanian-Maastrichtian marine climates and oceans. In: Barrera E, Johnson CC (eds) Evolution of the Cretaceous Ocean-Climate System. Geological Society of America, Special Paper 332, pp 245-282

Beckmann JP (1960) Distribution of benthonic foraminifera at the Cretaceous-Tertiary boundary of Trinidad (West Indies). In: Rosenkrantz A, Brotzen F (eds) 21st International Geological Congress, Part V, Proceedings of Section 5: The Cretaceous-Tertiary Boundary. Det Berlingske Bogtrykkeri, Copenhagen, pp 57-69

Berner RA (1994) Geocarb II: A revised model of atmospheric CO_2 over Phanerozoic time. American Journal of Science 294: 56-91

Bertels-Psotka A (1995) The Cretaceous-Tertiary boundary in Argentina and its ostracodes. In: Ríha J (ed) Ostracoda and Biostratigraphy. Balkema, Rotterdam, pp 163-170

Birkelund T (1993) Ammonites from the Maastrichtian White Chalk of Denmark. Bull Geol Soc Denmark 40: 33-81

Bohor BF (1990) Shocked quartz and more; Impact signatures in Cretaceous/Tertiary boundary clays. In: Sharpton VL, Ward PD (eds) Global catastrophes in Earth History. Geological Society of America, Special Paper 247, pp 335-342

Bohor BF, Glass BP (1995) Origin and diagenesis of K/T impact spherules—From Haiti to Wyoming and beyond. Meteoritics 30: 182-198
Bohor BF, Izett GA (1986) Worldwide size distribution of shocked quartz at the K/T boundary: evidence for a North American impact site [abs]. Lunar and Planetary Science 17: 68-69
Bohor BF, Foord EE, Modreski PJ, Triplehorn DM (1984) Mineralogic evidence for an impact event at the Cretaceous-Tertiary boundary. Science 224: 867-869
Bohor BF, Modreski PJ, Foord EE (1987) Shocked quartz in the Cretaceous-Tertiary boundary clays: Evidence for a global distribution. Science 236: 705-709
Bostwick JA, Kyte FT (1996) The size and abundance of shocked quartz in Cretaceous-Tertiary boundary sediments from the Pacific basin. In: Ryder G, Fastovski D, Gartner S (eds) The Cretaceous-Tertiary Event and Other Catastrophes in Earth History. Geological Society of America, Special Paper 307, pp 403-416
Braccini E, Peypouquet JP (1995) A paleoceanological reconstruction of the Djebel-Dyr outcrop (Algeria) based on ostracodes from Paleocene to early Eocene. In: Riha J (ed) Ostracoda and Biostratigraphy. Balkema, Rotterdam, pp 171-181
Braga G, Bignot G (1986) Les bryozoaires de la formation d'age paléocène (danien probable) du Mont Aimé (Marne, Bassin Parisien). Geobios 19: 279-298
Bramlette MN (1965) Massive extinctions in biota at the end of Mesozoic time. Science 148: 1696-1699
Brett R (1992) The Cretaceous-Tertiary extinction: a lethal mechanism involving anhydrite target rocks. Geochimica et Cosmochimica Acta 56: 3603-3606
Brinkhuis H, Schioler P (1996) Palynology of the Geulhemmerberg Cretaceous-Tertiary boundary section (Limburg, SE Netherlands). Geologie en Mijnbouw 75: 193-213
Brinkhuis H, Bujak JP, Smit J, Versteegh GJM, Visscher H (1998) Dinoflagellate-based sea surface temperature reconstructions across the Cretaceous--Tertiary boundary. Palaeogeography, Palaeoclimatology, Palaeoecology 141: 67-83
Brood K (1976) Cyclostomatous bryozoa from the Paleocene and Maastrichtian of Majunga basin Madagascar. Geobios 9: 393-408
Brouwers EM, De Deckker P (1993) Late Maastrichtian and Danian ostracode faunas from northern Alaska: Reconstructions of environment and paleogeography. Palaios 8: 140-154
Brouwers EM, De Deckker P (1996) Earliest origins of Northern Hemisphere temperate nonmarine ostracode taxa: evolutionary development and survival through the Cretaceous-Tertiary boundary mass-extinction event. In: MacLeod N, Keller G (eds) Cretaceous-Tertiary Mass Extinctions. Norton, New York, pp 205-229
Bryan JR (1991) A Paleocene coral-algal-sponge reef from southwestern Alabama and the ecology of Early Tertiary reefs. Lethaia 24: 423-438
Bubík M, Bak M, Svábenická L (1999) Biostratigraphy of the Maastrichtian to Paleocene distal flysch sediments of the Raca unit in the Uzgrun section (Magura group of Nappes, Czech Republic). Geologica Carpathica 50: 33-48
Buck BJ, Mack GH (1995) Latest Cretaceous (Maastrichtian) aridity indicated by paleosols in the McRae Formation, south-central New Mexico. Cretaceous Research 16: 559-572
Canudo JI, Keller G, Molina E (1991) Cretaceous-Tertiary boundary extinction pattern and faunal turnover at Agost and Caravaca, SE Spain. Marine Micropaleontology 17: 319-341
Carozzi AV, Castro JC, Beltrami CV, Spadini AR, Wolff B (1981) Microfacies, depositional model and porosity control of Amap carbonates, Paleogene, Foz do Amazonas Basin, offshore NW Brazil. VIII Congreso Geol. Argentino Actas 2: 651-693
Casadío S (1998) Las ostras del límite Cretácico-Paleógeno de la cuenca Neuquina (Argentina). Su importancia bioestratigráfica y paleobiogeográfica. Ameghiniana 35: 449-471
Chauris H, Lerousseau J, Beaudoin B, Propson S, Montanari A (1998) Inoceramid extinction in the Gubbio basin (northeastern Apennines of Italy) and relations with mid-Maastrichtian environmental changes. Palaeogeography, Palaeoclimatology, Palaeoecology 139: 177-193

Cheetham AH (1971) Functional morphology and biofacies distribution of cheilostome Bryozoa in the Danian stage (Paleocene) of southern Scandinavia. Smithsonian Contributions to Paleobiology 6: 1-87

Cherchi A, Fantozzi P-L, Abdirahman H (1993) Micropaleontological data on the Jurassic-Cretaceous sequences and the Cretaceous-Paleocene boundary in northern Somalia (Bosaso region). Comptes Rendus de l'Academie des Sciences Serie II Mecanique Physique Chimie Sciences de l'Univers Sciences de la Terre 316: 1179-1185

Christensen WK (1997a) The Late Cretaceous belemnite family Belemnitellidae: Taxonomy and evolutionary history. Bull Geol Soc Denmark 44: 59–88

Christensen WK (1997b) Palaeobiogeography and migration in the Late Cretaceous belemnite family Belemnitellidae. Acta Palaeont Polonica 42: 457-495

Claeys P, Montanari A, Smit J, Asaro F, Alvarez W, Vega F, Bermudez J (1995) KT boundary sections in Southern Mexico: Closing in on Chicxulub [abs]. 4th Int. Workshop of the ESF Scientific Network on Impact Cratering and Evolution of Planet Earth: 58-59, Ancona, Italy.

Claeys P, Montanari A, Smit J, Alvarez W, Vega F (1996) KT boundary megabreccias in southern Mexico [abs]. Geol Soc America, 28th annual meeting, Abstracts with Programs 28: 181

Claeys P, Smit J, Montanari A, Alvarez W (1998) L'impact de Chicxulub et la limite Crétacé-Tertiaire dans la région du golfe du Mexique. Bull Soc Geol France 169: 3-9

Coccioni R, Galeotti S (1994) K-T boundary extinction; geologically instantaneous or gradual event? Evidence from deep-sea benthic foraminifera. Geology 22: 779-782

Coccioni R, Galeotti S (1998) What happened to small benthic foraminifera at the Cretaceous/Tertiary boundary? Bull Soc Geol France 169: 271-279

Copper P (1994) Ancient reef ecosystem expansion and collapse. Coral Reefs 13: 3-11

Courtillot V (1990) Deccan volcanism at the Cretaceous Tertiary boundary - past climatic crises as a key to the future. Global and Planetary Change 89: 291-299

Courtillot V (1999) Evolutionary Catastrophes: The Science of Mass Extinction. Cambridge University Press, Cambridge, 173 pp

Crame JA, Luther A (1997) The last inoceramid bivalves in Antarctica. Cretaceous Research 18: 179-195

Cros P, Lucas G (1982) Le recif coralligene a algues de Vigny (Danien, environs de Paris). Sciences de la terre (Nancy) 25: 3-37

Crowley TJ, North GR (1991) Paleoclimatology. Oxford University Press, Oxford, 349 pp

D'Hondt S, Pilson MEQ, Sigurdsson H, Hanson AK, Jr., Carey S (1994) Surface-water acidification and extinction at the Cretaceous-Tertiary boundary. Geology 22: 983-986

D'Hondt S, Herbert TD, King J, Gibson C (1996) Planktic foraminifera, asteroids, and marine production: Death and recovery at the Cretaceous-Tertiary boundary. In: Ryder G, Fastovski D, Gartner S (eds) The Cretaceous-Tertiary Event and Other Catastrophes in Earth History. Geological Society of America, Special Paper 307, pp 303-317

D'Hondt S, Donaghay P, Zachos JC, Luttenberg D, Lindinger M (1998) Organic carbon fluxes and ecological recovery from the Cretaceous-Tertiary mass extinction. Science 282: 276-279

Dailey DH (1983) Late Cretaceous and Paleocene benthic foraminifers from Deep Sea Drilling Project Site 516, Rio Grande Rise, western South Atlantic. Initial Reports of the Deep Sea Drilling Project 72: 757-782

Deloffre R, Snea P (1980) Dasycladales (algues vertes) du Danien recifal d'Aquitaine occidental (France SW). Bulletin des Centres de Recherches Exploration - Production Elf-Aquitaine, Memoires 2: 609-631

Deloffre R, Fleury J-J, Fourcade E, Michaud F (1989) Dasycladales (algues vertes) du Paléocène-Èocène inférieur du Mexique. Revue de Micropaléontologie 32: 3-15

Donovan AD, Baum GR, Blechschmidt GL, Loutit TS, Pflum CE, Vail PR (1988) Sequence stratigraphic setting of the Cretaceous-Tertiary boundary in central Alabama. In: Wilgus CK, Hastings BS, Kendall CGSC, Posamentier HW, Ross CA (eds) Sea-Level Changes - An Integrated Approach. SEPM Special Publications 42, pp 299-308

Drobne K, Barattolo F (1995) Algae and foraminifera along the K/T boundary from the Karst area [abs]. 4th Int. Workshop of the ESF Scientific Network on Impact Cratering and Evolution of Planet Earth: 70-71, Ancona, Italy

Drobne K, Ogorelec B, Plenicar M, Zucchi-Stolfa MK, Turnsek D (1988) Maastrichtian, Danian and Thanetian beds in Dolenja vas (NW Dinarides, Yugoslavia). Microfacies, foraminifers, rudists and corals. Razprave IV, Razreda Sazu 29: 147-224

Drobne K, Ogorelec B, Dolenec T, Marton E, Pugliese N (1997) Biota and abiota in the subtropic carbonate platform conditions during the K/T transition [abs]. Recoveries '97 Conference Abstract Book: 35-36.

Duncan PM (1880) Sind fossil corals and Alcyonaria. Memoirs of the Geological Survey of India, Palaeontologica Indica, Ser. 7 and 14, 1: 1-110

Evans NJ, Gregoire DC, Goodfellow WD, McInnes BI, Miles N, Veizer J (1993) Ru/Ir ratios at the Cretaceous-Tertiary boundary: Implications for PGE source and fractionation within the ejecta cloud. Geochimica et Cosmochimica Acta 57: 3149-3158

Fleming RF, Nichols DJ (1990) The fern-spore abundance anomaly at the Cretaceous-Tertiary boundary: a regional bioevent in western North America. In: Kauffman EG, Walliser OH (eds) Extinction Events in Earth History. Springer, Berlin, Lecture Notes in Earth Sciences, 30, pp 347-349

Foreman HP (1968) Upper Maestrichtian Radiolaria of California. Spec Pap Paleontol Assoc London 3: 1-82

Fourcade E, Alonzo M, Barrillas M, Bellier J-P, Bonneau M, Cosillo A, Cros P, Debrabant P, Gardin S, Masure E, Philip J, Renard M, Rocchia R, Romero J (1997) The K-T boundary in Southwestern Peten (Guatemala). Comptes Rendus de l'Academie des Sciences Serie II A325: 57-64

Fourcade E, Piccioni L, Escribá J, Rosselo E (1999) Cretaceous stratigraphy and palaeoenvironments of the Southern Petén Basin, Guatemala. Cretaceous Research 20: 793-811

Frakes LA, Probst J-L, Ludwig W (1994) Latitudinal distribution of paleotemperature on land and sea from early Cretaceous to middle Miocene. Comptes Rendus de l'Academie des Sciences Paris, série II 318: 1209-1218

Frank TD, Arthur MA (1999) Tectonic forcings of Maastrichtian ocean-climate evolution. Paleoceanography 14: 103-117

Fütterer DK (1990) Distribution of calcareous dinoflagellates at the Cretaceous-Tertiary boundary of Queen Maude Rise, eastern Weddell Sea, Antarctica (ODP Leg 113). Proceedings of the Ocean Drilling Program, Scientific Results 113: 533-548

Gaffin S (1987) Ridge volume dependence on seafloor generation rate and inversion using long term sea-level change. American Journal of Science 287: 596-611

Gardin S, Monechi S (1998) Palaeoecological change in middle to low latitude calcareous nannoplankton at the Cretaceous/Tertiary boundary. Bull Soc Geol France 169: 709-723

Gartner S (1996) Calcareous nannofossils at the Cretaceous-Tertiary boundary. In: MacLeod N, Keller G (eds) Cretaceous-Tertiary Mass Extinctions. Norton, New York, pp 27-47

Gilliland PM (1993) The skeletal morphology, systematics and evolutionary history of holothurians. Special Papers in Palaeontology 47: 1-147

Gilmour I, Russell SS, Arden JW, Lee MR, Franchi IA, Pillinger CT (1992) Terrestrial carbon and nitrogen isotopic ratios from Cretaceous-Tertiary boundary nanodiamonds. Science 258: 1624-1626

Ginsburg RN (1997) An attempt to resolve the controversy over the end-Cretaceous extinction of planktic foraminifera at El Kef, Tunisia using a blind test Introduction: background and procedures. Marine Micropaleontology 29: 67-68

Glasby GP, Kunzendorf H (1996) Multiple factors in the origin of the Cretaceous/Tertiary boundary: the role of environmental stress and Deccan Trap volcanism. Geologische Rundschau 85: 191-210

Glass BP, Burns CA (1988) Microkrystites: A new term for impact-produced glassy spherules containing primary crystallites. Proceedings, Lunar and Planetary Science Conference 18: 455-458

Habib D, Moshkovitz S, Kramer C (1992) Dinoflagellate and calcareous nannofossil response to sea-level change in Cretaceous-Tertiary boundary sections. Geology 20: 165-168

Habib D, Olsson RK, Liu C, Moshkovitz S (1996) High-resolution biostratigraphy of sea-level low, biotic extinction, and chaotic sedimentation at the Cretaceous-Tertiary boundary in Alabama, north of Chicxulub crater. In: Ryder G, Fastovski D, Gartner S (eds) The Cretaceous-Tertiary Event and Other Catastrophes in Earth History. Geological Society of America Special Paper 307, pp 243-262

Håkansson E, Thomsen E (1979) Distribution and types of bryozoan communities at the boundary in Denmark. In: Birkelund T, Bromley RD (eds) Cretaceous-Tertiary boundary events. I. The Maastrichtian and Danian of Denmark. University of Copenhagen, Copenhagen (vol 1, pp 78-91)

Håkansson E, Thomsen E (1999) Benthic extinction and recovery patterns at the K/T boundary in shallow water carbonates, Denmark. Palaeogeography, Palaeoclimatology, Palaeoecology 154: 67-85

Hallam A, Wignall PB (1997) Mass extinctions and their aftermath. Oxford University Press, Oxford, 320 pp

Hanna RK (1995) Some macrofossils from the Aqra Limestone Formation (Maastrichtian), Aqra, northern Iraq. Neues Jahrbuch für Geologie und Paläontologie, Monatshefte 1995: 295-304

Hansen T, Farrand RB, Montgomery HA, Billman HG, Blechschmidt G (1987) Sedimentology and extinction patterns across the Cretaceous-Tertiary boundary interval in east Texas, USA. Cretaceous Research 8: 229-252

Hansen TA, Farrell BR, Upshaw B (1993) The first 2 million years after the Cretaceous-Tertiary boundary in east Texas - rate and paleoecology of the molluscan recovery. Paleobiology 19: 251-265

Haq BU, Hardenbol J, Vail PR (1988) Mesozoic and Cenozoic chronostratigraphy and cycles of sea-level change. In: Wilgus CK, Hastings BS, Kendall CGSC, Posamentier HW, Ross CA (eds) Sea-Level Changes - An Integrated Approach, Tulsa, SEPM Special Publication 42, pp 71-108

Hardenbol J, Thierry J, Farley MB, Jacquin T, de Graciansky P-C, Vail PR (1998) Mesozoic and Cenozoic sequence chronostratigraphic framework of European basins. In: de Graciansky P-C, Hardenbol J, Jacquin T, Vail PR (eds) Mesozoic and Cenozoic Sequence Stratigraphy of European Basins. SEPM Special Publications, Tulsa, vol 60, pp 3-13

Hardie LA (1996) Secular variation in seawater chemistry: An explanation for the coupled secular variation in the mineralogies of marine limestones and potash evaporites over the past 600 m.y. Geology 24: 279-283

Hayami I (1997) Size changes of bivalves and a hypothesis about the cause of mass extinction. Fossils 62: 24-36

Heinberg C (1979) Bivalves from the latest Maastrichtian of Stevns Klint and their stratigraphic affinities. In: Birkelund T, Bromley RG (eds) Cretaceous-Tertiary boundary events. I. The Maastrichtian and Danian of Denmark. University of Copenhagen, Copenhagen, vol 1, pp 58-64

Heinberg C (1999) Lower Danian bivalves, Stevns Klint, Denmark: continuity across the K/T boundary. Palaeogeography, Palaeoclimatology, Palaeoecology 154: 87-106

Henderson RA, McNamara KJ (1985) Maastrichtian non-heteromorph ammonites from the Miria formation, Western Australia. Palaeontology 28: 35-88

Henriksson AS (1996) Calcareous nannoplankton productivity and succession across the Cretaceous-Tertiary boundary in the Pacific (DSDP Site 465) and Atlantic (DSDP Site 527) Oceans. Cretaceous Research 17: 451-477

Herm D, Hillebrandt A, Perch-Nielsen K (1981) Die Kreide/Tertiär Grenze im Lattengebirge (Nördliche Kalkalpen) in mikropaläontologischer Sicht. Geologica Bavarica 82: 319-344

Herrig E, Nestler H, Frenzel P, Reich M (1996) Discontinuity surfaces in the high Upper Cretaceous of northeastern Germany and their reflection by fossil associations. Göttinger Arbeiten zur Geologie und Paläontologie Sb 3: 107-111

Hildebrand AR, Boynton WV (1990) Proximal Cretaceous-Tertiary boundary impact deposits in the Caribbean. Science 248: 843-847

Hildebrand AR, Penfield GT, Kring DA, Pilkington M, Camargo A, Jacobsen SB, Boynton WV (1991) Chicxulub crater: a possible Cretaceous/Tertiary boundary impact crater on the Yucatan Peninsula, Mexico. Geology 19: 867-871

Hollis CJ, Rodgers KA, Parker RJ (1995) Siliceous plankton bloom in the earliest Tertiary of Marlborough, New Zealand. Geology 23: 835-838

Hollis CJ (1996) Radiolarian faunal change through the Cretaceous-Tertiary transition of eastern Marlborough, New Zealand. In: MacLeod N, Keller G (eds) Cretaceous-Tertiary mass extinctions: Biotic and environmental changes. Norton and Company, New York, pp 173-204

Hollis CJ (1997) Cretaceous-Paleocene radiolaria from eastern Marlborough, New Zealand. Institute of Geological and Nuclear Sciences, Monograph 17: 1-152

Hottinger L (1997) Recovery of K-strategy and subsequent evolutionary events in shallow benthic foraminifera after the K-T boundary [abs]. Recoveries '97 Conference Abstract Book: 30-31.

Hough RM, Gilmour I, Pillinger CT, Langenhorst F, Montanari A (1997) Diamonds from the iridium-rich K-T boundary layer at Arroyo el Mimbral, Tamaulipas, Mexico. Geology 25: 1019-1022

Huber BT (1996) Evidence for planktonic foraminifer reworking versus survivorship across the Cretaceous-Tertiary boundary at high latitudes. In: Ryder G, Fastovski D, Gartner S (eds) The Cretaceous-Tertiary Event and Other Catastrophes in Earth History. Geological Society of America, Special Paper 307, pp 319-334

Hudson JD (1997) Discussion on the Cretaceous-Tertiary biotic transition. Journal of the Geological Society 155: 413-415

Ivanov M (1993) Uppermost Maastrichtian ammonites from the uninterrupted Upper Cretaceous: Paleogene section at Bjala (East Bulgaria). Geologica Balcanica 23: 54

Izett GA (1990) The Cretaceous/Tertiary boundary interval, Raton Basin, Colorado and New Mexico, and its content of shock-metamorphosed minerals; evidence relevant to the K/T boundary impact-extinction theory. Geological Society of America Special Paper 249: 1-100

Izett GA, Maurrasse FJ-MR, Lichte FE, Meeker GP, Bates R (1990) Tektites in Cretaceous-Tertiary boundary rocks on Haiti. U.S. Geological Survey Open-File Report 90-635: 1-31

Jablonski D (1986) Background and mass extinctions: the alternation of macroevolutionary regimes. Science 231: 129-133

Jablonski D (1991) Extinctions: A paleontological perspective. Science 253: 754-757

Jablonski D (1996) Mass extinctions: persistent problems and new directions. In: Ryder G, Fastovski D, Gartner S (eds) The Cretaceous-Tertiary Event and Other Catastrophes in Earth History. Geological Society of America, Special Paper 307, pp 1-10

Jablonski D (1997) Progress at the K-T boundary. Nature 387: 354-355

Jablonski D (1998) Geographic variation in the molluscan recovery from the end-Cretaceous extinction. Science 279: 1327-1330

Jablonski D, Raup DM (1995) Selectivity of end-Cretaceous marine bivalve extinctions. Science 268: 389-391

Jagt JWM (1995) Late Cretaceous and early Cainozoic crinoid assemblages from northeast Belgium and the southeast Netherlands. In: Emson RH, Smith AB, Campbell AC (eds) Echinoderm Research 1995. Balkema, Rotterdam, pp 185-196

Jansa L (1993) Cometary impacts into ocean: their recognition and the threshold constraint for biological extinctions. Palaeogeography, Palaeoclimatology, Palaeoecology 104: 271-286

Jeffery CH (1997a) All change at the Cretaceous-Tertiary boundary? Echinoids from the Maastrichtian and Danian of the Mangyshlak Peninsula, Kazakhstan. Palaeontology 40: 659-712

Jeffery CH (1997b) Dawn of echinoid nonplanktotrophy: Coordinated shifts in development indicate environmental instability prior to the K-T boundary. Geology 25: 991-994

Jeffery CH (2001) Heart urchins at the Cretaceous/Tertiary boundary: a tale of two clades. Paleobiology 27: 140–158

Jeffery CH, Smith AB (1998) Estimating extinction levels for echinoids across the Cretaceous-Tertiary boundary. In: Mooi R, Telford M (eds) Echinoderms. A. A. Balkema, Rotterdam, pp 695-701

Jelaska V, Bulic J (1975) Paleogeografska razmatranja gornjokrednih i paleogenskih klastita sjeverne Bosne i njihovo moguce naftnogeolosko znacenje. Nafta 26: 371-385

Johansen MB (1987) Brachiopods from the Maastrichtian-Danian boundary sequence at Nye Klov, Jylland, Denmark. Fossils and Strata 20: 1-99

Johansen MB (1988) Brachiopod extinction in the Upper Cretaceous to lowermost Tertiary chalk of Northwest Europe. Revista Espanola de Paleontología Extraordinario: 41-56

Johansen MB (1989) Background extinction and mass extinction of the brachiopods from the chalk of northwest Europe. Palaios 4: 243-250

Johansen MB (1989) Adaptive radiation, survival and extinction of brachiopods in the northwest European Upper Cretaceous-Lower Paleocene chalk. Palaeogeography, Palaeoclimatology, Palaeoecology 74: 147-204

Johnson KR, Hickey LJ (1990) Megafloral change across the Cretaceous/Tertiary boundary in the northern Great Plains and Rocky Mountains, U.S.A. In: Sharpton VL, Ward PD (eds) Global Catastrophes in Earth History. Geological Society of America Special Paper 247, pp 433-443

Johnson CC, Kauffman EG (1990) Originations, radiations and extinctions of Cretaceous rudistid bivalve species in the Caribbean province. In: Kauffman EG, Walliser OH (eds) Extinction Events in Earth History. Springer, Berlin, Lecture Notes in Earth Sciences, 30, pp 305-324

Johnson CC, Kauffman EG (1996) Maastrichtian extinction patterns of Caribbean Province rudistids. In: MacLeod N, Keller G (eds) Cretaceous-Tertiary Mass Extinctions. Norton, New York, pp 231-273

Jones DS, Nicol D (1986) Origination, survivorship and extinction of rudist taxa. Journal of Paleontology 60: 107-115

Jones TP, Lim B (2000) Extraterrestrial impacts and wildfires. Palaeogeography, Palaeoclimatology, Palaeoecology 164: 57-66

Jorgensen NO (1979) Maastrichtian ostracods from Denmark. In: Birkelund T, Bromley RG (eds) Cretaceous-Tertiary boundary events. I. The Maastrichtian and Danian of Denmark. University of Copenhagen, pp 95-100

Kaiho K (1992) A low extinction rate of intermediate-water benthic foraminifera at the Cretaceous/Tertiary boundary. Marine Micropaleontology 18: 229-259

Kaiho K, Kajiwara Y, Tazaki K, Ueshima M, Takeda N, Kawahata H, Arinobu T, Ishiwatari R, Hirai A, Lamolda MA (1999) Oceanic primary productivity and dissolved oxygen levels at the Cretaceous/Tertiary boundary: Their decrease, subsequent warming, and recovery. Paleoceanography 14: 511-524

Kaminsiki MA, Kuhnt W, Moullade M (1999) The evolution and paleobiogeography of abyssal agglutinated foraminifera since the Early Cretaceous: A tale of four faunas. Neues Jahrbuch für Geologie und Paläontologie, Abhandlungen 212: 401-439

Keller G (1988a) Extinction, survivorship and evolution of planktic foraminifera across the Cretaceous/Tertiary boundary at El Kef, Tunisia. Marine Micropaleontology 13: 239-263

Keller G (1988b) Biotic turnover in benthic foraminifera across the Cretaceous/Tertiary boundary at El Kef, Tunisia. Palaeogeography, Palaeoclimatology, Palaeoecology 66: 153-171

Keller G (1989) Extended period of extinctions across the Cretaceous/Tertiary boundary in planktonic foraminifera of continental shelf sections: Implications for impact and volcanism theories. Geol Soc America Bull 101: 1408-1419

Keller G, Barrera E, Schmitz B, Mattson E (1993) Gradual mass extinction, species survivorship, and long-term environmental changes across the Cretaceous-Tertiary boundary in high latitudes. Geol Soc America Bull 105: 979-997

Keller G, Stinnesbeck W, Guadelupe Lopez-Oliva J (1994) Age, deposition and biotic effects of the Cretaceous-Tertiary boundary event at Mimbral, NE Mexico. Palaios 9: 144-157

Keller G, Stinnesbeck W (1996) Sea-level changes, clastic deposits, and megatsunamis across the Cretaceous-Tertiary boundary. In: MacLeod N, Keller G (eds) Cretaceous-Tertiary Mass Extinctions. Norton, New York, pp 415-449

Keller G, Adatte T, Hollis C, Ordóñez M, Zambrano I, Jiménez N, Stinnesbeck W, Aleman A, Hale-Erlich W (1997) The Cretaceous/Tertiary boundary event in Ecuador: reduced biotic effects due to eastern boundary current setting. Marine Micropaleontology 31: 97-133

Keller G, Adatte T, Stinnesbeck W, Stüben D, Kramar U, Berner Z, Li L, von Salis Perch-Nielsen K (1998) The Cretaceous-Tertiary transition on the shallow Saharan platform of southern Tunisia. Geobios 30: 951-975

Kennedy WJ (1986) The ammonite fauna of the Calcaire à *Baculites* (Upper Maastrichtian) of the Cotentin Peninsula (Manche, France). Palaeontology 29: 25-83

Kennedy WJ (1989) Thoughts on the evolution and extinction of Cretaceous ammonites. Proceedings of the Geological Association 100: 251-280

Kennedy WJ, Henderson RA (1992) Heteromorph ammonites from the Upper Maastrichtian of Pondicherry, South India. Palaeontology 35: 693-731

Kennedy WJ, Klinger HC (1975) Cretaceous faunas from Zululand and Natal South Africa. Introduction: stratigraphy. Bulletin of the British Museum 25: 263-315

Kennedy WJ, Bilotte M, Lepicard B, Segura F (1986) Upper Campanian and Maastrichtian ammonites from the Petites-Pyrenees (Southern France). Eclogae Geologicae Helvetiae 79: 1001-1037

Kennedy WJ, Cobban WA, Landman NH (1997) Maastrichtian ammonites from the Severn Formation of Maryland. American Museum Novitates 0: 1-30

Kennedy WJ, Landman NH, Christensen WK, Cobban WA, Hancock JM (1998) Marine connections in North America during the late Maastrichtian: Palaeogeographic and palaeobiogeographic significance of Jeletzkytes nebrascensis Zone cephalopod fauna from the Elk Butte Member of the Pierre Shale, SE South Dakota and NE Nebraska. Cretaceous Research 19: 745-775

Khadkikar AS (1999) The influence of Deccan volcanism on climate: insights from lacustrine intertrappean deposits, Anjar, western India. Palaeogeography, Palaeoclimatology, Palaeoecology 147: 141-149

Kienel U (1994) Die Entwicklung der kalkigen Nannofossilien und der kalkigen Dinoflagellaten-Zysten an der Kreide/Tertiär-Grenze in Westbrandenburg im Vergleich mit Profilen in Nordjütland und Seeland (Dänemark). Berliner geowissenschaftliche Abhandlungen E12: 1-87

Kiessling W, Flügel E, Golonka J (1999) Paleo Reef Maps: Evaluation of a comprehensive database on Phanerozoic reefs. AAPG Bulletin 83: 1552-1587

Klaver GT, van Kempen TMG, Bianchi FR, van der Gaast SJ (1987) Green spherules as indicators of the Cretaceous/Tertiary boundary in Deep Sea Drilling Project Hole 603B: Initial Reports of the Deep Sea Drilling Project 93: 1039-1055

Koch CF (1996) Latest Cretaceous mollusc species 'fabric' of the US Atlantic and Gulf Coastal Plain: a baseline form measuring biotic recovery. In: Hart MB (ed) Biotic recovery from mass extinction events. Geological Society of London Special Publication 102: 309-317

Koeberl C, Armstrong RA, Reimold WU (1997) Morokweng, South Africa: a large impact structure of Jurassic-Cretaceous boundary age. Geology 25: 731-734

Köhler E, Borza K (1984) Oberkreide mit Orbitoiden in den Kleinen Karpaten. Geologica Carpathica 35: 195-204

Kucera M, Malmgren BA (1997) Foraminiferal dissolution at shallow depths of the Walvis Ridge and Rio Grande Rise during the latest Cretaceous: Inferences for deep-water circulation in the South Atlantic. Palaeogeography, Palaeoclimatology, Palaeoecology 129: 195-212

Kucera M, Malmgren BA (1998) Terminal Cretaceous warming event in the mid-latitude South Atlantic Ocean: evidence from poleward migration of Contusotruncana contusa (planktonic foraminifera) morphotypes. Palaeogeography, Palaeoclimatology, Palaeoecology 138: 1-15

Kuehn O (1960) Neue Untersuchungen über die Dänische Stufe in Österreich. In: Rosenkrantz A, Brotzen F (eds) 21st International Geological Congress, Part V, Proceedings of Section 5: The Cretaceous-Tertiary Boundary. Det Berlingske Bogtrykkeri, Copenhagen, pp 162-169

Kuhn W, Kirsch K-H (1992) Ein Kreide/Tertiär-Grenzprofil aus dem Helvetikum nördlich von Salzburg (Österreich). Mitteilungen der Bayerischen Staatssammlung für Paläontologie und Historische Geologie 32: 23-35

Kuhnt W, Kaminski MA (1993) Changes in the community structure of deep water agglutinated foraminifers across the K/T boundary in the Basque basin (northern Spain). Revista Española de Micropaleontología 25: 57-92

Kuhnt W, Moullade M, Kaminski MA (1998) Upper Cretaceous, K/T boundary, and Paleocene agglutinated foraminifera from Hole 959D (ODP Leg 159, Cote d'Ivorie - Ghana Transform Margin). Proceedings of the Ocean Drilling Program, Scientific Results 159: 389-441

Kyte FT, Bohor BF (1995) Nickel-rich magnesiowüstite in Cretaceous/Tertiary boundary spherules crystallized from ultramafic, refractory silicate liquids. Geochimica et Cosmochimica Acta 59: 4967-4974

Kyte FT, Zhou Z, Wasson JT (1980) Siderophile-enriched sediments from the Cretaceous-Tertiary boundary. Nature 288: 651-656

Kyte FT, Bostwick JA, Zhou L (1996) The Cretaceous-Tertiary boundary on the Pacific plate: composition and distribution of impact debris. In: Ryder G, Fastovski D, Gartner S (eds) The Cretaceous-Tertiary Event and Other Catastrophes in Earth History. Geological Society of America, Special Paper 307, pp 389-402

Li L, Keller G (1998) Abrupt deep-sea warming at the end of the Cretaceous. Geology 26: 995-998

Liebau A (1982) Faunengeschichte epineritischer Ostrakoden an der Kreide-Tertiär-Grenze in Mitteleuropa. Neues Jahrbuch für Geologie und Paläontologie, Abhandlungen 164: 280-288

Lipps JH, Hickman CS (1982) Origin, age, and evolution of Antarctic and deep-sea faunas. In: Ernst WG, Morris JG (eds). Prentice-Hall, Englewood Cliffs, pp 324-356

Luterbacher H, Premoli-Silva I (1964) Biostratigrafia del limite Cretaceo-Terziario nell Apennino centrale. Rivista Italiana di Paleontologia e Stratigrafia 70: 67-128

Macdougall JD (1988) Seawater strontium isotopes acid rain and the Cretaceous-Tertiary boundary. Science 239: 485-487

Machalski M (1996) Scaphitid ammonite correlation of the late Maastrichtian deposits in Poland and Denmark. Acta Palaeontologica Polonica 41: 369-383

MacLeod N (1996) Nature of the Cretaceous-Tertiary planktonic foraminiferal record: Stratigraphic confidence intervals, Signor-Lipps effect, and patterns of survivorship. In: MacLeod N, Keller. G (eds) Cretaceous-Tertiary mass extinctions: Biotic and environmental changes. W. W. Norton and Company, New York, pp 85-138

MacLeod N, Keller G (1991) How complete are Cretaceous/Tertiary boundary sections? A chronostratigraphic estimate based on graphic correlation. Geological Society of America Bulletin 103: 1439-1457

MacLeod N, Keller G (1994) Comparative biogeographic analysis of planktic foraminiferal survivorship across the Cretaceous-Tertiary (K-T) boundary. Paleobiology 20: 143-177

MacLeod KG, Huber BT, Ward PD (1996) The biostratigraphy and paleobiogeography of Maastrichtian inoceramids. In: Ryder G, Fastovski D, Gartner S (eds) The Cretaceous-Tertiary Event and Other Catastrophes in Earth History. Geological Society of America, Special Paper 307, pp 361-374

MacLeod N, Rawson PF, Forey PL, Banner FT, Boudagher-Fadel MK, Bown PR, Burnett JA, Chambers P, Culver S, Evans SE, Jeffery C, Kaminski MA, Lord AR, Milner AC, Milner AR, Morris N, Owen E, Rosen BR, Smith AB, Taylor PD, Urquhart E, Young JR (1997) The Cretaceous-Tertiary biotic transition. Journal of the Geological Society of London 154: 265-292

MacLeod KG, Huber BT, Fullagar PD (2001) Evidence for a small (~0.000 030) but resolvable increase in seawater $^{87}Sr/^{86}Sr$ ratios across the Cretaceous-Tertiary boundary. Geology 29: 303-306

Maddocks RF (1985) Ostracoda of the Cretaceous-Tertiary contact sections in central Texas. Gulf Coast Association of Geological Societies Transactions 35: 445-456

Mai H (1999) Paleocene coccoliths and coccospheres in deposits of the Maastrichtian stage at the type locality' and type area in SE Limburg, The Netherlands. Marine Micropaleontology 36: 1-12

Majoran S, Widmark JGV (1998) Response of deep-sea ostracod assemblages to Late Cretaceous palaeoceanographical changes: ODP site 689 in the Southern Ocean. Cretaceous Research 19: 843-872

Majoran S (1999) Palaeoenvironment of Maastrichtian ostracods from ODP Holes 1049B, 1050C and 1052E in the Western North Atlantic. Journal of Micropalaeontology 18: 125-136

Marincovich L, Jr. (1993) Danian molluscs from the Prince Creek Formation, northern Alaska, and implications for Arctic Ocean paleogeography. Paleontological Society Memoir 35: 1-35

Markov AV, Solovjev AN (1998) Echinoids at the Cretaceous/Paleogene boundary. In: Mooi R, Telford M (eds) Echinoderms. A. A. Balkema, Rotterdam, pp 733-733

McArthur JM, Thirlwall MF, Engkilde M, Zinsmeister WJ, Howarth RJ (1998) Strontium isotope profiles across K/T boundary sequences in Denmark and Antarctica. Earth Planet Sci Letters 160: 179-192

McKinney FK, Lidgard S, Jr. JJS, Taylor PD (1998) Decoupled temporal patterns of evolution and ecology in two post-Paleozoic clades. Science 281: 807-809

Miller KG (1982) Cenozoic benthic foraminifera: case histories of paleoceanographic and sea-level changes. In: Broadhead TW (ed) Foraminifera, notes for a short course. University of Tennessee, Studies in Geology 6, pp 107-126

Miller KG, Fairbanks RG, Mountain GS (1987) Tertiary oxygen isotope synthesis, sea level history, and continental margin erosion. Paleoceanography 2: 1-19

Miller KG, Barrera E, Olsson RK, Sugarman PJ, Savin SM (1999) Does ice drive early Maastrichtian eustasy? Geology 27: 783-786

Montanari A (1991) Authigenesis of impact spheroids in the K/T boundary clay from Italy: New constraints for high-resolution stratigraphy of terminal Cretaceous events. Journal of Sedimentary Petrology 61: 315-339

Montanari A, Claeys P, Asaro F, Bermudez J, Smit J (1994) Preliminary stratigraphy and iridium and other geochemical anomalies across the KT boundary in the Bochil section (Chiapas, Southeastern Mexico) [abs]. Lunar and Planetary Institute Contribution 825: 84-85

Moshkovitz S, Habib D (1993) Calcareous nannofossil and dinoflagellate stratigraphy of the Cretaceous-Tertiary boundary, Alabama and Georgia. Micropaleontology 39: 167-191

Moussavian E (1992) On Cretaceous bioconstructions: composition and evolutionary trends of crust-building associations. Facies 26: 117-144

Moussavian E, Vecsei A (1995) Paleocene reef sediments from the Maiella carbonate platform, Italy. Facies 32: 213-222

Nomura R (1991) Paleoceanography of upper Maestrichtian to Eocene benthic foraminiferal assemblages at Sites 752, 753 and 754, eastern Indian Ocean. Proceedings of the Ocean Drilling Program, Scientific Results 121: 3-29

Norris RD, Huber BT, Self-Trail J (1999) Synchronicity of the K-T oceanic mass extinction and meteorite impact: Blake Nose, western North Atlantic. Geology 27: 419-422

O'Keefe JD, Ahrens TJ (1982) Impact mechanics of the Cretaceous-Tertiary extinction bolide. Nature 298: 123-127

Olsson RK, Liu C (1993) Controversies on the placement of Cretaceous-Paleogene boundary and the K/P mass extinction of planktonic foraminifera. Palaios 8: 127-139

Olsson RK, Miller KG, Browning JV, Habib D, Sugarman PJ (1997) Ejecta layer at the Cretaceous-Tertiary boundary, Bass River, New Jersey (Ocean Drilling Program Leg 174AX): Geology, 25: 759-762

Pardo A, Adatte T, Keller G, Oberhänsli H (1999) Paleoenvironmental changes across the Cretaceous--Tertiary boundary at Koshak, Kazakhstan, based on planktic foraminifera and clay mineralogy. Palaeogeography, Palaeoclimatology, Palaeoecology 154: 247-273

Perch-Nielsen K, McKenzie J, He Q (1982) Biostratigraphy and isotope stratigraphy and the 'catastrophic' extinction of calcareous nannoplankton at the Cretaceous/Tertiary boundary. In: Silver LT, Schultz PH (eds) Geological Implications of Impacts of Large Asteroids and Comets on the Earth. Geological Society of America, Special Paper 190, pp 353-371

Peybernes B, Fondecave-Wallez M-J, Eichene P, Robin E, Rocchia R (1998) The Cretaceous-Tertiary boundary in marine environments from Central Pyrenees (sub-Pyrenean zone, France). Comptes Rendus de l'Academie des Sciences Serie II 326: 647-653

Philip J, Masse J-P, Camoin G (1995) Tethyan carbonate platforms. In: Nairn AEM, Ricou L-E, Vrielynck B, Dercourt J (eds) The Ocean Basins and Margins. Volume 8: TheTethys Ocean. Plenum Press, New York (vol 8, pp 239-265)

Phillips J (1860) Life on the Earth: Its Origin and Succession. Macmillan, Cambridge

Pillmore CL, Flores RM (1987) Stratigraphy and depositional environments of the Cretaceous - Tertiary boundary clay and associated rocks, Raton basin, New Mexico and Colorado. In: Fassett JE, J. K. Rigby J (eds) The Cretaceous-Tertiary boundary in the San Juan and Raton Basins, New Mexico and Colorado. Geological Society of America Special Paper 209, pp 111-130

Pindell JL (1994) Evolution of the Gulf of Mexico and the Caribbean. In: Donovan SK, Jackson TA (eds) Caribbean Geology: an introduction. University of the West Indies Publishers, Kingston, Jamaica, pp 13-39

Plenicar M, Caffau M, Jurkovsek B (1995) Morphologic changes of rudist shell as reflection of the environment at the terminal Cretaceous layers [abs]. 4th Int. Workshop of the ESF Scientific Network on Impact Cratering and Evolution of Panet Earth. p 133

Poag CW (1997) Roadblocks on the Killcurve: Testing the Raup hypothesis. Palaios 12: 582-590

Polsak A (1985) The boundary between the Cretaceous and Tertiary in terms of the stratigraphy and sedimentology of the biolithitic complex in Mount Medvednica northern Croatia, Yugoslavia. Prirodoslovna Istrazivanja Acta Geologica 15: 1-23

Pons JM, Gallemi J, Höfling R, Moussavian E (1994) Los Hippurites del Barranc del Racó, microfacies y fauna asociada (Maastrichtiense Superior, sur de la provincia de Valencia). Cuadernos de geologia iberica 18: 271-307

Pope KO, Baines KH, Ocampo AC, Ivanov BA (1994) Impact winter and the Cretaceous-Tertiary extinctions: Results of a Chicxulub asteroid impact model. Earth Planet Sci Letters 128: 719-725

Pospichal JJ (1994) Calcareous nannofossils at the K-T boundary, El Kef: no evidence for stepwise, gradual, or sequential extinction. Geology 22: 99-102

Pospichal JJ (1996a) Calcareous nannofossils and clastic sediments of the Cretaceous-Tertiary boundary of northeastern Mexico. Geology 24: 255-258

Pospichal JJ (1996b) Calcareous nannoplankton mass-extinctions at the Cretaceous-Tertiary boundary: an update. In: Ryder G, Fastovski D, Gartner S (eds) The Cretaceous-Tertiary Event and Other Catastrophes in Earth History. Geological Society of America, Special Paper 307, pp 335-360

Premoli Silva I, Luterbacher HP (1966) The Cretaceous-Tertiary boundary in the Southern Alps (Italy). Rivista Italiana di Paleontologia 72: 1183-1266

Premovic PI, Nikolic ND, Tonsa IR, Pavlovic MS, Premovic MP, Dulanovic DT (2000) Copper and copper(II) porphyrins of the Cretaceous-Tertiary boundary at Stevns Klint (Denmark). Earth and Planetary Science Letters 177: 105-118

Rao RL, Pia J (1936) Fossil algae from the uppermost Cretaceous beds (Niniyur group) of the Trichinopoly district South India. Palaeontologia Indica 21: 1-49

Rasmussen HW (1979) Crinoids, asteroids and ophiuroids in relation to the boundary. In: Birkelund T, Bromley RG (eds) Cretaceous-Tertiary boundary events. I. The Maastrichtian and Danian of Denmark. University of Copenhagen, Copenhagen, pp 65-71

Reiskind J (1973) Marine concretionary faunas of the uppermost Bearpaw shale (Maastrichtian) in eastern Montana and southwestern Saskatchewan. In: Caldwell WGE (ed) The Cretaceous system in the western interior of North America. Geological Association of Canada, Waterloo, Ontario, pp 235-252

Rhodes MC, Thayer CW (1991) Mass extinctions: ecological selectivity and primary production. Geology 19: 877-880

Robin E, Boclet D, Bonte P, Froget L, Jehanno C, Rocchia R (1991) The stratigraphic distribution of Ni-rich spinels in Cretaceous-Tertiary boundary rocks at El-Kef (Tunisia), Caravaca (Spain) and Hole-761C (Leg-122). Earth Planet Sci Letters 107: 715-721

Robin E, Bonté P, Froget L, Jéhanno C, Rocchia R (1992) Formation of spinels in cosmic objects during atmospheric entry: a clue to the Cretaceous-Tertiary boundary event. Earth Planet Sci Letters 108: 181-190

Robin E, Froget L, Jéhanno C, Rocchia R (1993) Evidence for a KT impact event in the Pacific Ocean. Nature 363: 615-617

Robin E, Froget L, Jehanno C, Rocchia R (1994) Evidence for a K/T impact event in the Pacific Ocean. Nature 363: 615-617

Rocchia R, Robin E, Smit J, Pierrard O, Lefevre I (1999) Iridium and Ni-rich spinel in an ammonite from the uppermost Maastrichtian of the K/T section of Bidart (French Basque Country) [abs]. In: Buffetaut E, Le Loeuff J (eds) Workshop on Geological and Biological Evidence for Global Catastrophes. Esperaza/Quillan (Aude, France), 26-30 Sept. 1999, Programme, Abstracts and Field Guide, p 62

Rosen BR (1988) From fossils to earth history: applied historical biogeography. In: Myers AA, Giller PS (eds) Analytical biogeography; an integrated approach to the study of animal and plant distributions. Chapman and Hall, London, pp 437-481

Rosen BR (2000) Algal symbiosis, and the collapse and recovery of reef communities: Lazarus corals across the K-T boundary. In: Culver SJ, Rwason PF (eds.) Biotic response to global change: The last 145 million years. Cambridge University Press, Cambridge, pp 164-180

Rosen BR, Turnsek D (1989) Extinction patterns and biogeography of scleractinian corals across the Cretaceous/Tertiary boundary. Memoirs of the Association of Australasian Palaeontologists 8: 355-370

Rosenkrantz A (1960) Danian Mollusca from Denmark. In: Rosenkrantz A, Brotzen F (eds) 21st International Geological Congress, Part V, Proceedings of Section 5: The Cretaceous-Tertiary Boundary. Det Berlingske Bogtrykkeri, Copenhagen, pp 193-198

Roy JM, McMenamin MAS, Alderman SE (1990) Trophic differences, originations and extinctions during the Cenomanian and Maastrichtian stages of the Cretaceous. In: Kauffman EG, Walliser OH (eds) Extinction Events in Earth History. Springer, Berlin, Lecture Notes in Earth Sciences 30, pp 299-303

Roy K (1994) Effects of Mesozoic marine revolution on the taxonomic, morphologic and biogeographic evolution of a group: Aporrhaid gastropods during the Mesozoic. Paleobiology 20: 274-296

Roy K (1996) The roles of mass extinction and biotic interaction in large-scale replacements: a reexamination sing the fossil record of stromboidean gastropods. Paleobiology 22: 436-452

Ryder G (1996) The unique significance and origin of the Cretaceous/Tertiary boundary: Historical context and burdens of proof. In: Ryder G, Fastovski D, Gartner S (eds) The Cretaceous-Tertiary Event and Other Catastrophes in Earth History. Geological Society of America, Special Paper 307, pp 31-38

Saito T, Yamanoi T, Kaiho K (1986) End-Cretaceous devastation of terrestrial flora in the Boreal Far East. Nature 323: 253-255

Saltzman ES, Barron EJ (1982) Deep circulation in the Late Cretaceous: Oxygen isotope paleotemperatures from Inoceramus remains in DSDP cores. Palaeogeography, Palaeoclimatology, Palaeoecology 40: 167-181

Scheibner E (1968) Contribution to the knowledge of the Palaeogene reef-complexes of the Myjava-Hricov-Haligovka Zone (West Carpathians). Mitteilungen der Bayerischen Staatssammlung für Paläontologie und historische Geologie 8: 67-97

Schmitz B, Keller G, Stenvall O (1992) Stable isotope and foraminiferal changes across the Cretaceous-Tertiary boundary at Stevns Klint, Denmark: arguments for long-term oceanic instability before and after bolide-impact event. Palaeogeography, Palaeoclimatology, Palaeoecology 96: 233-260

Schultz PH, D'Hondt S (1996) Cretaceous-Tertiary (Chicxulub) impact angle and its consequences. Geology 24: 963-966

Schuster F (1996) Paleocene coral reefs and related facies associations, Kharga Oasis, Western Desert, Egypt. In: Reitner J, Neuweiler F, Gunkel F (eds) Global and Regional Controls on Biogenic Sedimentation. I. Reef Evolution. Göttinger Arbeiten zu Geologie und Paläontologie, Sonderband, Göttingen, pp 169-174

Scott AC, Lomax BH, Collinson ME, Upchurch GR, Beerling DJ (2000) Fire across the K-T boundary: initial results from the Sugarite Coal, New Mexico, USA. Palaeogeography, Palaeoclimatology, Palaeoecology 164: 151-165

Sheehan PM, Hansen TA (1986) Detritus feeding as a buffer to extinction at the end of the Cretaceous. Geology 14: 868-870

Sheehan PM, Fastovsky DE, Hoffmann RG, Berghaus CB, Gabriel DL (1991) Sudden extinction of the dinosaurs: Latest Cretaceous, Upper Great Plains, U.S.A. Science 254: 835-839

Sheehan PM, Coorough PJ, Fastovski DE (1996) Biotic selectivity during the K/T and Late Ordovician extinction events. In: Ryder G, Fastovski D, Gartner S (eds) The Cretaceous-Tertiary Event and Other Catastrophes in Earth History. Geological Society of America, Special Paper 307, pp 477-490

Sheehan PM, Fastovsky DE, Barreto C, Hoffmann RG (2000) Dinosaur abundance was not declining in a "3 m gap" at the top of the Hell Creek Formation, Montana and North Dakota. Geology 28: 523-526

Shigeta Y (1989) Systematics of the ammonite genus tetragonites from the upper cretaceous of Hokkaido Japan. Transactions and Proceedings of the Palaeontological Society of Japan, New Series 156: 319-342

Sigurdsson H, D'Hondt S, Arthur MA, Bralower TJ, Zachos JC, Van Fossen M, Channell JET (1991) Glass from the Cretaceous-Tertiary boundary in Haiti. Nature 349: 482-487

Simon E (1998) Maastrichtian brachiopods from Ciply: Palaeoecological and stratigraphical significance. Bulletin de l'Institut Royal des Sciences Naturelles de Belgique Sciences de la Terre 68: 181-232

Sirel E (1996) Description and geographic, stratigraphic distribution of the species of Laffitteina Marie from the Maastrichtian and Paleocene of Turkey. Revue de Paléobiologie 15: 9-35

Sirel E, Dager Z, Sözeri B (1986) Some biostratigraphic and paleogeographic observations on the Cretaceous/Tertiary boundary in the Haymana Polatli region (Central Turkey). In: Walliser OH (ed) Global bio-events, Lecture Notes in Earth Sciences 8, pp 385-396

Sloss LL (1963) Sequences in the cratonic interior of North America. Geological Society of America Bulletin 74: 93-113

Sloss LL (1988) Tectonic evolution of the craton in Phanerozoic time. In: Sloss LL (ed) Sedimentary cover - North American craton; U.S. Geological Society of America, Boulder, The Geology of North America D-2, pp 25-51

Smit J (1982) Extinction and evolution of planktonic foraminifera after a major impact at the Cretaceous/Tertiary boundary. In: Silver LT, Schultz PH (eds) Geological Implications of Impacts of Large Asteroids and Comets on the Earth. Geological Society of America, Special Paper 190, pp 329-352

Smit J (1990) Meteorite impact, extinctions and the Cretaceous-Tertiary boundary. Geologie en Mijnbouw 69: 187-204

Smit J (1999) The global stratigraphy of the Cretaceous-Tertiary boundary impact ejecta. Annual Reviews of Earth and Planetary Science 27: 75-113

Smit J, Romein AJT (1985) A sequence of events across the Cretaceous-Tertiary boundary. Earth Planet Sci Letters 74: 155-170

Smit J, Montanari A, Swinburne NHM, Alvarez W, Hildebrand AR, Margolis SV, Claeys P, Lowrie W, Asaro F (1992) Tektite-bearing, deep-water clastic unit at the Cretaceous-Tertiary boundary in northeastern Mexico. Geology 20: 99-103

Smit J, Roep TB, Alvarez W, Montanari A, Claeys P, Grajales-Nishimura JM, Bermudez J (1996) Coarse-grained, clastic sandstone complex at the KT boundary around the Gulf of Mexico: Deposition by tsunami waves induced by the Chicxulub impact. In: Ryder G, Fastovsky D, Gartner S (eds) The Cretaceous-Tertiary Boundary Event and other Catastrophes in Earth History. Geological Society of America, Special Paper 307, pp 151-182

Smith AB, Jeffery CH (1998) Selectivity of extinction among sea urchins at the end of the Cretaceous period. Nature 392: 69-71

Smith AB, Morris NJ, Gale AS, Rosen BR (1995) Late Cretaceous (Maastrichtian) echinoid--mollusc--coral assemblages and palaeoenvironments from a Tethyan carbonate platform succession, northern Oman Mountains. Palaeogeography, Palaeoclimatology, Palaeoecology 119: 155-168

Smith AB, Gallemí J, Jeffery CH, Ernst G, Ward PD (1999) Late Cretaceous-early Tertiary echinoids from northern Spain: implications for the Cretaceous-Tertiary boundary. Bulletin of the Natural History Museum of London (Geology) 55: 81-137

Sohl NF (1987) Cretaceous gastropods: contrasts between Tethys and the temperate provinces. J Paleont 61: 1085-1111

Speijer RP, van der Zwann GT (1996) Extinction and survivorship of southern Tethyan benthic foraminifera across the Cretaceous/Palaeogene boundary. In: Hart MB (ed) Biotic recovery from mass extinction events. Geological Society Special Publication 102, pp 343-371

Steneck RS (1983) Escalating herbivory and resulting adaptive trends in calcareous algal crusts. Paleobiology 9: 44-61

Stinnesbeck W (1996) Ammonite extinctions and environmental changes across the Cretaceous-Tertiary boundary in central Chile. In: MacLeod N, Keller G (eds) Cretaceous-Tertiary Mass Extinctions. Norton, New York, pp 289-302

Surlyk F, Johansen MB (1984) End Cretaceous brachiopod extinctions in the chalk of Denmark. Science 223: 1174-1177

Surlyk F (1997) A cool-water carbonate ramp with bryozoan mounds: Late Cretaceous- Danian of the Danish Basin. In: James NP, Clarke JAD (eds) Cool-Water Carbonates. SEPM Special Publication 56, pp 293-307

Tambareau Y, Hottinger L, Rodriguez-Lazaro J, Villatte J, Babinot J-F, Colin J-P, Garcia-Zarraga E, Rocchia R, Guerrero N (1997) Communautés fossiles benthiques aux alentours de la limite Crétacé/Tertiaire dans les Pyrénées. Bulletin de la Societe géologique de France 168: 795-804

Taylor JD, Morris NJ, Taylor CN (1980) Foos specialization and the evolution of predatory prosobranch gastropods. Palaeontology 23: 375-409

Thierstein HR (1982) Terminal Cretaceous plankton extinctions: A critical assessment. In: Silver LT, Schultz PH (eds) Geological Implications of Impacts of Large Asteroids and Comets on the Earth. Geological Society of America, Special Paper 190, pp 385-399

Thierstein HR, Okada H (1979) The Cretaceous-Tertiary boundary event in the North Atlantic. Initial Reports of the Deep Sea Drilling Project 43: 601-616

Thomas E (1990) Late Cretaceous through Neogene deep-sea benthic foraminifers (Maud Rise, Weddell Sea, Antarctica). Proceedings of the Ocean Drilling Program Scientific Results 113B: 571-594

Toon OB, Zahnle K, Morrison D, Turco RP, Covey C (1997) Environmental perturbations caused by the impact of asteroids and comets. Reviews of Geophysics 35: 41-78
Toulmin LD (1977) Stratigraphic distribution of Paleocene and Eocene fossils in the eastern Gulf Coast region. Geological Survey of Alabama - Monograph 13: 1-601
Tragelehn H (1996) Maastricht und Paläozän am Südrand der Nördlichen Kalkalpen (Niederösterreich, Steiermark) - Fazies, Stratigraphie und Fossilführung des 'Kambühelkalkes' und assoziierter Sedimente, Erlangen, 216 pp
Turekian KK (1982) Potential of $^{187}Os/^{186}Os$ as a cosmic versus terrestrial indicator in high iridium layers of sedimentary strata. In: Silver LT, Schultz PH (eds) Geological Implications of Impacts of Large Asteroids and Comets on the Earth. Geological Society of America, Special Paper 190, pp 243-249
Valentine JW, Jablonski D (1986) Mass extinctions sensitivity of marine larval types. Proceedings of the National Academy of Science, USA 83: 6912-6914
Van Andel TH (1975) Mesozoic/Cenozoic calcite compensation depth and the global distribution of calcareous sediments. Earth Planet Sci Letters 26: 187-194
Vecsei A, Moussavian E (1997) Paleocene reefs on the Maiella Platform margin, Italy: An example of the effects of the Cretaceous/Tertiary boundary events on reefs and carbonate platforms. Facies 36: 123-140
Veizer J, Ala D, Azmy K, Bruckschen P, Buhl D, Bruhn F, Carden GAF, Diener A, Ebneth S, Godderis Y, Jasper T, Korte C, Pawellek F, Podlaha OG, Strauss H (1999) $^{87}Sr/^{86}Sr$, $\delta^{13}C$ and $\delta^{18}O$ evolution of Phanerozoic seawater. Chemical Geology 161: 59-88
Voigt E (1960) Zur Frage der stratigraphischen Selbständigkeit der Danienstufe. In: Rosenkrantz A, Brotzen F (eds) 21st International Geological Congress, Part V, Proceedings of Section 5: The Cretaceous-Tertiary Boundary. Det Berlingske Bogtrykkeri, Copenhagen, pp 193-198
Voigt E (1987) On new cyclostomate Bryozoa from the upper Maastrichtian chalk-tuff near Maastricht, Netherlands. Paläontologische Zeitschrift 61: 41-56
Voigt E (1995) Septocea n.g. (Bryozoa Cyclostomata), a new bryozoan genus from Ruegen and Maastricht. Paläontologische Zeitschrift 69: 173-179
Voigt E (1999) Neue Bryozoen aus dem Baltischen Danium (I. Cheilostomata). Greifswalder Geowissenschaftliche Beiträge 6: 301-325
von Blanckenburg F (1999) Tracing past ocean circulation? Science 286: 1862-1863
Vonhof HB, Smit J (1997) High-resolution late Maastrichtian–early Danian oceanic $^{87}Sr/^{86}Sr$ record: Implications for Cretaceous-Tertiary boundary events. Geology 25: 275-282
Ward P (1983) The extinction of the ammonites. Scientific American 249: 136-147
Wells JW (1956) Scleractinia. In: Moore RC (ed) Treatise on invertebrate paleontology, Part F, Coelenterata. Geological Society of America and University of Kansas Press, Lawrence, pp F328-F444
Whatley R (1990) Ostracoda and global events. In: Whatley R, Maybury C (eds) Ostracoda and Global Events. Chapman and Hall, London, pp 3-24
Widmark JGV, Malmgren BA (1992) Benthic foraminiferal changes across the Cretaceous-Tertiary boundary in the deep sea; DSDP Sites 525, 527, 465. Journal of Foraminiferal Research 22: 81-113
Wiedmann J (1973) Evolution or revolution of ammonoids at Mesozoic system boundaries. Biology Review 48: 159-194
Wiedmann J (1988a) Ammonoid extinction and the "Cretaceous-Tertiary Boundary Event". In: Wiedmann J, Kullmann J (eds) Cephalopods - Present and Past. Schweizerbart, Stuttgart, pp 117-140
Wiedmann J (1988b) The Basque coastal sections of the K/T boundary - a key to understanding "mass extinction" in the fossil record. Revista Española de Paleontología Extraordinario: 127-140
Wiedmann J, Kullmann J (1996) Crises in ammonoid evolution. In: Landman N, Tanabe K, Davis RA (eds) Ammonoid Paleobiology. Plenum Press, New York, Topics in Geobiology 13, pp 795-813

Willems H (1996) Calcareous dinocysts from the Geulhemmerberg K-T boundary section (Limburg, SE Netherlands). Geologie en Mijnbouw 75: 215-230

Witte L, Schuurman H (1996) Calcareous benthic foraminifera across the Cretaceous-Tertiary boundary in the Geulhemmerberg (SE Netherlands). Geologie en Mijnbouw 75: 173-185

Wolbach WS, Gilmour I, Anders E, Orth CJ, Brooks RR (1988) Global fire at the Cretaceous-Tertiary boundary. Nature 334: 665-669

Wolfe JA, Upchurch GR (1987) Leaf assemblages across the Cretaceous-Tertiary boundary in the Raton Basin, New Mexico and Colorado. Proceedings of the National Academy of Science, USA 84: 5096-5100

Zachos JC, Arthur MA, Thunell RC, Williams DF, Tappa EJ (1985) Stable isotope and trace element geochemistry of carbonate sediments across the Cretaceous/Tertiary boundary at Deep Sea Drilling Project Hole 577, Leg 86. Initial Reports of the Deep Sea Drilling Project 86: 513-532

Zachos JC, Arthur MA, Dean WE (1989) Geochemical evidence for suppression of pelagic marine productivity at the Cretaceous/Tertiary boundary. Nature 337: 61-64

Zakharov YD, Ignatyev AV, Ukhaneva NG, Afanaseva TB (1996) Cretaceous ammonoid succession in the Far East (south Sakhalin) and the basic factors of syngenesis. Bulletin de l'Institut Royal des Sciences Naturelles de Belgique Sciences de la Terre 66: 109-127

Zinsmeister WJ, Feldmann RM, Woodburne MO, Elliot DH (1989) Latest Cretaceous/Tertiary transition on Seymour Island, Antarctica. Journal of Paleontology 63: 731-738

Effects of the Cretaceous-Tertiary Boundary Event on Bony Fishes

Lionel Cavin

GIS PalSédCo (Toulouse-Espéraza), Musée des dinosaures, 11260 Espéraza, France, (lionel.cavin@dinosauria.org)

Abstract. The effects of the Cretaceous-Tertiary boundary (KTB) event on the bony fishes are explored. The data are compiled by the analysis of the literature concerning bony fishes occurrences ranging from the Late Jurassic to the Paleocene, and resting on more than 750 references. The quality of the fossil record of bony fishes during the Campanian-Danian span is checked by calculation of the simple completeness metric (SCM). The analysis shows that (1) ten families (19% of the total) became extinct during or at the end of the Maastrichtian; (2) most of the victims (80%) were families exclusively restricted to marine environments; (3) most of the families (94%) with freshwater and/or brackish representatives during the Late Cretaceous crossed the KTB; (4) most of the victims (90%) were fast swimming and piscivorous predators. Extinction of this type of fishes is more easily explainable by a collapse of their food chain based on plankton. No extinction of planktivorous families, the intermediate level between plankton and piscivores, is observed. This fact is explain by ecological reasons (representatives of these families were living in brackish or freshwater) and by taxonomical reasons (phylogenetic relationships are unresolved and the families are not included in the analysis). These results are discussed in the light of some possible bias inherent to the fossil record and its study. The general trophic dependent pattern of extinctions agrees better with the expected consequences of a short-term catastrophic event than with a long term environmental change. It fits well the expected consequences of an impact: a break in the primary production by darkening of the atmosphere due to the dust generated by the impact, and followed by a chain of effects on the higher levels of the food chain, while detritus feeders and most of the following organisms survived, because there never was a lack of food at the base of the trophic chain.

1
Introduction

Investigation of a global catastrophe based on biological evidence should rest on the study of organisms strongly affected by the global event in question (extinctions), as well as on the study of organisms weakly or not affected by it. The comparison between both kinds of response to the event provides information about the modalities of the global catastrophe, and eventually about its nature. The Cretaceous-Tertiary boundary (KTB) is well known for the complete extinction of some groups of large reptiles (dinosaurs, pterosaurs, sauropterygians and mosasaurs), of invertebrates (ammonites), as well as of most of the plankton. Investigation of the response of vertebrate taxa which were affected, but did not disappear as a whole at the KT boundary, are not numerous. One can mention among others the works of Thomson (1977) on the fishes in general, of Cappetta (1987) on the elasmobranchs, of Cavin and Martin (1995) on the actinopterygians, of Vasse and Hua (1998) on the crocodiles, of Bardet (1995) on the marine reptiles, of Feduccia (1999) on the birds, and of MacLeod et al. (1997) for a global review.

Bony fishes form a paraphyletic group containing about half of all known vertebrates. It comprises the monophyletic actinopterygians, and part of the sarcopterygians: the Dipnoi (lungfishes) and Actinistia (coelacanths) for the Upper Cretaceous – Present time span. Bony fish remains are relatively abundant in both freshwater and marine deposits from the Mesozoic and Cenozoic, and fossil material commonly occurs as complete skeletons, allowing to draw up a rather comprehensive image of the history of the group. The relatively good qualitative and quantitative record of bony fishes through time compared to other vertebrate groups and the fact that the classification of many groups of fishes is more refined than that of most other animal groups (MacLeod et al. 1997) provide a high potential for studies of earth history, in particular to observe the biological consequences of the KTB event.

Now that the extinction event of the KTB is well established, disagreements occur about the duration of the extinction process itself (Buffetaut 2000), and the drawing up of a precise scenario for the pattern of extinction (a short-term catastrophic event, or one or two short-term events superimposed upon a long term environmental change; Keller 2000). The fossil record of bony fishes does not possess a sufficiently precise temporal resolution to provide an answer to the first question. But the qualitative information it contains, due to the various paleoenvironments it reflects, may serve to solve the second problem.

2 Methods

To study the response of bony fishes to the KTB event, the data may be dealt with quantitative (rates of extinction for instance) or qualitative (groupings according to paleoenvironments for instance) analysis. I favour herein the second approach, because it is less dependent on the variations of the fossil record and more informative about the modalities of the phenomenon. The taxonomic unit of analysis is the family. I used as much as possible families which have been shown to be monophyletic. The use of non-monophyletic taxa in analyse of variations of biodiversity has been criticized, especially for studies of supposed periodic extinctions (Patterson and Smith 1987, 1989). Monophyletic families of bony fishes include species, which share derived morphological characters and generally close ecological features. Species grouped into non-monophyletic families share plesiomorphic morphological characters and, as the monophyletic families, share close ecological features (general morphological resemblance and similarities in the mode of life were reasons which previously led to gather species into unnatural taxa, or grades). Some non-monophyletic families have probably been involuntarily inserted in this study but, insofar as taxonomic entities show strong ecological identity, they also provide qualitative information on the effect of the KTB event on biodiversity (the effects of environment on organisms act on entities sharing close ecological features, not on clades *a priori*). Moreover, recent work (Robeck et al. 2000) showed that the number of groups in the classification and the distribution of the sizes of those groups have much more important effects on the recovery of diversity information that the use of non-monophyletic taxa. Although our data do not allow us to put a figure on the different parameters they used in their models, one of them, the sampling rate, is probably low for bony fishes (as for other vertebrate groups) for the Cretaceous and Paleocene time span (presumably lower than the lowest value used in their study, 0.05). Robeck et al. (2000) showed that when sampling is poor, better results for temporal diversity patterns can be obtained with a range of taxon sizes including some large groups (generally paraphyletic in their model). Thus the various size of the bony fish taxa and their possible non-monophyly should not be regard as a disadvantage in our study.

The database used here is compiled by the analysis of the literature about bony fishes from the Upper Jurassic to the Palaeocene. The starting points are the compilations of the fossil record of basal actinopterygians made by Gardiner (1993) and of Teleostei made by Patterson (1993a), which were checked, criticised and updated on the basis of subsequent works (description of new forms, of new fish assemblages, and phylogenetic works). I included some of my own observations and decisions about bones of contention. Most of the data on extant families are extracted from the online FishBase (www.fishbase.org). My database now contains 1623 occurrences comprising 486 genera spread over 249 different localities. The compilation rests on more than 750 references.

We indicate for each family:

1. The diet of species included in the family. The range of diets and of modes of feeding among bony fishes is extremely diverse. I gathered the families into five main groups: (1) the planktivorous fishes feed on phyto- and/or zooplankton; (2) the piscivorous fishes feed on other fishes only, by pursuing their preys or stalking; (3) the carnivores include fishes feeding mainly on benthic invertebrates and the insectivores; (4) the durophagous fishes are the coral-browsers, the grinding and crushing fishes feeding on hard molluscs and crustaceans, and are recognisable by their dentition; (5) the omnivores are euryphagic fishes (this group may include strict herbivorous fishes hardly recognisable on fossil specimens). Most fishes, especially from groups 3 to 5, show a considerable amount of plasticity in the foods used (Lowe-McConnell 1991). The diet of fossil forms can be determined by analysis of gut contents or by inference from the morphology of teeth, fin insertion, general shape of the body and comparison with extant representatives when available. Diet is generally homogenous in a given family, although notable exceptions are observed in extant characids and cichlids for instance.
2. The first and last occurrence. Several of the KTB surviving families are not known in Maastrichtian and/or Danian deposits; however, these Lazarus taxa are included in the analysis. For the present study, the time slice used is the geological stage. Unfortunately, it is now impossible to get a more precise time resolution with the available data found in literature.
3. Environments. In order to avoid circular reasoning, we attempted to gather information on the paleoenvironments of the fossiliferous localities from published results resting on data other than ecological interpretations of fish assemblages (i.e. inferred from the sedimentology of localities in which the fossils were found, from the associated flora and invertebrate or tetrapod fauna, etc.). The information about the paleoenvironment are combined with data inferred from the general morphology and fin insertion on the body of fossil specimens to determine the habits of extinct forms (epi-, meso-, bathy-benthic or pelagic).

One way to assess the quality of the fossil record is to calculate the simple completeness metric (SCM) used by Benton (1987) and Fara and Benton (2000) among others. This is the ratio of observed fossil occurrences to total inferred fossil occurrences. The total fossil occurrences are calculated as observed fossils plus Lazarus taxa. The SCM is calculated for the Campanian, Maastrichtian and Danian stages for the whole families, as well as for both exclusively marine families and other families (exclusively freshwater and freshwater tolerant families).

3 Results

The data concerning the potential victims of the KTB event are summarized in Table 1, and in Tables 2 to 7 concerning the survivors. Data without references come from Gardiner (1993) and Patterson (1993a) for fossil forms, and from the FishBase for extant forms.

Table 1. Victims

Families	Diet	First occurrence	Last occurrence	Inferred and/or observed nvironments
Semionotidae	Durophagous	Middle Triassic	Maastrichtian. South America (Gayet and Meunier 1998) and India (Mohabey and Udhoji 1996)	Marine and freshwater
Pachycormidae	Piscivorous	Lower Jurassic	Maastrichtian. "Craie phosphatée de Ciply", Hennuyer Basin, Belgium (Leriche 1929)	Marine, epipelagic
Aspidorhynchidae	Piscivorous	Middle Jurassic	Maastrichtian. Heel Creek and Lance formations, US (Brito 1997)[1]	Marine and freshwater
Ichthyodectidae	Piscivorous	Upper Jurassic	Lower Maastrichtian. Mont Laurel, New Jersey, US (Gallagher 1986)	Marine, epipelagic
Saurodontidae	Piscivorous	Upper Cretaceous	Maastrichtian. Navesink formation, New Jersey, US; Gramame formation, Brazil and Maastricht, The Netherlands	Marine, epipelagic
Pachyrhizodontidae	Piscivorous	Lower Cretaceous	Maastrichtian. "Craie phosphatée de Ciply", Hennuyer Basin, Belgium (Leriche 1929)[2]	Marine, epipelagic
Apateopholidae	Piscivorous	Lower Cretaceous	Lower Maastrichtian. The Netherlands (Leriche 1929)	Marine, epipelagic
Cimolichthyidae	Piscivorous	Cenomanian	Maastrichtian. Zaire (Patterson, 1993a) and Niger (Cappetta 1972)	Marine, epipelagic
Dercetidae	Piscivorous	Cenomanian	Maastrichtian. Morocco (Arambourg 1952) and Israel (Patterson 1993a)	Marine, epipelagic
Enchodontidae	Piscivorous	Lower Cretaceous	Maastrichtian. Africa, Europe and US (numerous occurrence)[3]	Marine, epipelagic

[1] Aspidorhynchidae. The single potential Thanetian occurrence from the Tongue River Formation, US, mentioned by Bryant (1987) is likely due to reworking from the Upper Cretaceous (P. M. Brito, personal communication).
[2] Pachyrhizodontidae. *Platinx macropterus* from the Eocene of Monte Bolca, Italy, is regarded by some authors as a possible pachyrhizodontids (Taverne 1980; Patterson 1993a). We questioned this attribution in a recent work (Cavin, work in progress).
[3] Enchodontidae. The occurrence of *Enchodus* teeth from the Paleocene of the Ouled Abdoun phosphatic basin in Morocco (Arambourg 1952) is likely due to reworking from the Maastrichtian, as for numerous isolated shark teeth (Cavin et al. 2000).

Table 2. Survivors: sarcopterygians

Families	Diet	First occurrence	Last occurrence	Inferred and/or observed environments
Coelacanthidae	Carnivorous	Lower Triassic	Extant	Marine, mesobenthic (extant) Freshwater and marine (fossil record)
Neoceratodontidae	Omnivorous	Upper Jurassic	Extant	Freshwater (for the Jurassic)
Lepidosirenidae	Omnivorous	Upper Cretaceous	Extant	Freshwater
Protopteridae	Carnivorous	Upper Cretaceous	Extant	Freshwater

Table 3. Survivors: basal actinopterygians

Families	Diet	First occurrence	Last occurrence	Inferred and/or observed environments
Polypteridae	Carnivorous	Cenomanian of the Kem Kem beds, Morocco (Dutheil 1999)	Extant	Freshwater
Acipenseridae	Carnivorous	Campanian of the Oldman formation, Canada (Gardiner 1984)	Extant	Marine and freshwater
Polyodontidae	Carnivorous	Basal Cretaceous (Jin 1999)	Extant	Freshwater
Lepisosteidae	Piscivorous	Lower Cretaceous of the Santana formation, Brazil (Wenz and Brito 1992) and of the Kem Kem beds, Morocco (Cavin and Brito, in press)	Extant	Freshwater
Amiidae	Carnivorous	Upper Jurassic	Extant	Marine, brackish and freshwater, chiefly freshwater for the Terminal Cretaceous
Pycnodontidae	Durophagous	Middle Jurassic Cretaceous	Eocene of Monte Bolca, Italy	Marine and freshwater

Table 4. Survivors: osteoglossomorphs and elopomorphs

Families	**Diet**	**First occurrence**	**Last occurrence**	**Inferred and/or observed environments**
Notopteridae	Carnivorous	Cenomanian of the Kem Kem beds (Forey 1997)	Extant	Freshwater
Osteoglossidae	Carnivorous	Lower Senonien of Niger (Gayet and Meunier 1983)	Extant	Marine and freshwater
Arapaimidae	Piscivorous	Aptian of the Areado formation, Brazil (Lundberg 1998)	Extant	Freshwater
Elopidae	Carnivorous	Late Jurassic of Germany (Arratia 1997)	Extant	Marine, rarely brackish and freshwater
Megalopidae	Carnivorous	Albian of France (Wenz, 1965) and Wealdian of Belgium and England (Taverne 1999)	Extant	Marine and brackish waters
Albulidae	Carnivorous	Cenomanian of Lebanon (Forey 1973)	Extant	Marine, rarely brackish and freshwater
Pterothrissidae	Carnivorous	Cenomanian of Lebanon (Forey 1973)	Extant	Mesopelagic
Phyllodontidae	Durophagous	Lower Cretaceous (Forey 1973)	Priabonian of Egypt	Marine, brackish and freshwater
Halosauridae	Carnivorous	Campanian	Extant	Bathypelagic
Congridae	Carnivorous	Campanian. Otoliths from the Coffee Sand formation, US	Extant	Epi- and mesopelagic

Table 5. Survivors: Otocephala

Families	**Diet**	**First occurrence**	**Last occurrence**	**Inferred and/or observed environments**
Ellimmichthyidae	Planktivorous	Barremian of Ilhas Formation	Lutetian of Green River Formation, US	Marine and freshwater
Clupeidae	Planktivorous	Upper Santonian of Vonzo, Zaire (Taverne 1997)	Extant	Epipelagic, rarely freshwater
Chanidae	Planktivorous	Lower Cretaceous of Spain (Wenz and Poyato-Ariza 1994)	Extant	Marine, brackish and rarely freshwater
Gonorhynchidae	Omnivorous	Lower Cretaceous of Spain (Wenz and Poyato-Ariza 1994)	Extant	Marine, rarely brackish and freshwater

Table 5: Continued

Erythrinidae	Piscivorous	Terminal Cretaceous of El Molino formation, Bolivia (Gayet et al. 1991)	Extant	Freshwater
Characidae	Carnivorous Omnivorous Piscivorous	Terminal Cretaceous of El Molino formation, Bolivia (Gayet et al. 1991)	Extant	Freshwater
Diplomystidae	Omnivorous?	Campanian of Los Alamitos formation, Argentina	Extant	Freshwater
Ariidae	Omnivorous?	Campanian of Los Alamitos formation, Argentina	Extant	Chiefly marine, rarely freshwater

Table 6. Survivors: euteleosteans I

Families	**Diet**	**First occurrence**	**Last occurrence**	**Inferred and/or observed environments**
Esocidae	Predator	Campanian of Canada (Wilson et al. 1992)	Extant	Freshwater
Argentinidae	Planktivorous	Lower Cretaceous	Extant	Mesobentic
Aulopodidae	Carnivorous	Cenomanian of Lebanon	Extant	Epi- and mesobentic
Chlorophthalmidae	Carnivorous	Cenomanian of Lebanon	Extant	Epi-, meso- and bathybentic
Myctophidae	Planktivorous	Campanian of Germany	Extant	Epi-, meso- and bathypelagic (300-1200 m by day, 10-100 m by night)
Asineopidae	Carnivorous	Campanian of Nardo, Italy	Eocene of Green River formation, US	Marine and freshwater
Blochiidae	Carnivorous	Cenomanian of Lebanon and England	Bartonian of Barton Clays, England	Epipelagic
Polymixiidae	Carnivorous	Cenomanian of Morocco and England	Extant	Epi- and mesobentic

Four main observations may be drawn from the analysis of the table content:

1. Ten families (19% of the total) became extinct during or at the end of the Maastrichtian. This corresponds to a decrease of the diversity of families of 7.5% at the KTB (10 families became extinct in the Maastrichtian and 6 originated in the Danian).
2. Most of the victims (80%) were families exclusively restricted to marine environments.
3. Most of the families (94%) with freshwater and/or brackish representatives during the Late Cretaceous crossed the KTB.
4. Most of the victims (90%) were fast swimming and piscivore predators.

Table 7. Survivors: euteleosteans II

Families	**Diet**	**First occurrence**	**Last occurrence**	**Inferred and/or observed environments**
Ophidiidae	Carnivorous	Santonian. Otoliths from Bavaria, Germany	Extant	Epi-, meso- and bathybentic
Berycidae	Carnivorous	Campanian. Otoliths from Cöthen, Germany	Extant	Epi- and mesobentic
Apogonidae	Carnivorous	Coniacian. Otoliths from Austria	Extant	Epi- and Chiefly marine epipelagic, some in brackish and in streams
Percichthyidae	Carnivorous	Campanian-Maastrichtian. Formation El Molino, Bolivia (Gayet and Meunier 1998)	Extant	Freshwater, rarely brackish
Bathyclupeidae	Carnivorous	Campanian. Otoliths from the Coffee Sand formation, US	Extant	Marine and brackish waters
Veliferidae	Omnivorous	Maastrichtian. Otoliths from Bavaria, Germany (Schwartzans 1996)	Extant	Epibenthic
Gempylidae	Piscivorous	Campanian. Otoliths from the Coffee Sand formation, US (Nolf and Stinger 1996)	Extant	Bathypelagic

The SCM for the whole families shows a decrease from the Campanian to the Danian (Fig. 1). The values of the SCM of bony fishes are much lower than the SCM values of tetrapods for the same time span (Fig. 1B; Fara and Benton, 2000). The SCM of the Danian stage is especially low (36.7%). The SCM of the exclusively marine families shows the same pattern as for the whole families, a progressive decrease, while the SCM for other families (exclusively freshwater and freshwater tolerant families) is similar in Maastrichtian and Campanian.

If we plot the stratigraphic occurrences of exclusively marine families of bony fishes concerned by the KTB (Fig. 2), we observe that the fossil record is much better for the victims in the Late Cretaceous than for the survivors in the Late Cretaceous and Tertiary. This fact is illustrated by the high average SCM calculated for the exclusively marine victims for the Campanian-Maastrichtian time span (93.7%) compared to average SCM calculated for the exclusively marine survivors for the same time span (54.2%). Moreover, most of the surviving families have living representatives in deep sea environments. Considering this fact and the fact that the fossil record of vertebrates is poor in deep sea environments, we can postulate that most of the exclusively marine surviving families should have had representatives in deep sea environments at the KTB.

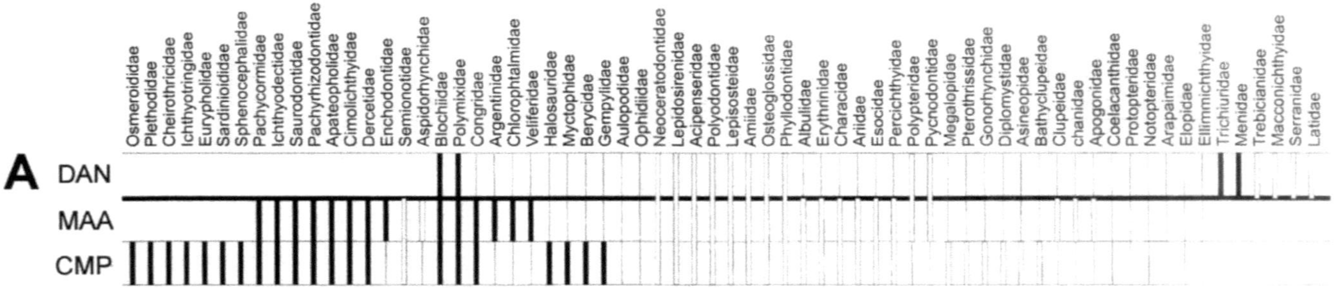

B	observed occurrences	total inferred occurrences	SCM	SCMmar.	SCMother	SCMtetra.
DAN	18 (4)	49 (14)	36.7%	28.6%	40%	74.7%
MAA	35 (14)	53 (20)	66%	70%	63.6%	89.3%
CMP	42 (21)	60 (27)	70%	77.8%	63.6%	78.1%

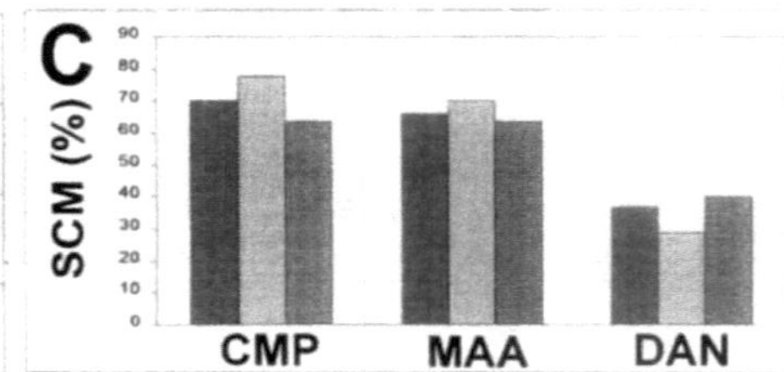

Figure 1. A: Stratigraphic occurrence of the families of bony fishes during the Campanian (CMP), Maastrichtian (MAA) and Danian (DAN). Black lines are the exclusively marine families, white lines are the other families (exclusively freshwater or freshwater tolerant), and thin lines are the Lazarus taxa. **B**: Table with the number of observed and of total inferred occurrences (number of exclusively marine families in brackets), and values of the SCM for all the bony fish families (SCM), for the exclusively marine families of bony fishes (SCMmar.), for the other families of bony fishes (SCMother) and for the tetrapods families (SCMtetra.) (from Fara and Benton, 2000). **C**: Histogram showing the SCM values during the Campanian (CMP), Maastrichtian (MAA) and Danian (DAN). Left: SCM, centre: SCMmar., Right: SCMother.

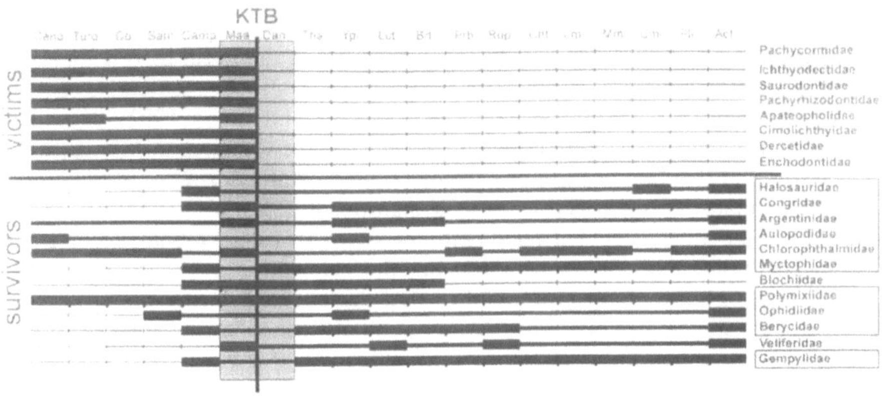

Figure 2. Stratigraphic occurrence of the exclusively marine families of bony fishes which became extinct at (above) and crossed (below) the KTB. Thick black lines: geological stages with observed fossil occurrence; thin black lines: geological stages with *lazarus* taxa. Families enclosed in frame have extant representatives living in deep-sea environments. The data for the Tertiary mainly come from Patterson (1993a).

4 Discussion

The extinction rate of families of bony fishes in the Maastrichtian (19%) is not very high. The decrease in the total diversity (7.5%) is much lower than for the amniote families (18%) and the non-marine terrestrial tetrapod families (14%) at the same time (Benton 1988). This decrease of the diversity should be linked to the effect of the KTB event, even if the temporal resolution of the fossil record of bony fishes is not good enough now to prove the exact concordance of both extinctions and KTB event. The standing diversity normally tends to increase for the Cretaceous, and decreases are generally link to mass extinction (Benton 1988).

These results should be discussed in the light of biases inherent to the fossil record and its study. The quality of the fossil record of bony fishes in the Campanian and Maastrichtian is quite good (SCM=70 and 66%), but is bad in the Danian (SCM=36.7%). This is probably due to the lack of Danian localities yielding rich assemblages of bony fishes (only the Santa Lucia formation in Bolivia contains more than 10 different taxa; Gayet and Meunier 1998). The low SCM value in the Danian could question the reality of the mass extinction at the KTB. But this assertion is lessened because of the selective pattern observed in the extinct families in the Maastrichtian: if the decrease of the diversity between the Maastrichtian and Danian was only due to the poor quality of the Danian fossil record, bony fishes from all types of environments and with all kinds of diet should have been affected (SCM of both exclusively marine and others families decrease). But the observed pattern is that mainly the epipelagic piscivorous predators were affected during the Maastrichtian: this pattern agrees with the consequences of an event independent of the quality of the fossil record, and with selective effects on life.

Another bias is the Signor-Lipps effect (Signor and Lipps 1982). I tried to assess its influence by examining what happened at the Campanian-Maastrichtian boundary: 8 (12%) families of bony fishes became extinct during or at the end of the Campanian. This extinction rate is significantly lower than at the KTB. Moreover, the difference between both increases if we take into account the estimated duration of the stages by using the calculation proposed by Benton (1988) (ratio of the number of families that disappeared during a stratigraphic stage to the estimated duration of that stage): total extinction rate is equal to 1.58 for the Maastrichtian and to 0.57 for the Campanian. All the families becoming extinct in the Campanian (Osmeroididae, Plethodidae, Cheirothricidae, Ichthyotringidae, Eurypholidae, Sardinioididae, Sphenocephalidae) were exclusively marine (Fig. 1a). On the other hand, the SCM of the exclusively marine families decreases between the Campanian and the Maastrichtian, while the SCM of other families is stable in the same time interval (Fig. 1c). It means that the quality of the marine fossil record is proportionally higher in the Campanian than in the Maastrichtian, while the situation remains constant for the continental fossil record. This proportional decrease of the quality of the marine fossil record in the Maastrichtian and the fact that only marine families seem to become extinct in the Campanian lead to suspect that we are faced with the Signor-Lipps effect: according to this hypothesis, I assume that a part or the totality of families cited above will be discovered in the future in marine Maastrichtian deposits.

In summary, because of the evolution of the SCM and because of the selectivity observed in the Maastrichtian victims, I propose that the decrease of the number of marine families observed between the Campanian and the Maastrichtian is due totally or in part to the Signor-Lipps effect, while the decrease of the number of marine families observed between the Maastrichtian and the Danian is mainly due to an event independent of the quality of the fossil record. We can notice that this decrease ended in the Maastrichtian, for no exclusively marine family became extinct during or at the end of the Danian. But new families of fast swimming and piscivorous bony fishes appeared in the fossil record in the Danian (Trichiuridae) and Thanetian (Scombridae, Carangidae).

The selective aspect of the bony fish extinction at the KTB is the more remarkable point resulting from the analysis. The ideal portrait of the victims may be illustrated by enchodontids, for instance. The body of these fishes is fusiform and somewhat laterally compressed, indicating fast-swimming pelagic fishes. The mouth is filled with needle-like teeth and large saber-toothed fangs are borne at the front of the snout by the dentary and palatine bones, indicating a diet based on other fishes and possibly on nectonic invertebrates. Contrary to most of the benthic and freshwater fishes, the diet of enchodontids and most of the other victims of the KTB event was probably quite specialized, with a weak plasticity (stenophagy), because food substitution of the necton in pelagic marine environments is rare (Lowe-McConnell 1991). Extinction of this type of fishes is more easily explainable by a collapse of their food chain. The hypothesis of the collapse at the KTB of food chains based on plant productivity is now widely accepted (Buffetaut 2000). Food chains rest mainly on phyto- and zooplankton in pelagic marine environments, while in freshwaters food chain rest essentially on detritus. The mass extinction of the plankton at the KTB is well established; this disappearance explains the extinction of the higher levels of the food chains, in particular the pelagic piscivorous bony fishes, some sharks (Cappetta 1987) and the mosasaurs (Bardet 1995). The weak point of this scenario is that we did not notice extinctions of the intermediate trophic levels at the KTB, such as the planktivorous fish families. Two reasons may explain this fact. (1) Some of the families of planktivorous fishes in the Late Cretaceous had representatives living in brackish or fresh water. In these environments, the species could have been more euryphagic, and thus, the families as a whole could have survived the KTB event. The Ellimmichthyidae for instance are known mainly from marine deposits in the Late Cretaceous, and mainly from continental deposits in the Paleogene (Grande 1985). (2) The phylogenetic relationships of several Late Cretaceous planktivorous fishes are not yet resolved and they are not gathered to form the familial units, which are studied in the present paper. The Cretaceous primitive acanthomorphs for instance contain numerous microphagous and planktivorous marine forms. But their relationships to one another and to extant taxa are equivocal or unsubstantiated (Patterson 1993b). Important changes in the acanthomorph faunas seem to occur between the Cretaceous and the Paleocene (Patterson 1993b) but, because of our poor understanding of the systematics of these groups (systematics should rests on phylogenetic relationships), they are not taken into account in this analysis.

Figure 3 illustrates what may have happened to bony fishes at the KTB: the extinction of bony fishes at the KTB was quite weak, very selective (epipelagic fishes) and trophic dependent (specialized piscivorous predators). The cause for such a selective extinction may be a collapse in the food chain. Marine planktonic organisms were strongly affected by the KTB event, and the consequence of such an extinction is the disruption of the higher levels of the food chain, i.e., planktivorous fishes, piscivorous predators and then mega-predators (for instance the mosasaurs and some sharks).

As previously observed (Buffetaut 1987), this trophic dependent extinction pattern disagrees with the expected consequences of a regression, but fits well the expected consequences of an impact: a break in the primary production by darkening of the atmosphere due to the dust generated by the impact, and followed by a chain of effects on the higher levels of the food chain, while detritus feeders and most of the following organisms survived because there never was a lack of food at the base of the trophic chain (at least for some large enough populations, which started again afterwards). The difference in the response to the KTB event between marine pelagic communities and the other ones is enhanced by the fact that the competition for food is more important than for space in pelagic environments, while the situation is reversed in environments with high spatial heterogeneity (rocky shores, for instance) (Lowe-McConnell 1991).

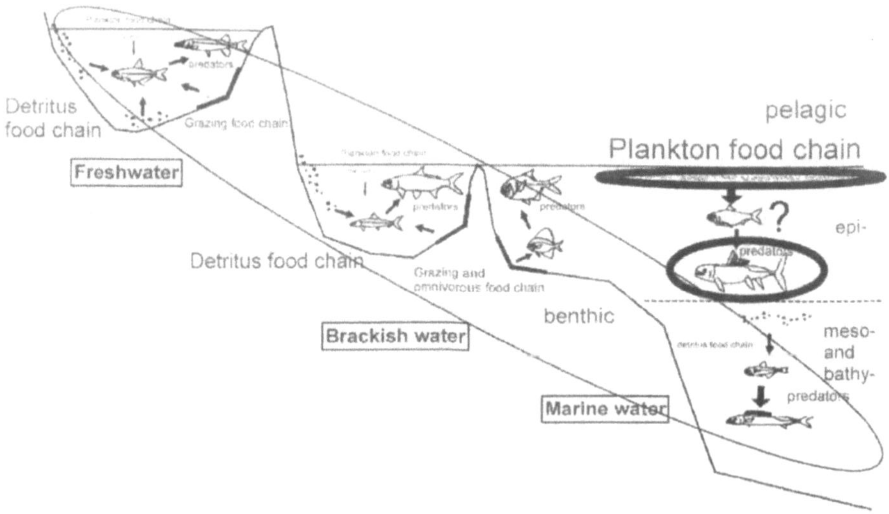

Figure 3. Diagrammatic drawing showing the consequences of the KTB event on the bony fishes according to the environment. Thick-line ellipses surround organisms strongly affected, the thin line ellipse surrounds organisms weakly affected by the KTB event.

Although the temporal resolution provided by the bony fish fossil record is not precise enough to appreciate the duration of the extinction process, the pattern of extinctions agrees better with the expected consequences a short-term catastrophic event than with a long term environmental change: the expected consequences on bony fishes of a global and long term environmental change are complete and profound renewals of faunal assemblages in various kind of environments. The settlement of freshwater fish communities started during the Late Cretaceous in North and South America and Africa (the situation is poorly known in Europe and Asia), and no important renewal is observed in the worldwide composition of freshwater fish families around the KTB. This is in contradiction with the expected consequences of a global environmental change, such as long term climatic changes due to an important volcanic activity spread over 500 ky, or to sea level fluctuations.

5 Conclusions

Since a previous study carried out on actinopterygians (Cavin and Martin 1995), the quantity of data has increased. But the general pattern obtained at that time was similar to the one observed here. This pattern partially agrees with what one observes at the KTB with the terrestrial vertebrates, i.e. important extinctions of organisms belonging to food chains resting on the primary production (Buffetaut 1990; Sheehan and Fastovsky 1992; Fara 2000). To the contrary, vertebrates totally or partially belonging to food chains based on detritus, which lived in terrestrial environments (squamates, small mammals, some birds, etc.), as well as in aquatic environments (most of lissamphibians and freshwater fishes, crocodiles, deep sea as well as some brackish and near-shore fishes, etc.) were not much affected by the KTB event. However we should keep in mind that some components of these surviving groups showed an important decrease or even became extinct at the KTB, for instance the marsupials among mammals in North America (Clemens 1984), and, possibly, birds (Feduccia 1999). Further studies of these differential responses of surviving groups to the KTB event will improve our understanding of the scenario of this mass extinction.

The trophic dependent extinction pattern observed in the fossil record of bony fishes fits well the expected consequences of an impact. To the contrary, this pattern disagrees with the consequences of a long and global environmental change, such as long term climatic modifications due to an important volcanic activity or to sea level fluctuations, because no complete and profound renewals of faunal assemblages in various kinds of environments are observed (in particular in freshwater environments).

Acknowledgements

I would like to thank Wolfgang Kiessling (Berlin), Eric Buffetaut (Paris), Jean Le Loeuff (Espéraza) and Michel Martin (Boulogne-sur-mer) for their critical reading of the manuscript and useful comments and suggestions. I also thank the participants of the *Impact* Meeting (Quillan, 1999), of the *Evolutionary Transformation and Mass Extinction Program* (Berlin, 2000), and especially Gloria Arratia and Hans-Peter Schultze (Berlin) for their fruitful comments and critics on this work. This research was supported by the Swiss National Science Foundation (grant n° 8220-56521).

References

Arambourg C (1952) Vertébrés fossiles des phosphates d'Afrique du Nord (Maroc, Algérie, Tunisie). Notes et Mémoires du Service Géologique du Maroc 92: 1-372

Arratia G (1997) Basal teleosts and teleostean phylogeny. Palaeo Ichthyologica 7: 5-168

Bardet N (1995) Evolution et extinction des reptiles marins au Cours du Mésozoïque.- Palaeovertebrata 24 (3-4): 177-283

Benton MJ (1987) Mass extinctions among families of non-marine tetrapods: the data. Mémoires de la Société Géologique de France, N. S. 150: 21-32

Benton MJ (1988) Mass extinctions in the fossil record of reptiles: Paraphyly, patchiness, and periodicity (?). In: Larwood GP (Ed) Extinction and survival in the fossil record. The Systematics Association. Oxford Science Publications, Special Volume 34: 269-294

Brito PM (1997) Révision des Aspidorhynchidae (Pisces, Actinopterygii) du Mésozoïque: Ostéologie, relations phylogénétiques, données environnementales et biogéographiques. Geodiversitas 19 (4): 681-772

Bryant LJ (1987) *Belonostomus* (Teleostei: Aspidorhynchidae) from the Late Paleocene of North Dakota. Paleobios 43: 1-3

Buffetaut E (1987) Why the Maastrichtian regression did not cause terminal Cretaceous mass extinctions ? Mémoire de la Société Géologique de France, N. S. 150: 75-80

Buffetaut E (1990) Vertebrate extinctions and survival across the Cretaceous-Tertiary boundary. Tectonophysics 171: 337-345

Buffetaut E (2000) Paleontological constraints on the duration of the K/T boundary extinction events. [abs] In: Catastrophic events and mass extinctions: Impacts and beyond, LPI Contribution No. 1053, Lunar and Planetary Institute, Houston. p 21-22

Cappetta H (1972) Les poissons Crétacés et Tertiaires du Bassin des Iullemmeden (République du Niger). Palaeovertebrata 5: 179-251

Cappetta H (1987) Exctinctions et renouvellement fauniques chez les sélaciens post-Jurassique. Mémoire de la Société Géologique de France, N. S. 150: 113-131

Cavin L (in progress) Osteology of *Goulmimichthys arambourgi* and phylogenetic relationships of the Pachyrhizodontidae.

Cavin L, Brito PM (in press) A new Lepisosteidae (Actinopterygii: Ginglymodi) from the Cretaceous of the Kem Kem Beds, Southern Morocco. Bulletin de la Société géologique de France

Cavin L, Martin M (1995) Les Actinoptérygiens et la limite Crétacé-Tertaire. Géobios Special Memoirs 19: 183-188

Cavin L, Bardet N, Cappetta H, Gheerbrant E, Iarochene SM, Sudre J (2000) A new Palaeocene albulid (Teleostei: Elopomorpha) from the Ouled Abdoun Phosphatic Bassin, Morocco. Geological Magazine 137 (5): 583-591

Clemens WA (1984) Evolution of Marsupials during the Cretaceous-Tertiary transition. Third Symposium on Mesozoic Terrestrial Ecosystems, Tübingen, Attempto Verlag: 47-52

Dutheil DB (1999) The first articulated fossil Cladistian: *Serenoichthys kemkemensis*, gen. et sp. nov., from the Cretaceous of Morocco. Journal of Vertebrate Paleontology 19: 243-246

Fara E (2000) Diversity of Callovian-Ypresian (Middle Jurassic-Eocene) tetrapod families and selectivity of extinctions at the K/T boundary. Géobios, 33: 387-396

Fara E, Benton MJ (2000) The fossil record of cretaceous tetrapods. Palaios, 15: 161-165.

Feduccia A (1999) The Origin and Evolution of Birds, Yale University Press, 465 p

Forey PL (1973) A revision of the Elopiform fishes, fossil and recent. Bulletin of the British Museum (Natural History) (Geol.) Suppl. 10: 1-222

Forey PL (1997) A Cretaceous Notopterid (Pisces: Osteoglossomorpha) from Morocco. South African Journal of Science 93: 564-569

Gallagher WB, Parris DC, Spamer EE (1986) Paleontology, biostratigraphy, and depositional environments of the Cretaceous-Tertiary transition in the New Jersey coastal plain. The Mosasaur 3: 1-35

Gardiner BG (1984) Sturgeons as living fossils. In: Eldredge N, Stanleys SM (Eds) Living Fossils, Springer-Verlag, New York, 148-152

Gardiner BG (1993) Osteichthyes: Basal Actinopterygians. In: Benton MJ (Ed) Fossils Record 2 London, Chapman and Hall: 611-619.

Gayet M, Meunier FJ (1983) Ecailles actuelles et fossiles d'ostéoglossiformes (Pisces, Teleostei). Comptes Rendus de l'Académie des Sciences, Paris, Série II, 297: 867-870

Gayet M, Meunier F (1998) Maastrichtian to Early Paleocene freshwater Osteichthyes of Bolivia: additions and comments. In: Malabara LRR, Vari RE, Lucena RP, Lucena CAS (Eds) Phylogeny and classification of neotropical fishes. Porto Alegre, Edipucrs: 85-110

Gayet M, Marshall LG, Sempere T (1991) The Mesozoic and Paleocene vertebrates of Bolivia and their stratigraphic context: A review. In: Suarez-Soruco R (Ed) Fosiles y facies de Bolivia Vol. I Vertebrados. Santa Cruz-Bolivia, Revista Técnica de Ypfb. 12: 393-433

Grande L (1985) Recent and fossil clupeomorph fishes with material for revision of the subgroups of Clupeoids. Bulletin of the American Mueum of Natural History 181: 231-272

Jin F (1999) Middle and Late Mesozoic Acipenseriforms from Northern Hebei and Western Liaoling, China. In: Chen PJ, Jin F (Eds) Palaeoworld, Hefei, China, Press of University of Science and Technology of China, 11: 188-261

Keller G (2000) Mass extinctions, catastrophes and environmental changes [abs]. In: Catastrophic Events and Mass Extinctions: Impacts and Beyond, LPI Contribution No. 1053, Lunar and Planetary Institute, Houston, pp 21-22

Leriche M (1929) Les poissons du Crétacé marin de la Belgique et du Limbourg Hollandais (Note Préliminaire). Bulletin de la Société belge de Géologie et d'Hydrologie, 37: 199-299

Lowe-McConnell RH (1991) Ecological studies in tropical fish communities. Cambridge University Press, Cambridge, 382 pp

Lundberg JG (1998) The temporal context for the diversification of neotropical fishes. In: LRR Malabara, RE, Vari, RP, Lucena, ZMS, Lucena, CAS (Eds) Phylogeny and classification of neotropical fishes. Porto Alegre, Edipucrs: 49-68

MacLeod N, Rawson PF, Forey PL, Banner FT, Boudagher-Fadel MK, Bown PR, Burnett JA, Chambers P, Culver S, Evans SE, Jeffery C, Kaminski MA, Lord AR, Milner AC, Milner AR, Morris N, Owen E, Rosen BR, Smith AB, Taylor PD, Urquhar E, Young JR (1997) The Cretaceous-Tertiary biotic transition. Journal of the Geological Society, London 154: 265-292

Mohabey DM, Udhoji SG (1996) Fauna and flora from Late Cretaceous (Maastrichtian) non-marine lameta sediments associated with Deccan volcanic episode, Maharashtra: Its relevance to the K-T boundary problem, paleoenvironment and paleogeography. Gondwana Geological Magazine, Spl. Vol. 2, Nat. Symp. Deccan Flood Basalts, India, 349-364

Nolf D, Stringer GL (1996) Cretaceous fish otoliths - A synthesis of the North American record. In: Arratia G, Viohl G (Eds) Mesozoic Fishes - Systematics and Paleoecology. Verlag Dr. Friedrich Pfeil, München, Germany, 433-459

Patterson C (1993a) Osteichthyes: Teleostei. In: Benton MJ (Ed) Fossils Record 2. London, Chapman and Hall: 621-656

Patterson C (1993b) An overview of the early fossil record of Acanthomorphs. Bulletin of Marine Science 52: 29-59

Patterson C, Smith AB (1987) Is the periodicity of extinctions a taxonomic artefact? Nature 330: 248-252

Patterson C, Smith AB (1989) Periodicity in extinction: The role of systematics. Ecology 70: 802-811

Robeck HE, Maley CC, Donoghue MJ (2000) Taxonomy and temporal diversity patterns. Paleobiology 26: 171-187

Schwarzhans W (1996) Otoliths from the Maastrichtian of Bavaria and their evolutionary significance. In: Arratia G, Viohl G (Eds) Mesozoic Fishes - Systematics and Paleoecology, Verlag Dr. Friedrich Pfeil, München, Germany: 417-431

Sheehan PM, Fastovsky DE (1992) Major extinctions of land-dwelling vertebrates at the Cretaceous-Tertiary boundary, Eastern Montana. Geology 20: 556-560

Signor PW, Lipps JH (1982) Sampling bias, gradual extinction patterns and catastrophes in the fossil record. In: Silver LT, Schultz PH (Eds) Geological implications of impacts of large asteroids and comets on the Earth. Geological Society of America, Special Paper 190: 291-296

Thomson KS (1977) The pattern of diversification among fishes. Patterns of evolution as illustrated by the fossil record. In: Hallam A (Ed) Developments in palaeontology and stratigraphy. New-York, Elsevier Scientific Publishing Company 5: 377-404

Taverne L (1980) Ostéologie et position systématique du genre *Platinx* (Pisces, Teleostei) de L'éocène du Monte Bolca (Italie). Bulletin de l'Académie Royale de Belgique (Classe des Sciences) 66: 873-889

Taverne L (1997) Ostéologie et position systématique d'*Audenaerdia casieri*, Téléostéen Clupéomorphe (Pisces) du Santonien (Crétacé) de Vonso, Bas-Zaïre. Musée Royal d'Afrique Centrale, Tervuren (Belgique), Département Géologie et Mines, Rapport Annuel 1995 et 1996: 203-213

Taverne L (1999) Ostéologie et position systématique d'A*rratiaelops vectensis* gen. nov., Téléostéen elopiforme du Wealdien (Crétacé Inférieur) d'Angleterre et de Belgique. Bulletin de L'institut Royal des Sciences Naturelles de Belgique, Sciences de la Terre 69: 77-96

Vasse D, Hua S (1998) Diversité des crocodiliens du Crétacé Supérieur et du Paléogène. Influences et limites de la crise Maastrichtien-Paléocène et des "Terminal Eocene Events". Oryctos 1: 65-77

Wenz S (1965) Les poissons Albiens de Vallentigny (Aube). Annales de Paléontologie, Vertébrés 51: 3-23

Wenz S, Brito PM (1992) Première découverte de Lepisosteidae (Pisces, Actinopterygii) dans le Crétacé Inférieur de la Chapada do Araripe (N-E du Brésil). Conséquence sur la phylogénie des Ginglymodi. Comptes Rendus de l'Académie des Sciences, Serie II, Paris 314: 1519-1525

Wenz S, Poyato-Ariza FJ (1994) Les Actinoptérygiens juvéniles du Crétacé Inférieur du Montsec et de Las Hoyas (Espagne). Géobios Special Memoires. 16: 203-212

Wilson MVH, Brinkman DB, Neuman AG (1992) Cretaceous Esocoidei (Teleostei): Early radiation of the pikes in North American fresh waters. Journal of Paleontology 66: 839-846

K/T Impact Remains in an Ammonite from the Uppermost Maastrichtian of Bidart Section (French Basque Country)

Robert Rocchia[1], Eric Robin[1], Jan Smit[2], Olivier Pierrard[1] and Irène Lefevre[1]

[1]LSCE, Laboratoire des Sciences du Climat et de l'Environnement, Bât. 12, Domaine du C.N.R.S. Avenue de la Terrasse. 91198 Gif-sur-Yvette cedex, France.
[2]Institute of Earth Sciences, Free University, de Boelelaan 1085, 1081HV Amsterdam, The Netherlands

Abstract. In two K/T sections of the French Basque Country, Hendaye and Bidart, ammonites have been found in the last few tens of centimetres of the Maastrichtian. In the Bidart section, one specimen was discovered less than five centimetres below the boundary clay layer. It contains a small, but significant amount of iridium and a few crystals of Ni-rich spinel, two markers of the K/T cosmic event. These findings support the idea that ammonites existed up to the very top of the Cretaceous and disappeared suddenly right at the K/T boundary.

1 Introduction

The extinction of ammonites is a persisting and interesting matter of debate. A rather common opinion, although derived from ancient observations, is that ammonites disappeared within the Cretaceous after a long period of decline. This position is based on two arguments: a loss of diversity observed from the Campanian through the Maastrichtian (Wiedmann 1969) and the stratigraphic position of the last ammonite shells well below the K/T boundary in supposed complete sections.

The first point was recently reconsidered by Ward (1990), who identified 35 ammonite taxa for the lower Maastrichtian and 27 for the upper Maastrichtian. According to this author, "Fifteen genera are known from the highest levels of the Cretaceous. The actual number still in the vicinity of the K/T boundary may be higher yet, since the total number of specimens collected from various boundary sections is still relatively small, and new discoveries will undoubtedly enlarge the ranges of some taxa". From his observations, Ward concluded that "the long-described, gradual decline in ammonite diversity was perhaps less marked than previously believed and that the purported long-term diversity reduction of ammonites near the end of the Cretaceous might be more a product of sea-level change than evidence for a slide toward extinction". A statistical study by Marshall and Ward (1996) confirms the sudden extinction at K/T preceded by a background decline of species.

The second point, the extinction levels of ammonite, should also be questioned. Fifteen-twenty years ago, ammonites were considered extinct more than 10 metres below the K/T boundary in sections of the Basque country (Wiedmann 1988). But in his review, Ward (1990) reported the presence (in the sections of Bidart and Hendaye) of ammonites a few tens of centimetres below the boundary. Birkelund (1969) also reported the existence of numerous juvenile forms in the uppermost Maastrichtian strata of Stevns Klint (Denmark). According to Ward, the Stevns Klint and Bidart sections are the best ones for the last ammonite occurrence. However, the exact level of ammonite extinction is not easily determined because, in both sites, there is some reason to believe that the uppermost Cretaceous strata are missing. Let us consider first the validity of this statement.

2 Geological Considerations Regarding K/T Sections in Denmark (Stevns Klint) and in the Basque Country (Hendaye, Bidart)

2.1 Stevns Klint Section

Stevns Klint is an important site for ammonite collection, but some observations suggest that this section might not be complete:

1• Hansen (1979a and b), from his study of dinoflagellates, concluded that the base of the Fishclay is marked by an unconformity.

2• *Micula prinsii* and *M. mura,* marker species for respectively the last and penultimate Cretaceous coccolith zones do not show at that site (Perch-Nielsen et al. 1982). This does not necessarily indicate the absence of uppermost Maastrichtian strata as these two species are typical tropical latest Maastrichtian markers, and do not occur in the Boreal realm (the latitude of Stevns Klint was around 45°N according to paleogeographic reconstruction). We cannot exclude either the possible role of differential dissolution.

3• Another argument derived from our own study deserves consideration: the absence of Ni-rich spinel in the Fishclay. This mineral is found in most K/T sections (Smit and Kyte 1984; Kyte and Smit 1986; Robin et al. 1991; Rocchia et al. 1996) at the very base of the boundary clay (Robin and Rocchia 1998). It is supposed to result from the worldwide dispersion of sizeable melted and oxidised K/T bolide debris (Robin et al. 1992). The deposition of these debris occurred shortly (hours or days) after the collision while the deposition of fine dust particles occurred much later (months or years). The absence of Ni-rich spinel does not necessarily imply that the uppermost Maastrichtian strata are missing. It might eventually result from the reducing conditions in the Fishclay producing the dissolution of spinel crystals. Dissolution could also explain the absence of spinel in some Italian sections (Fonte d'Olio near Ancona, for instance) apparently more reduced than the historical K/T sites near Gubbio, where spinel crystals are rather abundant. We have to note, however, that the absence of Ni-rich spinel in sections where reducing conditions prevailed is not a global feature: the Baie de Loya section, in spite of its apparently reduced character, contains numerous and beautifully preserved spinel crystals. Another explanation for the lack of spinel in some localities is the non-uniform or patchy dispersion of sizeable bolide debris on the surface of the Earth.

None of the preceding arguments demonstrate that the K/T section of Stevns Klint is not complete. But put together they suggest that this section it not the most appropriate one for a reliable evaluation of the time separating the deposition of the latest ammonites and the formation of the first Tertiary strata (Ward 1990).

2.2 Hendaye and Bidart Sections

The situation is different in the Basque country sections, at least in the two sections considered in this study, Hendaye and Bidart.

In Bidart, the most extensively studied section, the last Cretaceous nannofossil zone (*Micula prinsii* zone) is expanded - 7 m (Perch Nielsen 1979) to 20 m (Clauser 1987) - compared with other sections (4 m at Caravaca; 2 cm only at Zumaya, where dissolution was likely). On the other hand the cosmic markers are all there: high iridium concentration (5-6 ng/g), Ni-rich spinel (one of the highest abundances observed in K/T sections) and a few rare shocked quartz grains (H. Leroux, private communication). The only alleged argument against completeness was the occurrence of a normal magnetic polarity just below the boundary (Delacotte et al. 1985). Recently, the site was revisited for magnetostratigraphy by Galbrun (1997). This author found a weak but definitely reverse polarity component all over the several uppermost meters of the Maastrichtian. These results remove the single objection against completeness. On the basis of available data, Bidart should be considered as a complete section with a high sedimentation rate suitable for a precise stratigraphic study.

These characteristics should also apply to the nearby section of Hendaye (Baie de Loya) where the K/T event has been perfectly recorded (anomalous Ir concentrations and Ni-rich spinel). The upper Maastrichtian part of the Hendaye and Bidart sections can be tied bed by bed to the sections of Zumaya and Sopelana. Percival and Fisher (1975) already noted the orbitally driven marl/limestone alternations in the Maastrichtian. Kate and Sprenger (1993) have shown that these alternations indeed are best explained as eccentricity-precession bundled rhythms. Smit and Vonhof (in preparation) have shown that in the top Maastrichtian interval of 70 m (about 70 precession cycles) not a single interval-bed (half precession cycle) is missing in any of the four sections. Magnetostratigraphic data from Hendaye indicate a reverse polarity over at least

20 meters of the uppermost Maastrichtian, representing 350 000 years, indicating a rather high sedimentation rate (5.7 cm/kyr) comparable with the sedimentation of 5.2 cm/kyr derived from the Milankovitch precession cyclicity (about 1.1 m per 21000 years). The section of Hendaye has been little studied for microfossils but, on the basis of available data, it should be considered as the twin-sister of Bidart.

Below we report our findings in these two Basque country sections.

3 Observations and Discussion

3.1 Ammonite Imprints at Hendaye

Our recent findings confirm the report of Ward (1990). In the Baie de Loya section near Hendaye, we have found three specimen imprints over the last 20 cm of the Maastrichtian. The highest ammonite was located around 15 cm below the Ir-enriched clay layer (Fig. 1). We estimate that this uppermost ammonite was probably deposited there less than 3000 years before the K/T event. We also have to note that the three specimens were located in layers which, because of their distant position with respect to the K/T boundary, do not contain any marker of the K/T event (no significant Ir concentration, no Ni-rich spinel). At Bidart, the situation is different.

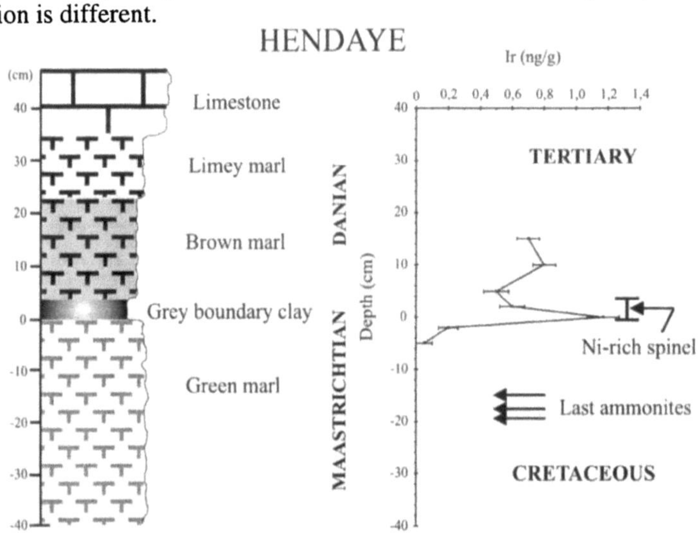

Fig. 1. Ammonite imprints in the Hendaye section (Baie de Loya). The figure shows a simplified stratigraphic log around the K/T boundary, together with the distribution of iridium. The stratigraphic distribution of Ni-rich spinel is less extended: this mineral is found essentially in the vicinity of the boundary clay (± 5 cm; max density: 110 crystals/g). Levels of ammonite findings, noted by arrows, are located below the strata containing iridium and Ni-rich spinel.

Fig. 2. Picture of the Bidart ammonite in situ. Note the dark burrow in the shell filled with boundary clay (Photo J. Smit).

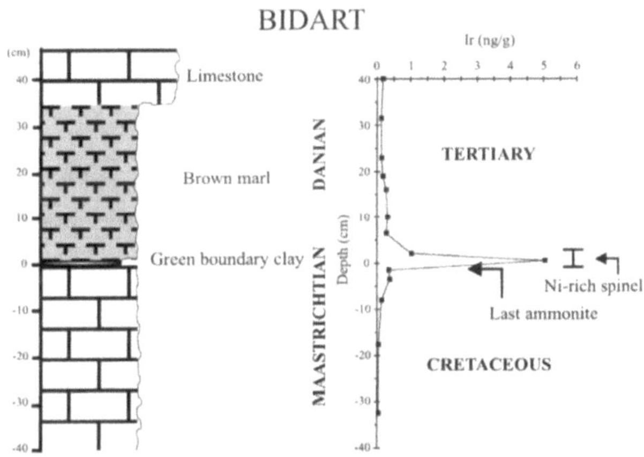

Fig. 3. Position of the ammonite shell found by J. Smit in the Bidart section. The figure shows a simplified stratigraphic log of the section together with the distribution of iridium. The Maastrichtian is the uppermost part of the *Micula prinsii* nannofossil zone. This marker species first occurred 7 to 20 m below the boundary. The ammonite was located less than 5 cm below the boundary clay, in the Cretaceous "tail" of the Ir distribution which is considered as resulting from bioturbation and diffusion. The presence of a few Ni-rich spinel crystals within the ammonite is consistent with the distribution of this mineral (width: ± 5 cm; max density: 3000 crystals/g) and the existence of burrows (visible on Fig. 2) filled with boundary clay.

3.2
Iridium and Ni-rich Spinel in an Ammonite from the Bidart K/T Section

In this section a specimen was found by one of us (J.S.) less than five cm below the boundary clay layer (Fig. 2), in Cretaceous sediments that contain a small amount of Ir (Rocchia et al. 1987). Such low but anomalous iridium concentrations in the uppermost layers of the Cretaceous probably result from diffusion and/or bioturbation from the overlying clay layer (Rocchia and Robin 1998). The specimen itself contains iridium, 0.3 ng/g in the upper part, 0.2 ng/g in the lower part, confirming its near K/T stratigraphic position. At Bidart, such values are usually observed in sediments collected less than 5 cm below the base of the K/T clay (Fig. 3). We have also found a few crystals of Ni-rich spinel *within* the ammonite filling material. One could argue that the ammonite is an old specimen (much older than the K/T event) subsequently reworked in the uppermost Maastrichtian, but the presence of Ni-rich spinel inside the shell does not support this idea. It rather suggests that the ammonite died shortly before the K/T boundary event and that its shell was deposited empty and partly filled (a very likely consequence of bioturbation) with debris of this event. Assuming a duration of 500,000 years for the *Micula prinsii* zone (Bralower 1995) and taking into account the revised thickness of this zone (20 m according to Clauser 1987) the deposition of Smit's specimen would have occurred less than 500-800 years before the K/T event. This is consistent with the sedimentation rate of 3.8 cm/kyr based on the precession cyclicity.

4
Conclusions

From the reported findings, limited in number and carried out in only two nearby localities, it would be hazardous to derive definite conclusions about the extinction of ammonites on a global scale. However, our observations illustrate a general trend in paleontological investigations since twenty years: ammonite remains are found closer and closer to the K/T boundary, but almost never (unless reworked) in the Tertiary. The Geulhemmerberg, in the Netherlands, is possibly the only exception (Smit and Brinkhuis 1996). The same trend is observed for dinosaur remains in continental sections. Our observations, especially the finding at Bidart of a specimen loaded with Ir and Ni-rich spinel, support the idea that at least some ammonite species existed up to the very top of the Cretaceous and disappeared suddenly right at the K/T boundary. It is clear that the accumulation of such observations on a wider, even global, scale is the only way to understand the link between the K/T event and the extinction pattern of ammonites at the end of the Cretaceous.

References

Birkelund T (1979) The last Maastrichtian ammonites. In: Birkelund T, Bromley RC (eds) Cretaceous-Tertiary boundary events. Copenhagen University, vol 1, pp 51-57

Bralower TJ, Leckie RM, Sliter WV, Thierstein HR (1995) An integrated Cretaceous microfossil biostratigraphy. In: Berggren WA, Kent DV, Hardenbol J (eds) Geochronology, Time Scales and Global Stratigraphic Correlations: A Unified Temporal Framework for an Historical Geology. Spec Publ Soc Econ Paleont Min 54, pp 65-79

Clauser S (1987) Evolution de la composition isotopique de l'oxygène des carbonates durant le Campanien-Maastrichtien; Données préliminaires de la série de Bidart (Pyrénées-Atlantiques). Compte-rendu de l'Académie des Sciences, série 2, 304: 579-584

Delacotte O, Renard M, Perch-Nielsen K, Premoli Silva I, Clauser S (1985) Magnétostratigraphie et biostratigraphie du passage Crétacé-Tertiaire de la coupe de Bidart (Pyrénées-Atlantiques). Géologie de la France 3: 243-254

Galbrun B (1997) Normal polarity magnetic overprint of chron C29r by diagenetic hematite growth in red marly limestones from Bidart and Baie de Loya sections (Pays basque, France) Abstract EUG9, Strasbourg, p 314

Hansen J (1979a) Dinoflagellate zonation around the boundary. In: Birkelund T, Bromley RC (eds) Cretaceous-Tertiary boundary events. Copenhagen University, vol 1, pp 136-140

Hansen J (1979b) A new dinoflagellate zone at the Maastrichtian-Danian boundary in Denmark. Danmarks geologie undersogelses, Arbog 1978: 131-140

Kate WGT, Sprenger A (1993) Orbital cyclicities above and below the Cretaceous/Paleogene boundary at Zumaya (N Spain), Agost and Relleu. Sedimentary Geology, 87: 69-101.

Kyte FT, Smit J (1986) Regional variations in spinel compositions: an important key to the Cretaceous-Tertiary event. Geology 14: 485-487

Marshall C, Ward P (1996) Sudden and gradual Molluscan extinctions in the latest Cretaceous of Western European Tethys. Science 274: 1360-1363

Perch-Nielsen K (1979) Cretaceous nannofossils at the Cretaceous-Tertiary boundary near Biarritz, France. In: Birkelund T, Bromley RC (eds) Cretaceous-Tertiary boundary events. Copenhagen University, vol 2, pp 151-155

Perch-Nielsen K, McKenzie J, He Q (1982) Biostratigraphy and isotope stratigraphy and the "catastrophic" extinction of calcareous nannoplankton at the Cretaceous-Tertiary boundary. In: Silver L, Schultz P (eds) Geological implications of impacts of large asteroids and comets on the Earth. Geological Society of America Special Paper 190, pp 353-371

Percival SF, Fischer AG (1977) Changes in calcareous nannoplankton in the Cretaceous-Tertiary biotic crisis at Zumaya, Spain. Evol Theory 2: 1-35

Robin E, Rocchia R (1998) Le spinelle nickélifère de la limite Crétacé-Tertiaire du site d'El Kef, Tunisie. Bull Soc géol Fr 169 (3): 365-372

Robin E, Boclet D, Bonté Ph, Froget L, Jéhanno C, Rocchia R (1991) The stratigraphic distribution of Ni-rich spinels in Cretaceous-Tertiary boundary rocks at El Kef (Tunisia), Caravaca (Spain) and Hole 761C (Leg 122). Earth Planet. Sci. Lett. 107: 715-721

Robin E, Bonté Ph, Froget L, Jéhanno C, Rocchia R (1992) Formation of spinels in cosmic objects during atmospheric entry: a clue to the Cretaceous-Tertiary boundary event. Earth Planet Sci Lett 108: 181-190

Rocchia R, Robin E (1998) L'iridium à la limite Crétacé-Tertiaire du site d'El Kef, Tunisie. Bull Soc géol Fr 169 (4): 515-526

Rocchia R, Boclet D, Bonté Ph, Devineau J, Jéhanno C, Renard M (1987) Comparaison des distributions de l'iridium observées à la limite Crétacé-Tertiaire dans divers sites européens. Mémoires Société géologique de France, nouvelle série 150, pp 95-103

Rocchia R, Robin E, Froget L, Gayraud J (1996) Stratigraphic distribution of extraterrestrial markers at the Cretaceous-Tertiary boundary in the Gulf of Mexico area: Implications for the temporal complexity of the event. In: Ryder G, Fastovsky D, Gartner S (eds) The Cretaceous-

Tertiary Event and other Catastrophes in Earth History. Geological Society of America Special Paper 307, pp 279-286

Smit J, Brinkhuis H (1996) The Geulhemmerberg Cretaceous/Tertiary boundary section (Maastrichtian type area, SE Netherlands); summary of results and a scenario of events. In: Brinkhuis H, Smit J (eds) The Geulhemmerberg Cretaceous/Tertiary boundary section (Maastrichtian type area, SE Netherlands). Geologie en Mijnbouw 75: 283-293

Smit J, Kyte FT (1984) Siderophile-rich magnetic spheroids from the Cretaceous-Tertiary boundary in Umbria, Italy. Nature 310: 403-405

Ward PD (1990) A review of Maastrichtian ammonite ranges. In: Sharpton VL, Ward PD (eds) Global Catastrophes in Earth History. Geological Society of America Special Paper 247, pp 519-530

Wiedmann J (1969) The heteromorphs and ammonoid extinction. Biology Review 48: 563-602

Wiedmann J (1988) The Basque coastal sections of the K/T boundary. A key to understanding "mass extinction" in the fossil record. Revista Espanola de Paleontologia, Num extraordinaria, pp 127-140

Petrographic and Geochemical Studies in the Cretaceous-Tertiary Boundary, Pernambuco – Paraíba Basin, Brazil

Gilberto A. Albertão[1] and Paulo P. Martins Jr.[2]

[1]PETROBRAS, Av. Elias Agostinho, 665, Ponta da Imbetiba, CEP 27913-350 Macaé, RJ, Brasil. (albertao@ep-bc.petrobras.com.br)
[2]Fundação CETEC, Av. J. C. da Silveira 2000, Horto, CEP 31170-000 Belo Horizonte, and Universidade Federal de Ouro Preto, Escola de Minas, DEGEO, Campus, CEP 35400-000 Ouro Preto, M.G., Brasil. (pmartin@cetec.br)

Abstract. The stratigraphic record of the Cretaceous-Tertiary (K-T) boundary (the Poty quarry section) in the Pernambuco-Paraíba coastal basin, Northeastern Brazil, provides evidence supporting the interpretation that the impact of a bolide has caused the widespread biotic extinction at the end of the Cretaceous. New considerations relating to the distribution of iridium and fluorine, as well the occurrence of shocked quartz, spherules, tsunamite bed, biotic extinction and other possible impact-related characteristics have been brought together in this report. The relative abundance of 35 chemical elements were analysed and are presented in geochemical profiles. A geochemical break for most of these elements is better expressed in the geological contact than in the K-T boundary, although iridium and fluorine anomalies have been determined exactly at the boundary. Palaeoenvironmental characterisation is determined by micropalaentological considerations, stable isotope data, and the presence of phosphatised fragments. Most of the particular characteristics described here, such as the geochemical anomalies (iridium, total organic carbon and fluorine) and the presence of a tsunami deposit, of possible shocked quartz grains, and of some impact-related spherules, give support to previous preliminary interpretations of this stratigraphic boundary as a sedimentary record of a catastrophic event marking the K-T boundary.

1 Introduction

The outcrops studied here are those at the Poty quarry, located in the Pernambuco-Paraíba coastal basin (the PE/PB basin) which span the Cretaceous-Tertiary (K-T) boundary in Brazil (Fig. 1). Preliminary studies in this area were, amongst others, reported by Beurlen (1967), Tinoco (1967), Mabesoone et al. (1968) and Stinnesbeck (1989). Albertão (1993) described the outcrops at the Poty quarry as the first K-T boundary location in the low southern latitudes in South America that exhibits an Ir anomaly.

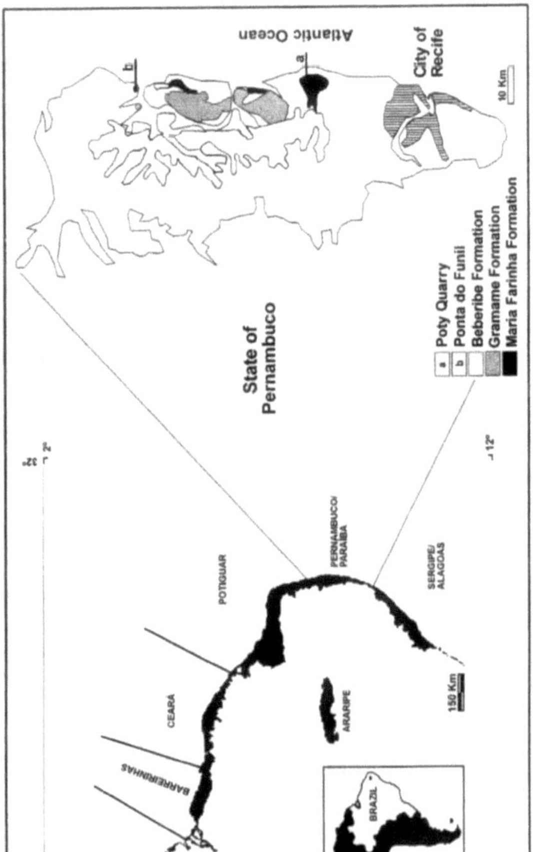

Fig. 1. Location map for the outcrops of the Poty quarry (*a*, UTM 9,152,000N / 300,000E) and Ponta do Funil area (*b*, UTM 9,117,000N / 296,000E), Pernambuco State. The two localities are about 30 km apart. Starting points for UTM coordinates: 10 x 10^6 m S of the equator, and 0.5 x 10^6 m W from the meridian 39°W of Greenwich, respectively.

The PE/PB coastal basin is a passive-margin rift basin. Its origin is related to the South Atlantic ocean opening. The studied sedimentary succession was deposited during Maastrichtian and Danian ages and characterises a marine regressive mega-sequence, as defined by Chang et al. (1988).

In the Poty quarry, the outcropping sequence is composed of two formations (Gramame and Maria Farinha), which exhibit an erosive lithological contact. The Gramame formation is mainly composed by marly biomicrites of deep neritic-upper bathyal environment. The Maria Farinha formation overlays the Gramame formation and consists of alternations between limestones (biomicrites, biosparites and calcilutites) and shales deposited in a middle-deep neritic environment. The sedimentary structures present at the transition from the uppermost Gramame formation to the basal portions of the Maria Farinha formation (such as hummocky cross-stratification, fining-upward and wavy bedding), as well as ichnofossil, geochemical, palaeontological and mineralogical data, characterise a carbonate ramp controlled by storms in a sedimentary process of progressive marine regression (Albertão 1993).

Slightly weathered outcrops from the quarry, and eventually, other proximal areas were selected for good-quality sampling.

2 Samples and Experimental Methods

Samples were collected for thin sections microscopic petrographic analysis, X-ray diffractometry (XRD), $\delta^{13}C$ and $\delta^{18}O$ total rock isotopic analysis (tr-IA), inorganic geochemistry (major, secondary and trace elements), micropalaeontology, scanning electron microscope analysis (SEM), qualitative/quantitative chemistry in energy disperse spectrometer (EDS), determination of phosphate mineral types and petrographic descriptions of hand samples with binocular lens. Sampled areas number 13; six in the Poty quarry, six in Ponta do Funil and one in Cruz de Rebouças. A total number of 99 samples were collected.

A total of 120 thin sections were studied by optical microscopy. XRD was performed at the PETROBRAS Centro de Pesquisas (CENPES) laboratory using a JEOL JDX 8030 instrument, with Cu K-alpha radiation, and followed a routine methodological procedure (Albertão 1993): "oriented" samples with a grain size diameter $< 2\mu m$ were studied for clay minerals determination; "non-oriented" samples with a grain size diameter $>2\mu m$ were checked for clay minerals and for total semi-quantitative mineralogical composition.

Stable isotope ratios of C and O ($\delta^{13}C$ and $\delta^{18}O$) were analysed on 46 samples at the Centro de Energia Nuclear na Agricultura, Universidade Estadual de São Paulo (CENA/UNESP), according to techniques described by Albertão (1993). As a by-product of the stable isotope analyses, the total organic carbon (TOC), insoluble residue (IR), and the $CaCO_3$ contents of the sediments were also determined.

The chemical determination of the abundance of 45 elements, including iridium (Ir), was conducted at Los Alamos National Laboratory by neutron activation analysis (NAA) in 70 samples of about 10 g each one (Albertão 1993). Instrumental NAA was done for most elements, but radiochemical NAA was done on larger sample aliquots in some selected samples to confirm Ir content values. Sample preparation and analysis were done accordingly to standard procedures. Ir was determined with a precision of 5 rel.% (M. Attrep Jr. 1992, personal communication). Albertão (1993) and Albertão et al. (1994b) give more details about the NAA analytical methods. Of the initial number of 45 elements, 30 to 36 elements were selected for interpretation, as they were detected in most samples.

Micropalaeontological studies (foraminifera and palynological ones) on all samples were conducted to determine the disruption of both fauna and ecosystems (Koutsoukos 1996). Sample preparation was done according to routine procedures of CENPES. The biostratigraphy team of CENPES conducted these studies (Albertão 1993; Albertão et al. 1993; Koutsoukos 1996).

Statistical treatment of the data was conducted with basic and multivariate statistics (analysis of variance, Pearson's correlation coefficient, "F" and "T" tests, normalisation of geochemical data, factor analysis and discriminant canonical analysis).

Spherules and quartz grains were manually separated from the rock samples and analysed (with binocular lens) according to the procedures described by Albertão et al. (1994b) and Delício et al. (2000). Scanning electron microscopy (SEM) with an attached semi-quantitative X-EDS system permitted the determination of the fluorine content in the spherules; a chemostratigraphic search for fluorine was performed later (Marini et al. 2000) on 11 whole-rock samples from beds in the neighbourhood of the K-T boundary, using ion-selective-electrode measurements, after complete removal of Al. This method has a detection limit of 20 ppm and a precision better than 5 rel.% for F >500 ppm.

3 Petrography and Geochemistry of the K-T Boundary at the Poty Quarry

3.1 Petrographic Description

The best and most complete sequence of outcrops across the sediments containing the K-T boundary in Brazil is located in the Poty quarry, in the PE/PB coastal basin (Fig. 1). The strata are continuous and perfectly preserved with small lateral-vertical facies variation. Strata are almost horizontal with a slight dip towards the East-Southeast.

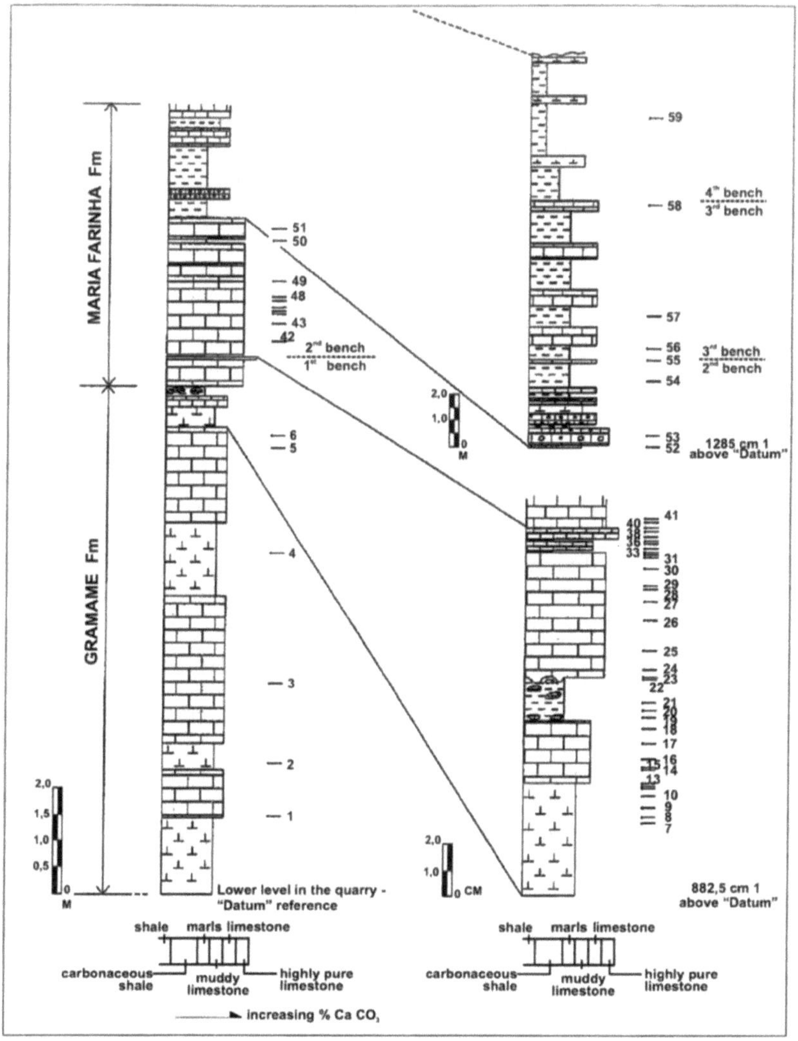

Fig. 2. Schematic geological section of the Poty quarry. The stratigraphic location of samples 1 to 59 are indicated. Details of the section (lower part, right side of the figure) are shown in Figure 5. Lithological symbols are the same as in Figure 5. See text for details.

Figure 2 shows the local sequence of outcropping beds, as well as the 59 main samples used in the present study. This profile is a composite of four different places of observation of the local stratigraphic record (named here points 1, 2, 3 and 4) and shows the reference level ("Datum" reference) for the profile determination; this "Datum" reference marks the lowest level of the quarry, situated in its western part, and it is here defined as the "zero" reference for all measurements. The interval from 0 cm up to 882.5 cm was measured and described in point 2; from level 882.5 cm up to 1022.5 cm in points 1 and 4, and from 1022.5 cm up to 2775.0 cm in point 3. The $CaCO_3$ content of the various beds was determined from average of XRD and calcimetry analyses performed on 46 samples and also from data obtained directly from the geological division of Poty quarry Company (Albertão 1993). Accordingly to Flügel's (1982) criteria (Table 1), it was possible to classify the stratigraphic sequence of beds (Fig. 2).

Table 1. Criteria for the characterisation of stratigraphic beds accordingly to the $CaCO_3$ content in the Poty quarry.

Content of $CaCO_3$ in wt%	Classification
(0-20)	shale (clay minerals)
(20-40)	carbonaceous shale
(40-65)	marls
(65-80)	muddy limestone
(80-95)	limestone
>95	high-purity limestone

The Poty quarry is divided into four mining benches for exploration purposes: one is in the Gramame formation (the first one from the bottom of the quarry), and the other three are in the Maria Farinha formation.

The complete outcropping exposure comprises an average thickness of 10 m for the Gramame formation and 18 m for the Maria Farinha formation. The lowest level of the quarry ("Datum" reference in Fig. 2) occurs in the Gramame Formation. Fluvial sediments of the Barreiras formation truncate the top of the Maria Farinha formation, but a discussion of this topic is beyond the scope of this paper.

The contact of both formations is thus identifiable at 964.5 cm (above the "Datum" reference). Rock types are described accordingly to the Dunham and/or Folk system of classification (Flügel 1982). The description below is devoted to both formations, to the contact within between them, and includes a discussion of the K-T boundary position.

3.1.1 The Gramame Formation

From the base, at the reference level ("Datum" reference), to the erosive top, sediments of this unit are relatively homogeneous in composition, alternating planktonic wackestones/mudstones and marls, with small variation of clays and bioclasts contents. This homogeneity is better represented in this formation than in the Maria Farinha formation. (Mabesoone et al. 1968; Stinnesbeck 1989). Primary sedimentary structures seem to be absent. In the horizontal direction the thickness of the great beds vary from a few centimetres to a few metres. Each great bed is composed of discontinuous laminae caused by bioturbation among alternating fine (millimetric) beds of limestones and marls.

Bioturbation structures are widespread in the contact between marls and carbonaceous rocks. Bioturbation structures are more conspicuous in marls. The sediments are constituted of alternating sequence of mudstones and planktonic wackestones with varying shale and bioclasts content. Micritic material is visible with a few cases of recrystallisation. Planktonic foraminifera predominate over benthonic ones. Calcispherulids, radiolaria and echinoderms are also present. Fragments of phosphatised vertebrates and grains of siliciclasts, mainly silt-size quartz, are rare.

The neighbourhood of the contact between the Gramame and the Maria Farinha formations is characterised by 14 beds that are named here by letters from **A** to **N**. Their petrographic and fossil content is presented in Table 2 and Figs 3, 4 and 5. Beds **A**, **B** and **C** are in the Gramame formation, whilst beds **D**, **E**, **F**, **G**, **H**, **I**, **J**, **K**, **L**, **M**, and **N** are in the Maria Farinha formation. Figure 3 shows the sequence of beds **B** to **H,** whereas Figure 4 shows the sequence **D** to **N** and other upper parts of the sequence. Details on the sedimentary structures are presented schematically in Figure 5.

In between beds **C** and **D**, a conspicuous break of the sedimentary depositional regime occurs, which characterises the lithological contact of both formations. Beds **A, B** and **C** have similar characteristics which are also similar to those of the underlying rock beds. From bed **D** upward there is an increasing lithological variation with complete distinction from the underlying beds (Table 2; Figs. 2, 3, 4, 5 and 6).

Fig. 3. Photograph showing the lithological contact (indicated by an arrow) between Gramame and Maria Farinha formations. (Poty quarry). From the base to the top, beds **B**, **C**, **D**, **F** and **H** are represented. Beds **E** and **G** are fine marls between respectively beds **D** and **F**, and **F** and **H**. Bed **I** overlies bed **H**, but is not visible in the photo. Bed **A** underlies bed **B**, but is covered by debris in the photo. It is possible to observe the erosive character of the contact between beds **D** and **C**, the interference ripples (similar to wavy bedding) structures above bed **D** and the breccia-aspect of bed **C** (scale bar in the photo: 1.0 m).

Fig. 4. Continuous and non-weathered portion of the section across bed **I** in the Poty quarry. Beds **D** (base) to **N** (top) are shown in the photograph. Marly beds **E**, **G** and **I** (the Ir-rich bed, designated "limit" in the figure) are indicated by arrows. It is possible to observe interference ripple structures in the couplets **E/F** and **G/H**. Total thickness of the sequence in the photograph is 2.5 m.

Table 2. Summary description of field and petrographic observations of the neighbourhood of the lithological contact between the Gramame formation (beds **A**, **B** and **C**) and the Maria Farinha formation (bed **D** up to bed **H**); see also Figures 2 and 5.

Beds	Local names	Width (cm)	Rock types and content description
A	Marga I	40	Wackestone/ packstone with planktonics / phosphatised grains / fossils / some worms' tubes *Hamulus* / bioturbation mainly *Thalassinoides* / pelecypods, echinoderms, foraminifera
B	Poty I	25	Doubtful K-T boundary / nodular limestone / wackestone – locally packstone / bioturbation / more planktonic foraminifera than benthonic ones / *Hamulus*, echinoderms, ostracods, calcispherulids / phosphatised fragments
C	without local name	15	Same as **B** though **C** is more "marly"/ spherule occurrences/ first observation of very rare and thinly Danian foraminifera
D	Capim	50	Erosive lithological contact (at the base) / Packstone upgrading to wackestone-mudstone / rare bioturbation / spherule occurrences / rare, possibly shocked quartz grains / bio and siliciclastic gross sands / phosphatised fragments (partially glauconitised – pyritised), of foraminifera, gastropods, pelecypods, worms' tubes, echinoderms, shark teeth, wood (rare)
E *	Without local name	2	Continuous marl / spherule occurrences / benthonic and planktonic foraminifera / echinoderms / phosphatised fragments / siliciclasts
F *	Topo do Capim	3	Similar to upper **D** bed / mudstone-wackestone / spherule occurrences
G *	without local name	2	Marl / spherule occurrences / planktonic foraminifera / echinoderms, fragments of worms tubes / less siliciclasts and phosphatised fragments than **E**
H *	Batentinho	4	Recrystallised limestone / mudstone / rare bioclasts, mainly foraminifera / spherule occurrences / bioturbation – *Chondrites*, *Planolites*, worm tubes / **E***, **F***, **G***. **H*** - alternating mudstone/marl thin beds with complex interference ripples structures throughout the quarry

Table 2. Continued. Summary description of field and petrographic observations of the Maria Farinha formation (from bed **I** up to bed **N**). In this table only beds of the Maria Farinha formation are described; see also Figures 2 and 5.

Beds	Local names	Width (cm)	Rock types and content description
I	without local name	2	"Marly" mudstone / most probable K-T boundary / spherule occurrences / rare, possibly shocked quartz grains / globigeriniform planktonic foraminifera, benthonic ones / few siliciclasts and phosphatised fragments / Ir, F, and TOC (total organic carbon) anomalies
J	Vidro	58	Apparently recrystallised micritic mudstone / tiny planktonic and benthonic foraminifera / echinoderms, rare calcispherulids / slightly bioturbated
K	without local name	20	Similar to **J** / bioturbation / bioclasts with glauconite grains and rare phosphatised grains
L	Topo do Vidro	23	Intensively bioturbated / almost brecciated / gastropods fragments / foraminifera / phosphatised and glauconitised grains
M	Enfornação do Vidro	35	With some elements of **L** / wackestone-packstone / fining-upward / abundant bioclasts / some phosphatised fragments / large (up to 7 mm) gastropods fragments/ worms tubes (serpulids), arthropods, mainly benthonic foraminifera, rare bryozoa / phosphatised pellets eventually from *Calianassa* arthropod /
N	Batente	28	Similar to **M** / thicker grains size at **N** basis than at the top of **M**

3.1.2 The Cretaceous-Tertiary (K-T) Boundary

The K-T boundary at the Poty quarry has been located in two different places (Albertão 1993; Albertão et al. 1994a; Koutsoukos 1996; see Figure 5 and Table 2).

The first location is a continuous and very thin marly bed (bed **I**). The boundary is defined by micropalaeontological analysis, with biotic extinction (Albertão 1993), geochemical anomalies, such as Ir, with a maximum value of 0.69 ppb, (Albertão 1993; Albertão and Martins Jr. 1996a; Figures 6, 7a, 7b) and TOC (Albertão 1993; Albertão and Martins Jr. 1996a; Figure 7c), spherules crystals (Albertão et al. 1994b; Albertão and Martins Jr. 1996a), some of them containing fluorite (Marini and Albertão 1999; Marini et al. 2000), and some rare, possibly shocked, quartz (Albertão et al. 1994b; Albertão and Martins Jr. 1996a). Despite the fact that fluorine is not generally accepted as an indicator of a K-T boundary, a distinct F anomaly of 5.57 wt.% (in stark contrast with F contents of less than 0.3 wt.% in all other samples) has also been observed in the same bed **I** (Marini and Albertão 1999).

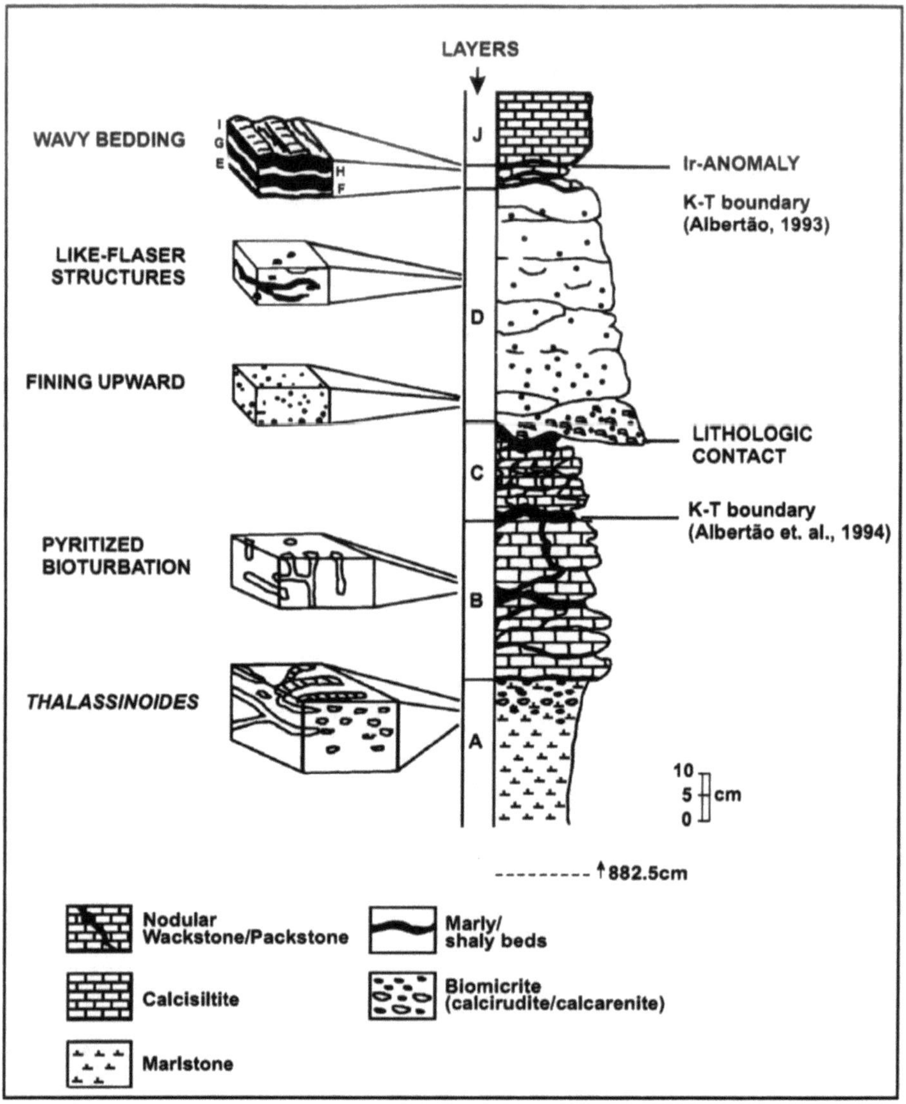

Fig. 5. Schematic lithological section at the Poty quarry, representing beds **A** to **J** with their main sedimentary structures. Two different possibilities for the location of the K/T boundary are considered; the upper one coincides with the Ir anomaly and the lower one with the first occurrence of very rare Danian foramifera species. The lithological contact defines the limit between Gramame formation (lower sequence, mostly Maastrichtian in age) and Maria Farinha formation (upper sequence, mostly Danian in age).

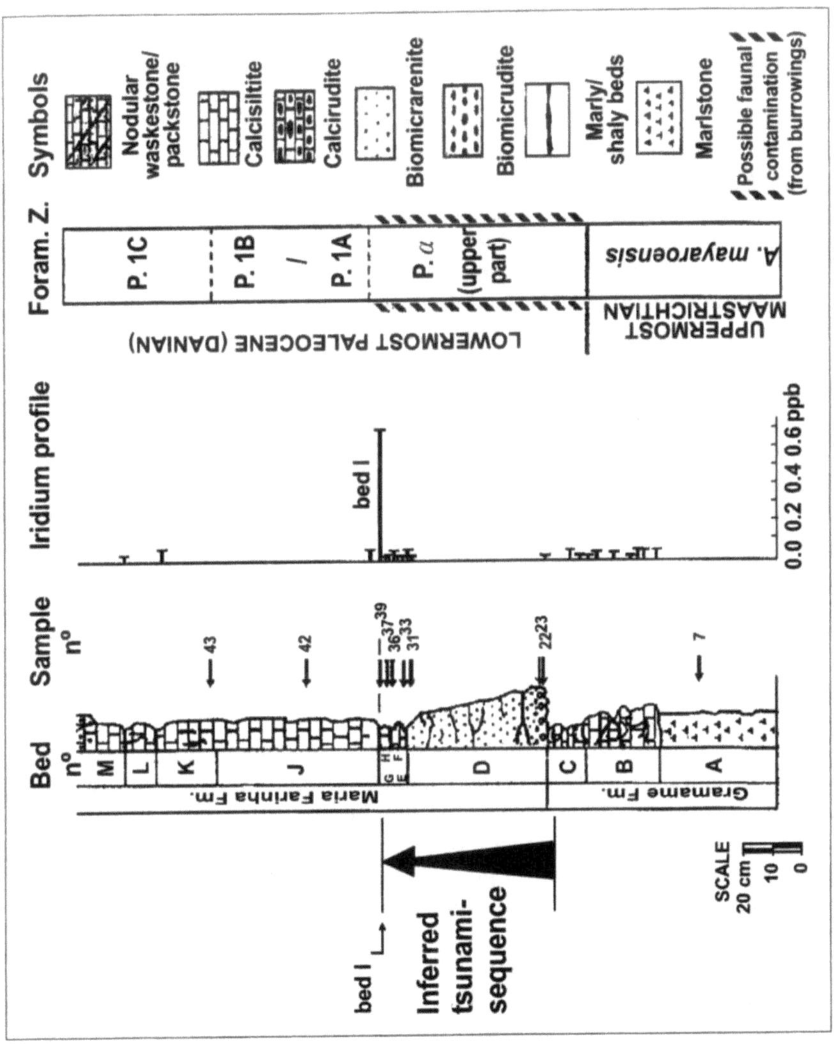

Fig. 6. Detailed lithostratigraphy, with Ir distribution and foraminifera zones, across the K-T boundary in Poty quarry. Beds **A** to **J** are shown.

The other location is at the top of a non-graded, nodular carbonate mudstone to wackestone (bed **B**), situated about 75 cm below bed **I** (Fig. 5). At this very position, after a very detailed foraminifera investigation, the first Danian planktonic taxa were determined (Albertão et al. 1994b; Koutsoukos 1996). Also more recent biostratigraphic analysis on marine ostracodes and again on foraminifera (Koutsoukos 2000, personal communication) indicate a boundary position at the top of **B**.

However, both the occurrence of a possible tsunami deposit (bed **D**), as is discussed below, and the location of the Ir anomaly, point to a more adequate K-T boundary position in bed **I** (Albertão 1993). Preliminary micropalaeontological analysis of foraminifera and palynomorphs, as mentioned above, supported this interpretation. Albertão and Martins Jr. (1996a) discussed this ambiguous and controversial situation, and suggested the possibility of contamination by post-depositional biogenic burrowing, which permitted microfossils to migrate from the more recent stratum to the older one. Although this hypothesis has neither been confirmed nor dismissed, it is possible to consider it as a working hypothesis. In this paper, the K-T boundary is thus proposed to be situated in bed **I**.

The same discussion may be applied to the spherules occurrence. Spherules are quite unusual in this outcrop, especially those of bed **I,** which contain the fluorine anomaly. Spherules occur only across the section between beds **C** and **I**, but their impact origin is not entirely evident (Marini et al. 2000). The fact that they do also occur even some tens of centimetres below bed **I** (at bed **C**) may be due to bioturbation, as was formerly discussed for the foraminifera occurrence.

Three beds were selected for more detailed analysis of the spherules (Marini et al. 2000): beds **E**, **G** and **I** (Figs. 3, 4, 5 and 6). The samples were chosen due to their proximity to the K-T boundary and one of them, from bed **I**, contains the Ir anomaly.

The vast majority of the spherules consist of small (80-200 μm), faintly translucent- to opaque- brownish, particles with a shiny luster. The spherules were divided into three types, based on their mechanical behaviour (Marini et al. 2000): [1] *brittle* class: fragile, with F-/Cl-rich apatite crust (0.5-5 μm thickness) enclosing an interior with an aggregate of minute (globular or rhomboidal) calcite and/or dolomite crystals (0.5-3 μm in size), which eventually forms concentric arrangements, reminiscent of oolitic textures; [2] *resistant* class: with dominant, concretionary Ca-/F- rich phosphates, with low Fe and Al content; and [3] *white* class: they are the largest ones, sometimes reaching 0.8 mm, and are in appearance similar to "sugar-coated almonds", due to the presence of "sugar"-like fluorite crystals. Some of the *resistant* class contains relict, slightly darker grains of K-feldspar resembling sanidine.

The majority of the sampled spherules in the PE/PB basin are composed mainly of F-rich apatites and strongly differ from Al- or Fe- rich phosphates described elsewhere (Marini et al. 2000) from altered K-T boundary impact glasses. On the other hand, spherules with fluorite crystals (*white* class), which are present only in bed **I,** although diagenetic, may have an indirect impact relation, considering the high F-abundance in this stratigraphic level (Marini et al. 2000). This hypothesis is supported by the following observations: [1] a F anomaly occurs in the same bed as the Ir anomaly, and [2] evaporitic rock sequences, such as those that were impacted at Chicxulub (Mexico), are usually F-bearing sediments. Abnormal and global fluorine release should be taken into consideration if further studies confirm increasing occurrence of diagenetic fluorite (and other F anomalies) close to the K-T boundary in other, more classical, sections (Marini et al. 2000).

3.1.3 The Maria Farinha Formation

Overlying the Gramame formation and underlying the Barreiras formation, the Maria Farinha formation is about 18 m thick in the quarry area. It is composed of alternating beds of limestones, marls and shales. From the base upwards, there is a general tendency of a progressive decrease of the $CaCO_3$ content and an increase of marly sediments. There is also both an increase of the abundance of siliciclasts and an increase in the frequency of dolomitised material. Structures such as swaley cross-stratification are of metre size.

Bioturbation with *Thalassinoides* (greater than 2 cm) and *Fugichinia* structures are common, especially around the horizon of contact between marls and limestones. Limestones show some fining-upward and wavy bedding structures. They are predominantly wackestones/packstones. Bioclasts are more common in sediments with gastropods, worm tubes, ostracods, and phosphatised fragments. Large foraminifera (*Nummulites* up to 2 mm) are common in clay-rich layers, although benthonic foraminifera dominate in the sediments.

Planktonic/benthonic ratios decrease from the bottom upwards, whereas the content of siliciclasts and clay increases. Locally, micro- and macro- biostructures, such as burrows and other trace fossils are also present. Tables 2.1 and 2.2 present a complete description of all beds across the K-T boundary and also of the lithological contact, as shown in Figures 2 and 5.

Bed **D**, the base of Maria Farinha formation, has been considered as a possible tsunami deposit (Albertão 1993; Albertão et al. 1994b; Albertão and Martins 1996a). Some further discussion on this matter is given below.

3.2 Geochemistry of the K-T Boundary in the Pe/Pb Basin

Neutron activation analysis provided data for a geochemical description of the relative abundance of 35 elements across K-T boundary at the Poty quarry. These elements were selected because they were determined in most of the analysed samples. The number of elements and samples permitted the statistical treatment of the analytical data, which aids in the description of the sedimentary processes. Figures 8a and 8b show the distribution of Na, Mg, Al, K, Ca, Sc, Ti, V, Cr, Mn, Fe, Co, Zn, As, Se, Br, Rb, Sr, Zr, Sb, Cs, Ba, La, Ce, Nd, Sm, Eu, Tb, Dy, Yb, Lu, Hf, Ta, Th and U.

It is noteworthy that the K-T boundary, considered here to be located in bed **I,** is characterised by small variations of the contents of some of the elements mentioned above as measured upwards from the lower bed to the upper bed (Fig. 8). Nevertheless, the contact between the Gramame and the Maria Farinha formations is definitely marked by a strong contrast in the abundance of most elements, characterising a net break of the geochemical content of both formations. Samples shown in Figures 6, 8a, and 8b are from a depth of 905 to 1055 cm. Figures 8a and 8b indicate also different average contents of the elements in each formation.

In fact, the abundances of Cr, Mn, Co, Rb, Zr, Th, U and the rare-earth elements (REEs) show the most distinct contrast in their distribution over the vertical profiles, across the Gramame formation to the Maria Farinha formation. From the REEs, La and Lu show the most distinct differences in average content in between the two formations.

Overall, the chemical data in the thin contact layer show that the K-T boundary (in bed **I**) is also marked by geochemical changes, which are, however, not as evident as in the lithological contact. Nevertheless, the most important geochemical anomaly in bed **I** is that of Ir, which will be discussed below.

The geochemical "break" that occurs at the contact between the two formations is in agreement with the erosive event that can be observed at the base of bed **D** (Figs. 3 and 5), and which also caused a sharp contrast in lithological composition

According to Huffman et al. (1990), elements can be grouped into three different categories in relation to their geochemical behaviour in calcite and clay minerals. Using this classification, the following grouping was obtained (values within parenthesis are correlation coefficients obtained in the present work).

1. elements which follow the behaviour of calcite: Sr (0.68),
2. elements with clay affinity: Cs (0.94), Al (0.93), Co (0.93), K (0.92), V (0.92), Sc (0.91), Th (0.90), Cr (0.89) and Hf (0.89),
3. elements with independent behaviour (which have abundances that are not correlated to the calcite or clay content): Ir, Sb, Ba, Fe, Se, U and Zr (these are the most important).

The previous relations were determined in accordance with the correlation coefficients for the following aspects: (1) high positive coefficients for [calcite *versus* the elements] and for [Ca *versus* the elements]; (2) high positive coefficients for [elements *versus* clay minerals] and for [elements *versus* insoluble residues], and high negative coefficients for [elements *versus* calcite]; (3) for the independent elements the criteria was of exclusion that neither (1) and (2) are valid.

Michel et al. (1985), Schmitz (1988), and Koeberl (1989) discussed various elements that show anomalies in K-T boundary samples, such as Ir, Ni, Co, Cr, Zn, Cu, As, Se and Sb. Nevertheless, Schmitz (1988) pointed out that, although certain types of meteorites are rich in Ir, Ni, Co and Cr, both Co and Ni, as well as As, Se and Sb, can behave as calcophile elements and in this case are very abundant in terrestrial sediments enriched in sulphur and organic matter. Nevertheless, Ni is very abundant in most meteorites. Preisinger et al. (1986) and Michel et al. (1990) discussed data for other K-T boundary locations, where there are only anomalies in the contents of the platinum group elements (PGEs). In the PE/PB basin only the content of Ir (and none of the other PGEs) was measured.

Table 3 presents some values for trace elements and their relative concentrations at some different K-T boundaries (of different depositional environments) and

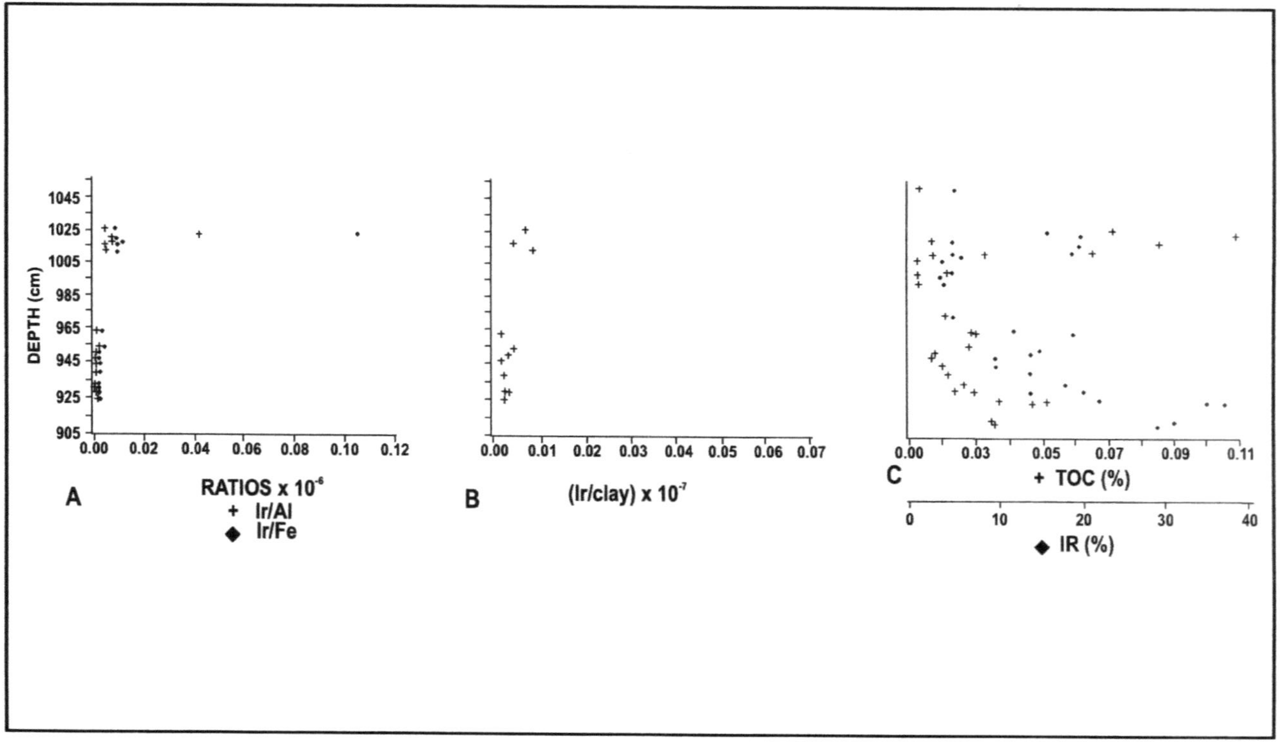

Fig. 7. Ir/Al and Ir/Fe ratios (A) and Ir/clay ratio (B) confirm the presence of an Ir anomaly in bed **I** (interval of 1021-1023 cm). (C) The total organic carbon (TOC, in wt%) anomaly in bed **I** is independent from the amount of the insoluble residue (IR, in wt%). Bed **I** represents the maximum values in all of the plots. " O cm" is the " Datum" reference at the base of the quarry (the lowest level of the quarry in Figure 2); " depth" means thickness of the section above the datum. Profile across the K-T boundary at the Poty quarry.

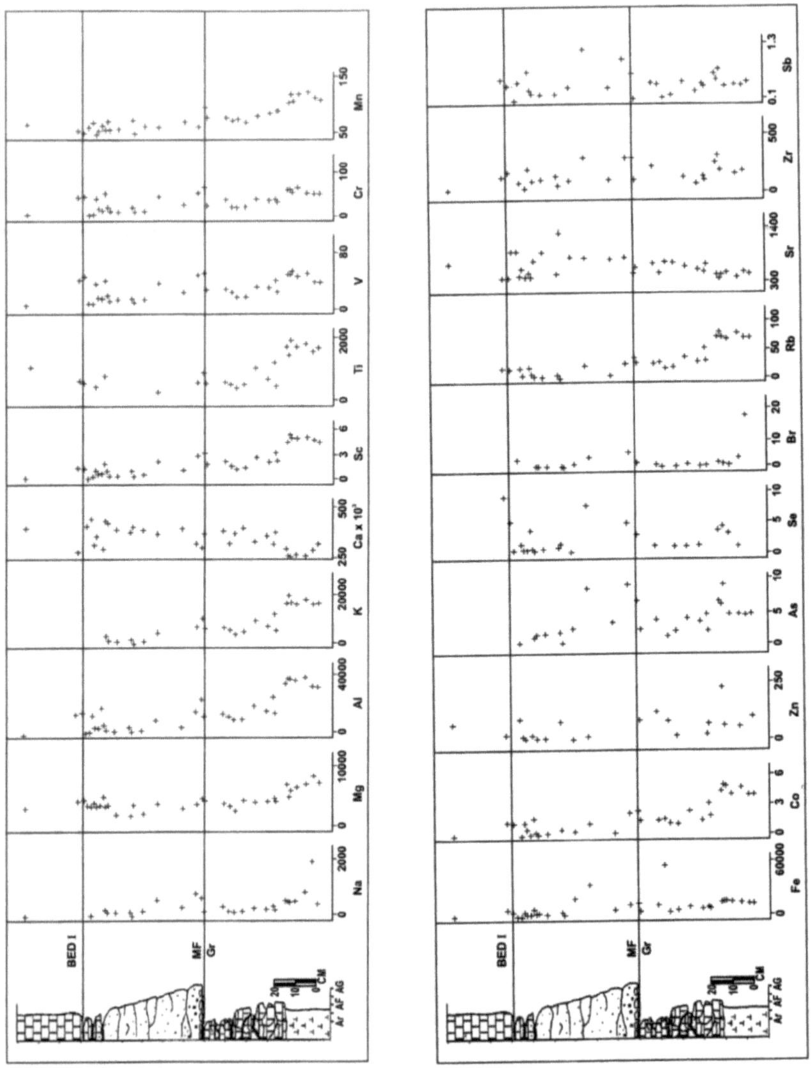

Fig. 8a. Geochemical profiles for the first 25 elements analysed in Poty quarry (concentrations in ppm). The lithological contact between Maria Farinha and Gramame formations (MF/Gr) and the Ir-rich bed I are indicated. Lithological symbols are the same as those in Figure 5.

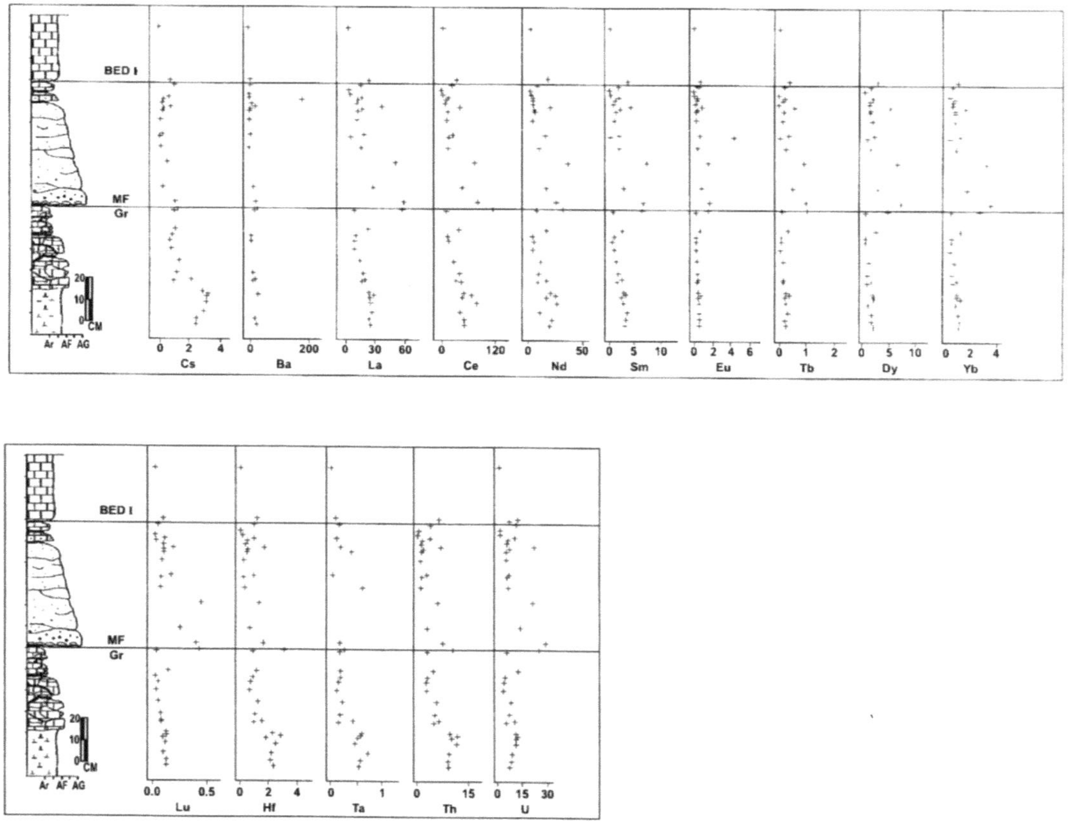

Fig. 8b. Geochemical profiles for the other 10 elements analysed in Poty quarry (concentrations in ppm). The lithological contact between Maria Farinha and Gramame formations (MF/Gr) and the Ir-rich bed **I** are indicated. Lithological symbols are the same as those in Figure 5.

Table 3. Contents of some elements in various K-T boundaries and rocks (adapted from Gilmore et al. 1984). The abundance of iridium is given in ppb. Contents of other elements are reported in ppm.

El.	PE/ PB	S.	Raton	G.	crust	chond.	US bas.	KI bas.
Se	5.3	4	-	-	0.05	1.8	-	-
La	17.9	44	16 – 80	32.4	30	0.24	28	17.5
Ce	22.1	74	-	-	60	0.62	-	-
Yb	0.56	2.4	0.7 - 2.2	2.4	3.4	0.16	3.53	1.6
Hf	1.0	8	3-6	2.5	3	0.12	5.0	4
Ta	0.24	1.3	-	-	2	0.02	-	-
Ir	0.69	3.2	3.2 – 15	9	$\ll 0.1$	480	35×10^{-4}	55×10^{-3}
LY	31.9	18	15 – 57	13.9	8.8	1.5	7.9	11

El. chemical elements; *LY* ratio *La/Yb* (dimensionless); *PE/PB* Pernambuco-Paraíba basin; *S.* Sugarite area in Raton basin, USA; *Raton* average data from Raton basin; *G.* data from Gubbio, Italy; *crust* crust average values; *chond.* average values from chondrites of type C1; *US bas.* average values of basalts from USA; *KI bas.* average values of basaltic lava from Kilauea Iki, Hawaii.

rocks. These values are presented to show the relative abundance of these elements at locations with different depositional conditions (crust and sedimentary rocks). In general, the contents of Se and Ir are very high in K-T boundary samples compared to average crustal rocks. Otherwise, Hf and Ta contents in the PE/PB coastal basin are less abundant than the mean value of the crust and basalts.

However, the ratios Cr/Al, Se/Al, Cr/Fe and Se/Fe (Figs. 9a, b, c and d) do not indicate the existence of Cr and Se anomalies in the K-T boundary at the Poty quarry, although Se shows high concentrations (Table 3). Otherwise, the Cr and Se ratios in relation to Al increase towards bed **D** and decrease from this level (bed **D**) upwards, which characterises a geochemical discontinuity in the lithological contact; low values of these ratios (especially for Cr/Al) occur in the vicinity of the Ir-rich bed **I** (Figs. 9a, b). The ratios with the element Fe do not show a similar behaviour, but support the existence of a discontinuity at the base of bed **D** (Figs. 9c, d).

3.3 Iridium Anomaly at the Poty Quarry

Iridium, as a representative of the platinum-group elements (PGEs), is an important indicator of the possible presence of extraterrestrial material in sediments, if its content is significantly higher than typical crustal values. Such significant enrichments in Ir and the other PGEs have been found in the K-T boundary samples from locations all over the world, including the PE/PB coastal basin. Sample 39 of bed **I** (Figs. 2, 5, 6; Table 2.2) has an elevated content of Ir compared to average crustal abundances (Fig. 6). The measured content (0.69 ppb)

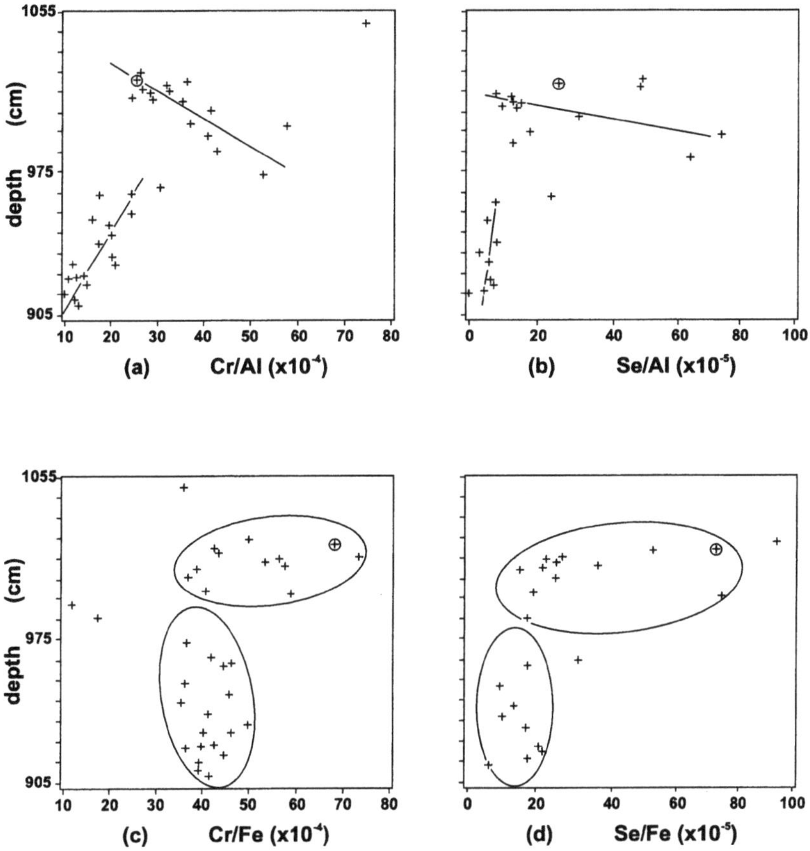

Fig. 9. Ratios of Cr/Al, Se/Al, Cr/Fe and Se/Fe *vs.* stratigraphic position in the Poty quarry. The cross with the circle represents sample 39 (bed **I**). The lithological contact between the Gramame and Maria Farinha formations (964.5 cm) is marked by a discontinuity in the ratios.

is about 26 times higher than the average measured for the other samples from the Poty quarry analysed in the present paper.

Michel et al. (1985 and 1990) observed that an increase in Ir content may be correlated with a decrease of the $CaCO_3$ content in sediments. These authors have suggested that this is the result of normal sedimentary processes, where sediments richer in clay minerals are also richer in Ir. It is well-known that the clay content is usually inversely proportional to the $CaCO_3$ content of sediments. Therefore, Michel et al. (1985 and 1990) suggested that it is important to determine the Ir/clay content ratio to determine if the high values of Ir present in sediments are really anomalous or if they are just "artifacts" of clay content influence.

According to Michel et al. (1985 and 1990), high contents of Fe and Al are representative of the presence of clay minerals. Thus, examining the ratios Ir/Fe and Ir/Al (Fig. 7a), it is possible to determine if indeed there is an enrichment in Ir compared to normal sedimentary values. Another way to verify the ratio Ir/clay content is directly through the ratio Ir/clay minerals (Fig. 7b), with the clay minerals content obtained from XRD analyses. In all of these examined cases (Ir/Fe and Ir/Al in Fig. 7a and Ir/clay minerals in Fig. 7b) it is evident that there is a strong anomaly of the ratio Ir/clay content in bed **I** of Poty quarry.

Bed **I** also has a TOC anomaly (Fig. 7c). Figure 7c shows that the TOC anomaly is not directly related to the insoluble residue (IR) content.

It seems that a fluorine anomaly also occurs at bed **I**. Marini and Albertão (1999) and Marini et al. (2000) discuss this characteristic and its possible implications.

4 Environmental Characterisation of Geological Processes Associated with the K-T Boundary

4.1 Micropalaeontological Considerations

Descriptions of fossils in these samples have been presented elsewhere by Albertão et al. (1993), Albertão et al. (1994b), Albertão and Martins Jr. (1996b) and Koutsoukos (1996). The Ir-rich bed **I** marks the limit between the Cretaceous and the Tertiary (Palaeocene) palaeofauna. The abundance of pollen showed a distinct difference between samples below bed **I** and above, whilst foraminifera showed a more gradual change from underlying beds to those immediately above this boundary. One possible reason for that may be the fact that foraminifera shells are more resistant to abrasion than pollen structure. Thus, foraminifera could have been contaminated during sampling. This might be the case for some types of Palaeocene foraminifera, which were collected in levels below bed **I**.

Palynological data give a good definition for level **I** as the last level deposited during the Cretaceous (Albertão 1993) and may also serve as a controlling factor for determining other extinction such as those of foraminifera or other taxa. In general, palaeontological data clearly show a period of crisis at the very end of Cretaceous, with the extinction of pollen and foraminifera. The relatively high abundance of smooth and ornate trilets spores of palms just above the K-T boundary gives the impression of an "explosion" of opportunist forms. The Ir anomaly is precisely at the boundary at which fossils become extinct, which confirms the crisis at the very end of the Cretaceous.

4.2 Stable Isotope Data

In many K-T boundary sections all over the world, a minimum for $\delta^{13}C$ has been observed to be associated directly with the boundary layer. This is not observed in the samples of the PE/PB coastal basin. A minimum of $\delta^{13}C$ occurs on the top of bed **D**, below the Ir-rich bed **I** (Fig. 10).

As a preliminary explanation for these observations, it is possible to consider the following hypotheses:

1. this minimum value for $\delta^{13}C$ ratio in the vicinity of the K-T boundary is only local and not of regional extent,
2. the decrease of the $\delta^{13}C$ values immediately below the Ir-rich bed **I** (975 cm up to 1022 cm) may indicate a reduction of primary oceanic productivity,
3. in this same interval the increasing in $\delta^{18}O$ values indicates a tendency of temperature reduction,
4. some peaks of $\delta^{13}C$ (high values) may indicate up-welling activities, or what was preserved of these events, mostly on the top of the Gramame formation and the base of the Maria Farinha formation, as well as on beds **M** and **N**.

These hypotheses must be reconsidered from a diagenetic point of view, given that the diagenetic processes may be important in the geochemical and petrographic characteristics of the sediments. In this study stable isotope ratios give results that differ from those obtained at K-T boundaries in other locations around the world, where isotope analyses were done not only for bulk rock, but also for carbonate shells of micro- and macro-organisms, in which case the diagenetic influence was determined through petrographic characteristics. In the present case, it was observed that many of the macroscopic and some of the microscopic shells, specially those of gastropods, are re-crystallised.

In discussing the present data from the PE/PB basin, two factors must be considered: [1] it is not always possible to identify diagenetic effects on rocks even with petrographic microscopy; [2] only bulk rock samples were analysed. In spite of these limitations some general trends are observed in the Maastrichtian sediments (Fig. 10), such as an increase of $\delta^{13}C$ isotopic ratios and a decrease of $\delta^{18}O$ isotopic ratios. Both of these trends indicate a general increase in temperature and salinity, and high oceanic primary productivity; this is in agreement with the observed increase of the $CaCO_3$ content. During early Palaeocene, opposite trends are observed (Fig. 10), indicating existence of more humid and temperate climates.

4.3 Presence of Phosphatised Fragments

Although phosphate occurrences are not commonly related to K-T boundary events, we present a discussion on the occurrence of phosphatised fragments at Poty quarry due to the controversy in opinions about their origin (Mabesoone et al. 1968; Albertão et al. 1993).

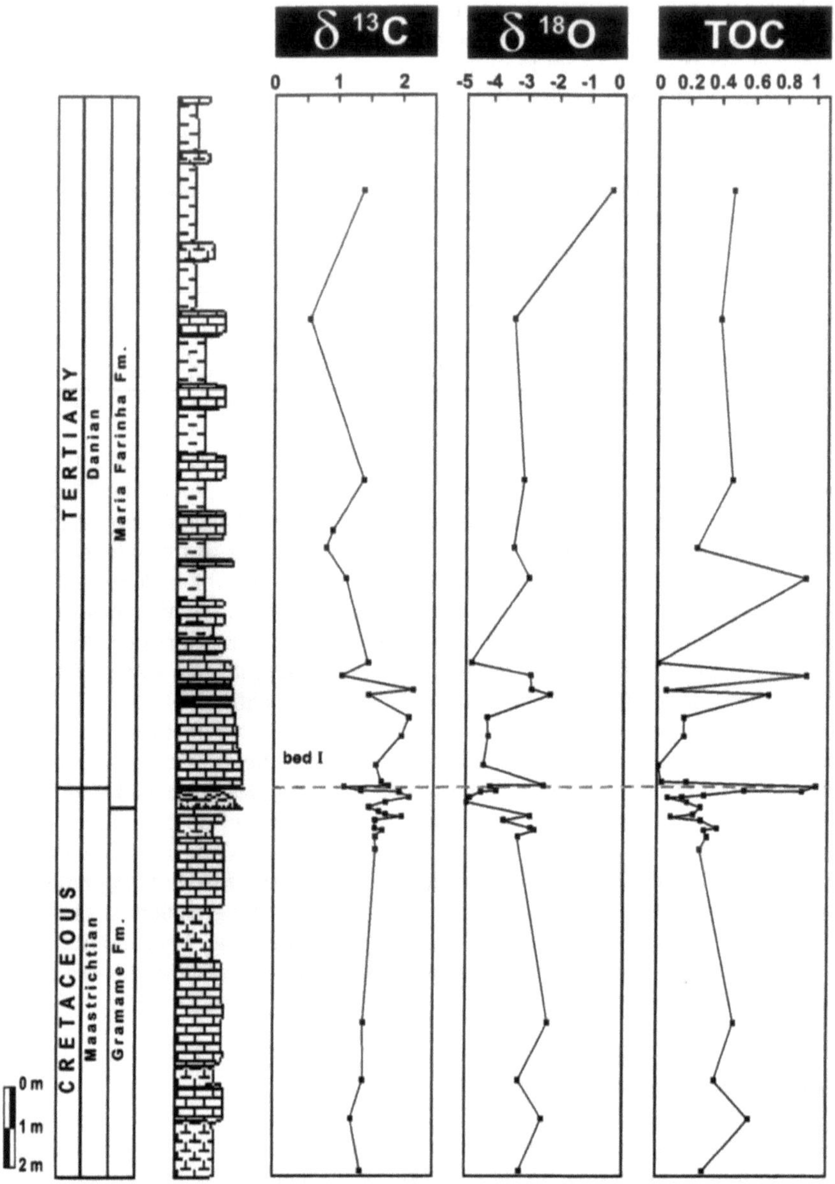

Fig. 10. Stable isotope data for $\delta^{13}C$ and $\delta^{18}O$, and total organic carbon (TOC) distribution in the sedimentary section of Poty quarry. See text for details. Isotope data $\delta^{13}C$ in ‰ relative to PDB and $\delta^{18}O$ in ‰ relative to SMOW; TOC data in wt%.

Phosphatised fragments (intraclasts and bioclasts, usually fine to coarse sand in size, but eventually reaching cobble size in bed **D**) occur in two parts of the main sequence: on the base of the Maria Farinha formation (base of bed **D**) and on beds

M and **N**, mostly on **M** (Table 2). The following observations may suggest that phosphatised fragments are the result of reworking from underlying beds.

1. round shape of phosphatised intraclasts (Fig. 11),
2. partially oxidised external borders of intraclasts (Fig. 11),
3. indiscriminate phosphatisation of many different materials, such as fecal pellets (mainly from *Calianassa*), pelloids, various bioclasts (molluscs, echinoderms, foraminifera, worm tubes, etc.) and intraclasts.
4. presence of fining-upward structures that occur not only in phosphatised fragments, but also in siliciclasts and non-phosphatised bioclasts, particularly in bed **D**.

Mabesoone et al. (1968) proposed a reworking process to explain the phosphatised grains from basal beds of the Gramame formation (the gradation contact with the underlying Beberibe formation, not exposed at the Poty quarry), where the phosphorite of the PE/PB coastal basin occurs.

It seems improbable that during this short time interval (late Maastrichtian to early Palaeocene) tectonic movements might have taken place that led to sufficient tilting of these strata to cause erosion of the phosphorites in proximal areas of the basin and subsequent deposition inwards of the basin, at stratigraphically higher locations. Besides, this process would have had to occur more than once, otherwise it would not be possible to explain the existence of phosphatised grains in two different stratigraphic levels (beds **D** and **M**).

It seems more likely that beds **D** and **M** represent deposits of sediments produced by reworking hardgrounds during storms. These hardgrounds were already rich in phosphates produced during "good weather" conditions. Beds **L** and the very base of **D** (in fact, the contact between beds **C** and **D**) can be considered the remaining records of these hardgrounds. Bed **L** shows intense bioturbation, glauconite and some phosphatised fragments, whereas bed **D** exhibits more continuous levels of phosphate and Fe-rich minerals, especially limonite, and shows intensive bioturbation and pyritisation at the contact with bed **C** (Table 1).

As already mentioned, high values of $\delta^{13}C$ coincide with the portions of the record that have higher phosphate contents. This could provide evidence for up-welling during recursive sea transgression, as suggested by Wilgus et al. (1988). Accordingly, the recurrence of up-welling can be expected in periods of rapid transgression, with the formation of hardgrounds, which would lately be eroded by storms. Phosphatisation events close to other K-T boundary locations were also described at the following sites: North Europe (Jarvis 1992), Greece (Pomoni-Papaiannou and Solakius 1991), Morocco (Parsons 1989) and Colombia (Föllmi et al. 1992).

Burst (1958) verified occurrences of selective glauconite formation inside certain organisms as evidence for early diagenetic processing; this is usually suggested to occur also with phosphate formation, which is a well-known process at least at the Campos basin, Brazil (Mundim UP 1992, personal communication).

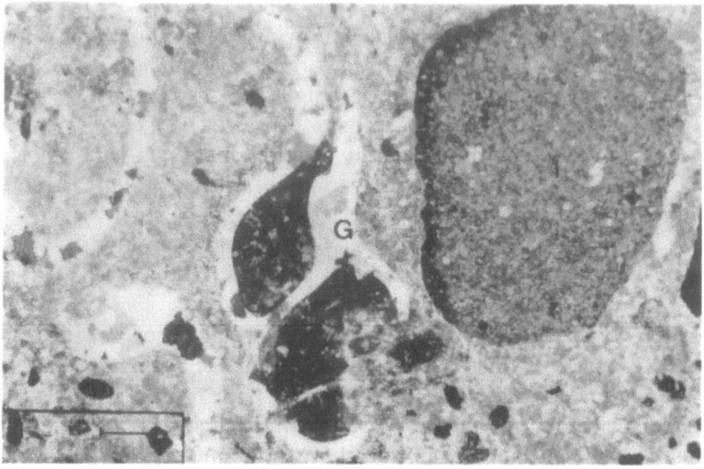

Fig. 11. Photomicrograph (sample 28, bed **D**, crossed polars, scale bar = 0.4 mm), showing the presence of phosphatised fragments. Larger fragments are phosphatised round intraclasts with partially oxidised external border (I), and gastropods (G) with re-crystallised shells and a phosphatised interior. Smaller fragments are mainly of foraminifera, pelloids, and worm tubes.

This process would characterise an early diagenesis of the recently deposited sediments. Such a hypothesis does not exclude the existence of phosphatisation events: phosphatisation would occur in up-welling periods, just after the deposition of detrital material, and would substitute the carbonates in adequate microenvironments.

Nevertheless, to differentiate reworked original phosphate materials from early diagenetic ones is an almost impossible task, as the substitution of material would occur inside reworked clasts; eventually a fining-upward structure would be a function of the original distribution of the material.

4.4 A Possible Tsunami Deposit

Bed **D** has been interpreted as a tsunami deposit (Albertão and Martins Jr. 1996a). The unique stratigraphic and sedimentological characteristics of bed **D** indicate rapid deposition:

1. a sharp erosive base (Figs. 3, 5),
2. following a fining-upward succession (Fig. 3) composed of shells, siliciclasts and abundant phosphatised fragments at the base,
3. extensive mixing and fragmentation of fossils derived from different palaeobathymetries and reworked from older strata,
4. coarse grain-size (intraclasts and bioclasts of up to 9 cm in diameter) and poor sorting (Fig. 11),
5. scattered presence of possibly impact-derived products (spherules and (possibly) shocked quartz grains),
6. interference ripples at the top of the bed (Fig. 4) and

7. great lateral continuity in the basin (for at least 30 km in Pernambuco).

The great lateral continuity of bed **D** is one of its most conspicuous characteristics, because it is possible to follow this bed from the Poty quarry to the Ponta do Funil areas, which are around 30 km far apart (Fig. 1). Between these two areas is Itamaracá Island, from where the same bed **D** was recovered in a core of the well 2-Ist-1-PE, drilled by PETROBRAS.

Semi-quantitative modelling performed for the particular characteristics of bed **D** (see Albertão and Martins 1996a, for details on the modelling parameters) has indicated the plausibility of the action of a tsunami as the responsible process for its deposition.

5 Conclusions

The data and arguments presented in this paper have led to the following conclusions:

1. Petrographic, geochemical, palaeontological and sedimentological data presented in this paper are consistent with the hypothesis of an impact event as the main cause for the observed anomalies in the K-T boundary,
2. Iridium shows a significant enrichment in the boundary layer, independent of the clay content; total organic carbon and fluorine anomalies are also observed at the boundary layer
3. The abundance of fossils, especially foraminifera and pollen, present non-controversial evidence for a sudden change in the ecological conditions, from the late Cretaceous (Maastrichtian) environment towards the early Tertiary (Palaeocene, Danian) environment,
4. The very particular sedimentological characteristics of bed **D** indicate a tsunami event,
5. Stable isotope data for $\delta^{13}C$ indicate a reduction in primary productivity corresponding to the boundary layer,
6. Stable isotope data for $\delta^{18}O$ indicate a temperature drop coincididing with the boundary layer,
7. Phosphatised layers are the possible product of a reworking of the hardgrounds of previous sediments by the tsunami event that is recorded in bed **D,** and, possibly, a lower-energy storm event in bed **M**,
8. Positive high values of $\delta^{13}C$ coincide with the portions of the sedimentary record that are richer in phosphates and may indicate recurrent up-welling events,
9. A geochemical break for most of the analysed elements is better characterised in the geological contact between the Gramame and the Maria Farinha formations than in the K-T boundary,
10. Geochemical data for alkalies, alkali-earth, rare-earth and other elements show some distinct changes in abundance at the boundary.

We propose additional geochemical and paleontological studies in this area of the PE/PB coastal basin to clarify the details concerning the precise location of the horizon of Cretaceous to Tertiary transition.

Acknowledgements

The authors thank Dr. Moses Attrep Jr. and the Los Alamos National Laboratory Reactor Group for performing the neutron activation analyses; Eng. Delmiro Paes de Lira Filho and the Fábrica de Cimento Poty of Grupo Votorantim for strong support during field work in the Poty quarry; Prof. Margareth M. Alheiros from Universidade Federal de Pernambuco, UFPE, for advice and support during the field work in the Pernambuco-Paraíba basin; Drs. Eduardo A. M. Koutsoukos and Marilia P. S. Regali for the micropalaeontological analyses; the Geochemistry Section of PETROBRAS/CENPES for the organic geochemistry analyses; Dr. Jorge C. Della Favera (Universidade Estadual do Rio de Janeiro-UERJ) and Adali R. Spadini (PETROBRAS) for the discussions and advice during the field work; Andréa da Silva Rosa (XEROX Co., drafting support section of PETROBRAS), for providing the final versions of the figures; and Drs. Christian Koeberl, Eric Buffetaut, and Philippe Claeys for the revision of the original manuscript.

References

Albertão GA (1993) Abordagem interdisciplinar e epistemológica sobre as evidências do limite Cretáceo-Terciário, com base em leituras efetuadas no registro sedimentar das bacias da costa leste brasileira. Escola de Minas de Ouro Preto, MG, Brasil: M.Sc. Thesis. vs. 1 and 2, 251 p

Albertão GA (1997) The Cretaceous-Tertiary Boundary in Brazil – State of the art and peculiarities. Sphaerula (the international journal of IGCP 384) 1: 101-114

Albertão GA, Martins Jr. PP (1996a) A possible tsunami deposit at the Creteceous-Tertiary boundary in Pernambuco, North-eastern Brazil. Sed Geol 104: 189-201

Albertão GA, Martins Jr. PP (1996b) Stratigraphic record and geochemistry of the Cretaceous-Tertiary (K-T) boundary in Pernambuco-Paraíba, North-eastern Brazil. In: Jardiné S, De Klasz I, Debenay J-P (eds) Géologie de l'Afrique et de l'Atlantique Sud, Elf Aquitaine Édition, Mémoire 16: 403-411

Albertão GA, Martins Jr. PP (1999) Petrographic and geochemical studies in the sedimentary succession spanning the Cretaceous-Tertiary boundary, Pernambuco basin, NE Brazil [abs.]. In: Buffetaut E, Le Loeuff J (eds) Abstracts, Workshop on geological and biological evidence for global catastrophes – Quillan (France), Impact Programme, European Science Foundation, p 11

Albertão GA, Koutsoukos EAM, Regali MPS, Martins Jr. PP (1993) O registro micropaleontológico com base em foraminíferos no limite Cretáceo-Terciário na bacia de Pernambuco, Nordeste do Brasil [abs.]. Bol Res Exp XIII Congresso Brasileiro de Paleontologia; Simpósio de Micropaleontologia, São Leopoldo, RS, Brasil, p 54

Albertão GA, Martins Jr. PP, Koutsoukos EAM (1994a) O limite Cretáceo-Terciário na bacia de Pernambuco-Paraiba: características que definem um marco estratigráfico relacionado a um evento catastrófico de proporções globais. Acta Geol Leopoldensia 17(39/1): 203-219

Albertão GA, Koutsoukos EAM, Regali MPS, Attrep Jr. M, Martins Jr. PP (1994b) The Cretaceous-Tertiary boundary in southern low-latitude regions: preliminary study in Pernambuco, north-eastern Brazil. Terra Nova 6: 366-375

Albertão GA, Grassi AA, Martins Jr. PP, De Ros LF (2001) The K-T boundary in Brazilian sedimentary basins and related spherules. Geochemical Journal (in press).

Beurlen K (1967) Estratigrafia da faixa sedimentar costeira Recife-João Pessoa. Boletim da Sociedade Brasileira de Geologia 16: 43-53.

Burst JF (1958) "Glauconite" pellets: their mineral nature and applications to stratigraphic interpretations. Bull AAPG 42(2): 310-327

Chang HK, Kowsmann RO, Figueiredo AMF (1988) New concepts on the development of East Brazilian marginal basins. Episodes 11: 194-202

Delício MP, Oliveira AD, Albertão GA, Martins Jr. PP (2000) Looking for spherules at the Cretaceous-Tertiary (K-T) Boundary in Pernambuco/Paraíba (PE/PB) Basin, NE Brazil. In: Detre CH (ed) Proc Annual Meet TECOS, 1998, Budapest, Hungarian Academy of Science, pp 35-43

Flügel E (1982) Microfacies Analysis of Limestones. Berlin, Springer-Verlag, 633 pp

Föllmi KB, Garrison RE, Ramirez PC, Zambrano-Ortiz F, Kennedy WJ, Lehner BL (1992) Cyclic phosphate-rich successions in the upper Cretaceous of Colombia. Palaeogeogr Palaeoclim Palaeoecol 93: 151-182

Gilmore JS, Knight JD, Orth, CJ, Pillmore CL, Tschudy RH (1984) Trace element patterns at a non-marine Cretaceous-Tertiary boundary. Nature 307: 224-228.

Huffman AR, Crocket JH, Carter NL, Borella PE, Officer CB (1990) Chemistry and mineralogy across the Cretaceous-Tertiary boundary at DSDP Site 527, Walvis Ridge, South Atlantic Ocean. In: Sharpton VL, Ward PD (eds) Global catastrophes in Earth history: an interdisciplinary conference on impacts, volcanism and mass mortality. Boulder: GSA Special Paper 247: 319-334.

Jarvis I (1992) Sedimentology, geochemistry and origin of phosphatic chalks: the Upper Cretaceous deposits of NW Europe. Sedimentology 39: 55-97

Koeberl C (1989) Iridium enrichment in volcanic dust from blue ice fields, Antarctica and possible relevance to the K-T boundary event. Earth Planet Sci Lett 92: 317-322

Koutsoukos EAM (1996) The Cretaceous-Tertiary Boundary at Poty, NE Brazil - Event stratigraphy and palaeoenvironments. In: Jardiné S, De Klasz I, Debenay J-P (eds) Géologie de L'Afrique et de L'Atlantique Sud, Elf Aquitaine Édition, Mémoire 16: 413-431

Mabesoone JM, Tinoco IM, Coutinho PN (1968) The Mesozoic-Tertiary boundary in Northeastern Brazil. Palaeogeogr Palaeoclim Palaeoecol 4: 161-185

Marini F, Albertão GA (1999) First report on KTB-related fluorine anomaly, Pernambuco area, Northeastern Brazil [abs.]. In: Buffetaut E, Le Loeuff J (eds) Abstracts, Workshop on Geological and Biological Evidence for Global Catastrophes - Quillan (France), Impact Programme, European Science Foundation, pp 56-57

Marini F, Albertão GA, Oliveira AD, Delício MP (2000) Preliminary SEM and EPMA investigations on KTB spherules from Pernambuco area (NE Brazil): diagenetic apatite and fluorite concretions, suspected fluorine anomalies. In: Detre CH (ed) Proc Annual Meet TECOS, 1998, Budapest, Hungarian Academy of Science, pp 109-117

Michel HV, Asaro FA, Alvarez W, Alvarez LW (1985) Elemental profile of iridium and other elements near the Cretaceous/Tertiary boundary in Hole 577 B. In: Heath GR, Burckle LH, D'Agostino AE, Bleil U, Horai K, Jacobi RD, Janecek TR, Koizumi I, Krissek LA, Monechi S, Lenotre N, Morley JJ, Schultheiss PJ, Wright AA, Turner K (eds) Init Repts DSDP 86: 533-538

Michel HV, Asaro FA, Alvarez W, Alvarez LW (1990) Geochemical studies of the Cretaceous-Tertiary boundary in ODP holes 689 B and 690 C. In: Barker PF, Kennet JP et al. Proc ODP, Sci Results 113: 159-168

Parsons DG (1989) Palaeecology of foraminiferid assemblages across late cretaceous phosphogenic cycles, South West Atlas, Morroco. Palaeogeogr Palaeoclim Palaeoecol 69: 193-212

Pomoni-Papaiannou F, Solakius N (1991) Phosphatic hardgrounds and stromatolites from the limestone/ shale boundary section at Prossilion (Maastrichtian Paleocene) in the Parnassus-Ghiona zone, Central Greece. Palaeogeogr Palaeoclim Palaeoecol 86: 243-254

Preisinger A, Zobets E, Gratz AJ, Lahodynsky R, Becke M, Mauritsch HJ, Eder G, Grass F, Rögl F, Stradner H, Surenian R (1986) The Cretaceous-Tertiary boundary in the Gosau Basin, Austria. Nature 322: 794-799

Schmitz B (1988) Origin of microbeding in worldwide distributed Ir-rich marine Cretaceous-Tertiary boundary clays. Geology 16: 1068-1072

Stinnesbeck W (1989) Fauna y microflora en el limite Cretacico-Terciario en el Estado de Pernambuco, Nordeste de Brasil. Contribuiciones de los Simposios sobre Cretácico de America Latina. Parte A: 215-230

Tinoco IM (1967) Micropaleontologia da faixa sedimentar costeira Recife-João Pessoa. Boletim da Sociedade Brasileira de Geologia 16: 81-85

Wilgus CK, Posamentier H, Hastings BS, Van Wagoner J, Ross CA, Kendall CGStC (1988) Sea-level changes: an integrated approach. Tulsa, SEPM Special Publication 42, 407 p

Effects of Bioturbation through the Late Eocene Impactoclastic Layer near Massignano, Italy

Heinz Huber[1], Christian Koeberl[1,3*], David T. King, Jr.[2], Lucille W. Petruny[2], and Alessandro Montanari[3]

[1]Institute of Geochemistry, University of Vienna, Althanstrasse 14, A-1090 Vienna, Austria. (*christian.koeberl@univie.ac.at)
[2]Department of Geology, Auburn University, Auburn, AL 36849-5305, USA.
[3]Osservatorio Geologico di Coldigioco, I-62020 Frontale di Apiro (MC), Italy.

Abstract. The purpose of this study was to investigate the variations in trace element composition resulting from bioturbation of a Late Eocene impactoclastic layer at Massignano, Italy. This layer has an age of 35.7 ± 0.4 Ma and is, therefore, coeval with both the North American tektite event (related to the Chesapeake Bay impact structure, USA) and the Popigai impact event (Siberia, Russia). The layer shows a distinct iridium anomaly of up to about 270 ppt. The stratigraphic leakage within *Planolites* and *Zoophycos* burrows was studied by using iridium coincidence spectrometry and conventional neutron activation analysis, on samples obtained by a micro drilling procedure. The iridium contents of the burrow fillings vary from 50 ppt about 20 cm below the impactoclastic layer to 280 ppt within the layer, depending upon the stratigraphic level where burrowing began. The stratigraphic distribution of Ir is the result of both bioturbation and chemical diffusion.

1 Introduction

The Late Eocene was a time of major changes in the history of the Earth. Separation of Antarctica from other southern continents caused a distinct change in temperature of the Earth's oceans and had long-lasting effects on the global ecosystem. According to $\delta^{18}O$ records of fine bulk carbonate and benthic foraminifers from middle and upper Eocene sediments, a major cooling trend progressively developed, which included a sharp temperature drop of about 2°C near the Eocene/Oligocene (E/O) boundary (Vonhof et al. 2000). Additionally, this time interval shows evidence of a cometary shower lasting some 2.5 million years, which caused highly increased accretion of extraterrestrial material in the sedimentary record. The ^{3}He record at Massignano (at Monte Cònero, near Ancona, Italy – see Fig. 1) described by Farley et al. (1998) supported the prediction of such a cometary shower during the Late Eocene made previously by Hut et al. (1987). At Massignano, the entire sequence of this long-lasting crisis can be documented in the Eocene/Oligocene Global Stratotype Section and Point (GSSP), which is the best exposure of the Eocene-Oligocene transition known in the pelagic basin of the Umbria-Marche Apennines of Italy. In this GSSP, an 8 cm thick impactoclastic layer is situated within marly limestones of the Scaglia Variegata Formation, between meter levels 5.64 and 5.72. This layer is well-dated at 35.7 ± 0.4 Ma (e.g., Montanari et al. 1988), using direct radioisotopic methods on volcanic material (i.e., biotite, zircons, and monazite). Biostratigraphically, the impactoclastic layer is located in the middle part of nannofossil zones NP19/20 and CP15b (Coccioni et al. 1988) and in the lower part of planktonic foraminiferal Zone P16 (the upper part of the *P. semiinvoluta* Zone – after Berggren et al. 1995). This layer is characterized by an iridium anomaly of up to 0.2 ppb (Montanari et al. 1993), impact-derived spinel (Pierrard et al. 1998), shocked quartz (Clymer et al. 1996; Langenhorst 1996), and anomalous extraterrestrial ^{3}He (Farley et al. 1998).

A similar iridium anomaly was also found, at the same stratigraphic level, at the Contessa Quarry near Gubbio, Italy (Clymer 1996), and it may correlate with a microkrystite (clinopyroxene-bearing spherule) layer of same age found in numerous deep-sea cores in the Caribbean Sea, North Pacific, and Indian Ocean (see, e.g., Alvarez et al. 1982; Glass et al. 1985; Keller et al. 1987; Whitehead et al. 2000). This Ir-rich microkrystite layer in Caribbean and North Atlantic cores is found a short distance (equivalent to 10-20 ka) below a microtektite- and shocked quartz-bearing layer (Glass et al. 1982, 1985), which represents the North America tektite strewn field, and which has been associated with the Chesapeake Bay impact structure (Koeberl et al. 1996). The absence of high-pressure silica phases in the layer at Massignano, which are present in the Chesapeake Bay-related microtektite layer of DSDP 612 (Bohor et al. 1988), led Langenhorst (1996) to suggest that Massignano shocked quartz was derived from the non-porous, crystalline target rock at Popigai. This has recently been supported by the isotopic data of Whitehead et al. (2000).

In the Massignano and Contessa sections, a change in oxidation conditions of the paleo-sea floor is manifested by a sudden color change of the marly limestones from pinkish to gray and back. The first change from oxidizing to reducing conditions at 5.64 m coincides with the base of the impactoclastic layer in the upper part of magnetic Chron 16 (Montanari and Koeberl 2000).

The impactoclastic layer at Massignano is penetrated by numerous burrows, which can be classified as either traces of *Planolites* or large flattened *Zoophycos*. In the present study, we investigate the extent to which biogenic activity (i.e., post depositional burrowing action of bottom dweller organisms) led to additional stratigraphic redistribution of iridium from the impactoclastic layer, besides chemical diffusion. Bioturbation has commonly been suggested as one agent in the dilution of extraterrestrial signals, such as stratigraphic distribution of siderophile elements.

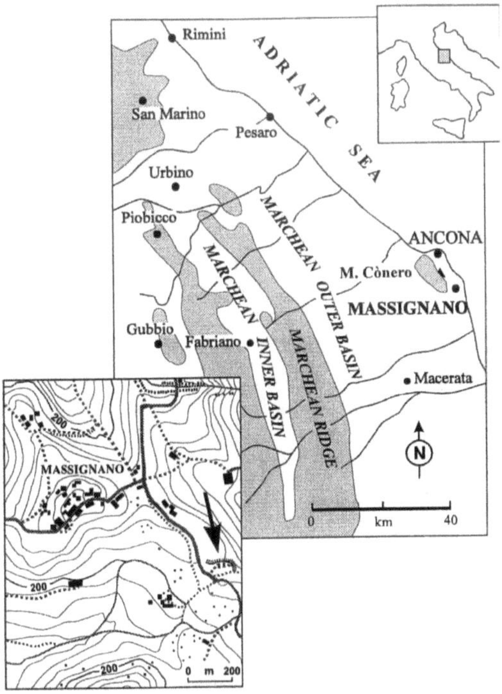

Fig. 1. Map of the Umbria-Marche region (after Coccioni et al. 1988). Mountain ridges (e.g., the Marchean- and the Umbro-Marchean Ridge), as well as the mountainous Monte Cònero Park, are shown in gray. The close-up shows the region around Massignano. The sampling site in an abandoned quarry, half a kilometer in the south-east of Massignano, is marked with a black arrow.

2 Geology and Stratigraphy of the Massignano Section

The abandoned quarry of Massignano is located along the provincial road of the Monte Cònero Park, about 4 km north of the resort town of Sirolo. It exposes a 23 m thick, continuous and complete sequence of pelagic marly limestones and calcareous marls, rich in well-preserved benthic and planktonic microfossils, and interbedded with several biotite-rich volcano-sedimentary layers, spanning the upper Eocene and lowermost Oligocene. Availability of this material yielded an ideal situation for the application of integrated studies aimed at the precise and accurate calibration of the litho-, bio-, magneto-, and chemostratigraphic records with direct radioisotopic dating.

Fig. 2. Panoramic view of the Late Eocene section at Massignano, Italy. (a) View of quarry including the Eocene/Oligocene GSSP (top left). The studied zone of Late Eocene limestones is marked by the rectangle. (b) Close-up of the impactoclastic layer, which is characterized by a faint color change, from pink below, to gray above, at 5.64 to 5.72 m.

The Massignano quarry was formally established by the IUGS-SPS at the 1993 IGC meeting in Kyoto (Premoli Silva and Jenkins 1993) as the Global Stratotype Section and Point (GSSP) for the E/O boundary. A number of studies conducted at this GSSP deal with the mineralogic and lithostratigraphic (Mattias et al. 1992), biostratigraphic and paleontologic (Gonzalvo and Molina 1992; Molina et al. 1993), magneto-stratigraphic (Lowrie and Lanci 1994), chemostratigraphic (Montanari et al. 1993; Vonhof et al. 1998), and geochronologic (Odin and Montanari 1989; Oberli and Meier 1991; Odin et al. 1991) aspects. In particular, high-resolution stratigraphic studies lead to the discovery, just above 5.6 m in the Massignano GSSP, of an impactoclastic layer, which is possibly of worldwide occurrence (Vonhof and Smit 1999), containing an iridium anomaly (Montanari et al. 1993), shocked quartz (Clymer et al. 1996; Langenhorst 1996), extraterrestrial spinel and altered microkrystites (Pierrard et al. 1998).

These discoveries encouraged further detailed studies on the paleontologic and paleobiologic records of the Massignano GSSP aimed at verification of the effects that the inferred cosmic events (impacts and comet shower) may have had on a terminal Eocene marine biota (Brinkhuis and Coccioni 1995; Gardin et al. 1999; Vonhof et al. 2000).

However, the discrete impactoclastic layer at meter level 5.64–5.72 in the Massignano GSSP is good evidence that invites a closer look at what may be sedimentologic and biologic responses to a "non-lethal" impact event. In terms of physical stratigraphy, this impactoclastic layer is located at the very top of a pinkish band which, in the Scaglia Variegata Formation of the Umbria-Marche Apennine, can be considered as a precise and consistent regional lithostratigraphic marker horizon. The impactoclastic layer is in the upper part of magnetic Chron 16n, although several magnetostratigraphic studies by different researchers in the CQ section at Contessa (Lowrie et al. 1982), Massignano section (Bice and Montanari 1988; Lowrie and Lanci 1994), and through a core drilled in the immediate vicinity of the exposed section (Lanci et al. 1996) have yielded slightly different results about the duration of Chron 16n. This is probably due to interlaboratory differences in demagnetization and analytical techniques, and ways of interpreting paleomagnetic data (for details, see Montanari and Koeberl 2000).

As for biostratigraphy, the impactoclastic layer is located in the upper part of planktonic foraminiferal *P. semiinvoluta* Zone (i.e., lower P16 Zone), and in the middle to lower part of nannofossil zones NP19/20 and CP15b (Montanari and Koeberl 2000). Therefore, the impactoclastic layer is not located at a biozonal boundary, across which a faunal or floral change could be readily recognized with a magnifying lens directly on hand samples (as for the K/T boundary) or even by looking at thin sections or smear slides with an optical microscope.

The numerical age of the impactoclastic layer has been calibrated with direct radioisotopic dating of volcanoclastic material, i.e., biotite, zircon, and monazite (see Montanari et al. 1988; Odin et al. 1991) and yielded in an interpolated age of 35.7 ± 0.4 Ma for the impactoclastic layer. This age is practically indistinguishable from the ages of the North American tektites and microtektites (Glass et al. 1986; Obradovich et al. 1989), the inferred age for the Chesapeake Bay impact structure (Poag and Aubry 1995; Koeberl et al. 1996), and the radioisotopic age of impact melt rocks from Popigai crater (Bottomley et al. 1997).

On the outcrop, there is a sharp change in lithology at 5.8 m, where a softer, more clay-rich layer, representing a biotite-rich tuff bed, is resting on top of a 50 cm thick marly limestone with a faint pinkish color. The pinkish limestone is relatively homogeneous, except for an 8 cm thick, greyish band at 5.64–5.72 m, which is the impactoclastic layer. In slabbed samples, scattered mineralized spherules are evident. In some instances, these spherules are concentrated within *Planolites* and *Zoophycos*. However, many of the through-going fossil traces, including those within the thin impactoclastic horizon, contain biotite flakes that were derived from the volcanoclastic layer at 5.8 m. This gives an indication of the degree of vertical mixing in these sediments caused by intense bioturbation activity. *Anellida* and other bottom-dweller organisms that were living at or above the 5.8 m biotite-rich level penetrated the subbottom sediment by as much as half a meter. On the other hand, the few burrows that contain spherules were probably dug by organisms extant during the impact event.

A polished rock slab of the top pinkish marly limestone reveals that the color change from pink to gray occurs right at the base of the impactoclastic horizon (see Fig. 2). Moreover, the intricate network of trace fossils seems not to exist between 5.61 and 5.63 m, as there exist only few burrow traces through this band of sediments. These observations suggest that there has been a very brief interruption or decrease in bioturbation during the deposition of the impactoclastic layer, which probably coincided also with a brief episode of relative reduction of the sea floor sediment. Normal oxidizing conditions were restored at around 5.67 m, as indicated by a return to a pinkish color. The arrival of volcanoclastic material on the sea floor at 5.83 m marks the definitive end of these slightly oxidizing conditions, and from here up to 9.5 m these marly limestones exhibit a gray color.

In all, the situation around the impactoclastic layer of Massignano, characterized by a decrease of bioturbation activity during a short episode of relatively low-Eh conditions, is somewhat similar to that at the K/T boundary (Lowrie et al. 1990; Montanari 1991); however the latter is more marked in appearance and records a much more severe event. Given this, the question remains whether there has been a significant response of the plankton and/or benthos to the distant impact represented by the impactoclastic layer. A strong abundance increase of cosmopolitan dinocyst *Thalassiphora pelagica* right after the deposition of the impactoclastic layer at Massignano was interpreted by Vonhof et al. (1998, 2000) as an indication of sudden cooling and/or increased productivity, resulting, perhaps, from the "impact winter" effect of the Popigai and Chesapeake impacts. A similar inference was reached by Gardin et al. (1999) after a high-resolution statistical analysis of foraminifera and calcareous nannofossil assemblages through the interval 4-8 m in the Massignano section. Gardin et al. (1999) proposed that during a time of general, gradual cooling, an environmental break took place right after the impactoclastic layer was deposited, leading to accelerated pulses of climatic deterioration, which may represent the relatively long-term effects of the terminal Eocene impact events.

3
Samples

Large rock blocks covering the complete interval between 5.0 m and 5.9 m of the GSSP at Massignano were cut into slabs perpendicular to bedding. Sampling started at 5.77 m, right above the impact layer and continued down to level 5.31 m, including *Zoophycos* and *Planolites* burrow fillings and matrix. Micro-drilling with a 2 mm drill of different parts of the slabs was used to isolate individual burrow fillings (9 samples) and compare their composition with matrix samples across the study interval (Fig. 3 a-d).

Table 1. Concentrations of Ir, Fe, Co, Ni, and Sc in matrix and burrow samples.

Matrix	Meter	Ir [ppt]	Fe [wt %]	Co [ppm]	Ni [ppm]
TF 25	5.77	186 ± 27	1.25	19.2	78.6
TF 24	5.63	271 ± 32	1.23	14.6	84.0
TF 21	5.565	217 ± 29	1.56	20.7	94.9
TF 2	5.53	119 ± 21	1.12	18.2	74.5
TF 4	5.52	38 ± 12	1.04	17.1	77.7
TF 7	5.495	106 ± 20	1.01	13.5	72.3
TF 9	5.47	101 ± 20	1.11	11.1	62.3
TF 11	5.44	119 ± 21	1.16	10.9	63.5
TF 13	5.42	132 ± 23	1.02	8.37	41.6
TF 15	5.385	63 ± 16	1.17	11.0	51.4
TF 16	5.375	50 ± 14	1.01	9.30	49.6
TF 17	5.36	83 ± 18	0.89	7.43	51.9
TF 20	5.315	55 ± 15	1.07	8.81	48.1
Burrows					
TF 23	5.66	287 ± 33	1.29	15.8	63.0
TF 22	5.58	260 ± 32	1.20	11.2	89.1
TF 1	5.54	73 ± 17	1.34	20.6	97.8
TF 3	5.53	47 ± 13	1.26	17.1	92.1
TF 5	5.51	221 ± 29	1.28	18.3	75.6
TF 6	5.505	102 ± 20	1.59	14.7	68.4
TF 8	5.46	124 ± 22	1.35	14.9	85.7
TF 10	5.449	114 ± 21	1.27	13.5	76.3
TF 12	5.43	155 ± 24	1.37	10.9	55.0
TF 14	5.41	21 ± 9	0.79	6.80	44.8
TF 18	5.355	86 ± 18	1.28	10.8	54.8
TF 19	5.33	112 ± 21	0.85	7.10	42.5

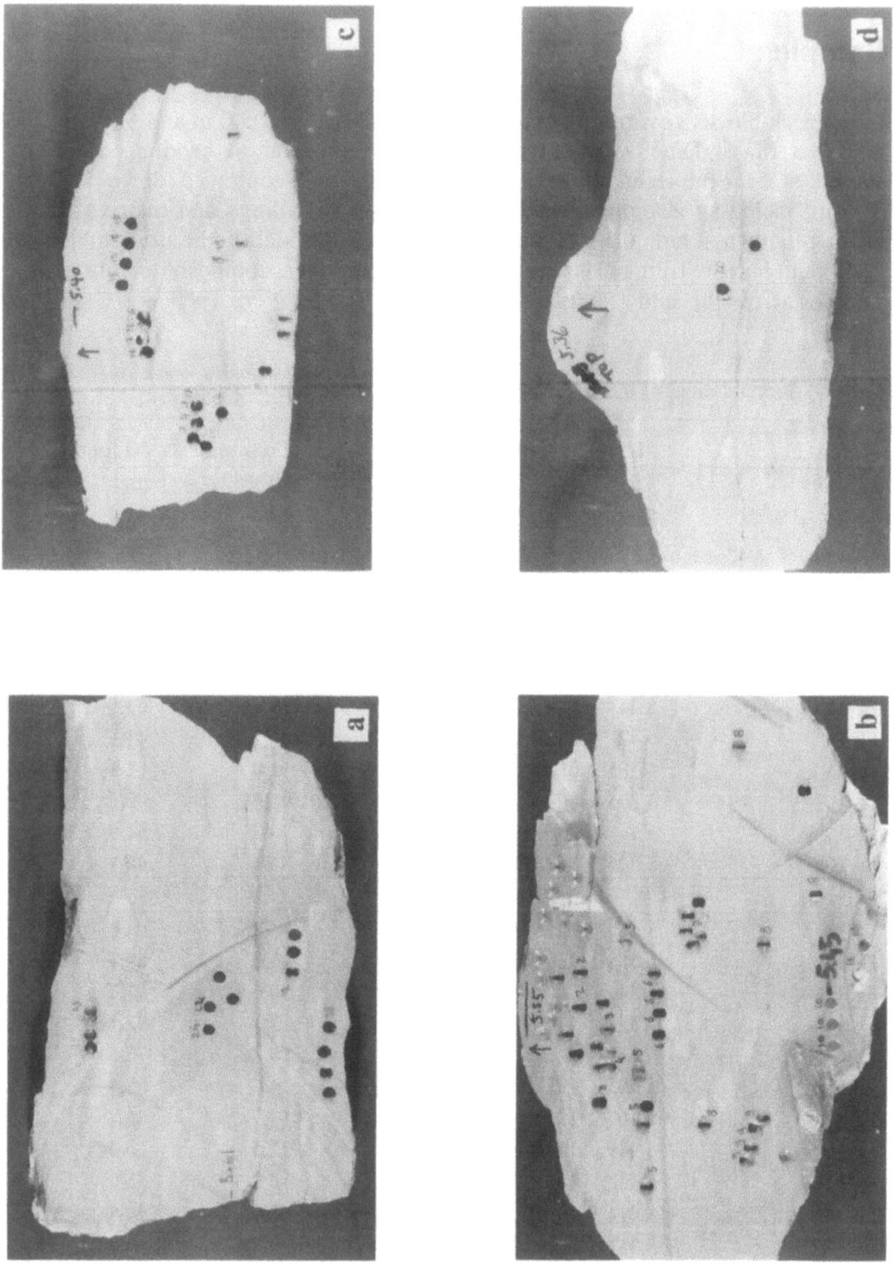

Fig. 3 a–d. Documentation of sample slabs that were micro-drilled (diameter 2 mm) to obtain the samples that were analyzed in the present study. Sample numbers on slabs refer to sample numbers given in Table 1.

The rock is a coccolith-foraminiferal micritic marly limestone, showing a distinct variation in color from pale red (10R6/2) to pinkish grey (5YR8/1) to light greenish grey (5GY8/1-5G8/1; all color terms and symbols from the Rock Color Chart Committee, 1991). The micrites are pale red from 4.98 to 5.39 m and from 5.42 to 5.45 m, light greenish gray from 5.39 to 5.42 and from 5.52 to 5.67 m, and pinkish gray from 5.45 to 5.52 and from 5.67 to 5.84 m. These colors are not specific to bedding, and commonly a single color bridges the contact between beds. Suites of tiered *Planolites* and *Zoophycos* are filled with sediments derived from overlying layers, as determined by comparing burrow-fill colors with the distinctive color shades at various levels. Volcanic biotite flakes, as well as impact spherules that were deposited between the 5.83 and 5.88 m and the 5.64 and 5.72 m level, respectively, may be found within the *Planolites* and *Zoophycos* fillings. The biotite flakes occur down to 5.02 m, whereas spherules are found only within *Planolites* down to 5.06 m.

4 Methods

The iridium content was measured with the iridium coincidence spectrometry (ICS) system at the Institute of Geochemistry at the University of Vienna (for details of the method see Koeberl and Huber 2000). The ICS method is well suited for the determination of iridium abundances at ultra-low levels. Compared to other methods for the determination of Ir concentrations in the sub-ppb range, such as radiochemical neutron activation analysis (RNAA) or inductively coupled plasma-source mass spectrometry (ICP-MS), ICS has a number of advantages. The ICS method does not require dissolution or any other chemical treatment of samples, and, therefore, possible contamination problems are avoided. Good sensitivity and precision can be obtained even for small samples (less than 200 mg). The main purpose of a coincidence setting is background reduction and removal of spectral interferences in the gamma spectrum. Coincidence counting of iridium was first used in the 1960s and 70s, mainly because of high detection limits with the then standard instrumental neutron activation analysis (INAA) methods (due to the low resolution of NaI detectors and low efficiency of Ge(Li) detectors available). Because highly sensitive high-purity Ge-detectors became available, coincidence counting was no longer useful for routine analyses and has only been used by a few laboratories for specific applications (Michel et al. 1991, Meyer et al. 1993). Nevertheless, ICS is still the only opportunity for the non-destructive determination of ultra-low iridium abundances in geological materials.

The ICS method is based on the counting of coincident events from two γ-rays from the decay cascade of ^{192}Ir. A coincidence system consists of two detectors (low energy planar HpGe-detectors), two preamplifiers, two spectroscopy amplifiers, and two analog-to-digital-converters (ADCs), connected to an MPA bus-box (containing the fast coincidence electronics). This external bus box connects each ADC to the computer for evaluation with multiparameter hard- and software. Only signals occurring within a certain time limit (the so-called coincidence window) from both detectors are accepted and plotted in a 1024*1024 matrix (Fig. 4). Then the regions of the iridium peaks (at 316.5 and 468.1 keV) are extracted, the resulting peak volumes corrected (live- and decay time correction, flux correction, background subtraction), and compared with the standards.

Rock samples were powdered during collection with a masonry drill. Aliquots of about 50 mg were sealed into Suprapur quartz glass tubes, and each sample vial was wrapped into aluminium foil with well-characterized reference material and packed into an aluminium capsule. For standard references, a series of five aliquots of the Allende meteorite reference sample (containing 740 ppb iridium), diluted with pure quartz powder, and containing 47 to 6930 ppt iridium, was used. Twenty-five crushed and powdered samples as well as the dilution standard series were packed and irradiated for 48 hours at the ASTRA reactor of the Forschungszentrum Seibersdorf (Austria) using a flux of about $7{\cdot}10^{13}$ $n{\cdot}cm^{-2}s^{-1}$. After a first cooling period of 7 days the samples were transferred to the Institute of Geochemistry at the University of Vienna for further processing. Counting started after a further cooling period of about two months.

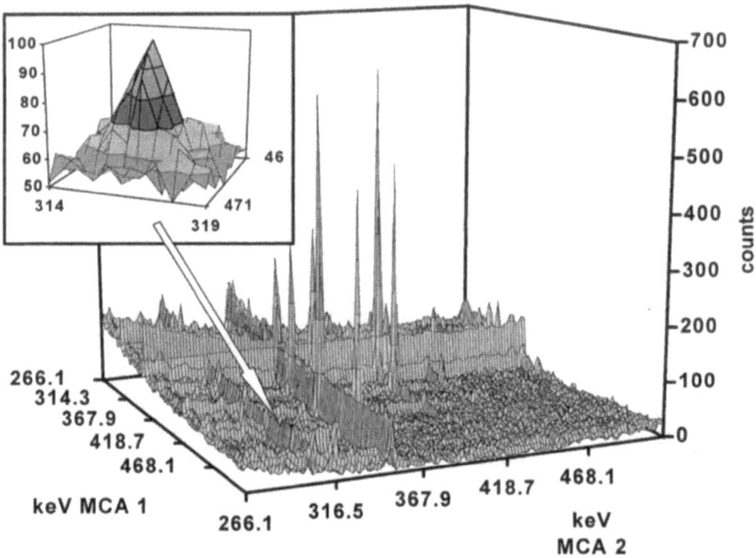

Fig. 4. Gamma-gamma multiparameter coincidence spectrum of the sample TF 22 containing 260 ppt iridium, showing the iridium peaks at 316x468 and 468x316 keV and the ridges of europium at 344 keV.

Routine INAA procedures were used to measure the contents of iron, cobalt, nickel, and scandium. After determining the iridium concentrations of the powdered samples using ICS, the quartz glass vials were counted with the INAA system at the Institute of Geochemistry, University Vienna, for at least five hours. For further details on our INAA-procedures, such as equipment, precision, and accuracy, see Koeberl (1993).

5 Results and Discussion

5.1 Geochemistry

The stratigraphic distribution of iridium is shown in Figure 5a. Maximum abundances of about 270 ± 32 ppt are found within the impactoclastic layer. This value agrees reasonably well with abundances of 199 ± 19 ppt reported by Montanari et al. (1993) and about 280 ± 20 ppt by Pierrard et al. (1998). A small peak of the iridium abundance at 5.4 m, previously found by Montanari et al. (1993), was also detected in our measurements. The bell-shaped distribution of iridium, which was found for iridium distributions at various K/T-boundary sites (Robin et al. 1991), can also be seen at Massignano (see Fig. 5b).

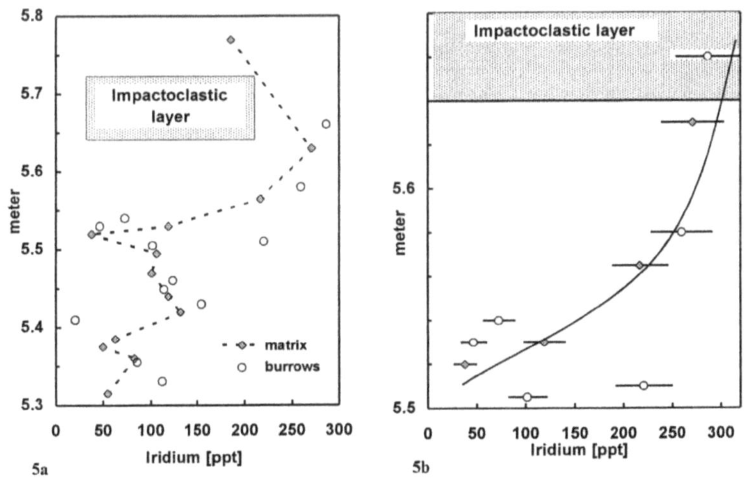

Fig. 5a/b. Distribution of Ir abundances across the Late Eocene impactoclastic layer at Massignano. Fig. 5a shows the distribution through the studied sequence. The close-up (Fig. 5b) includes an interpolation line for matrix abundances. Post-impact chemical diffusion can be followed in the close-up down to about meter level 5.50, whereas the distinct enrichments in burrows at level 5.53 are due to bioturbation of bottom-dwelling organisms.

Some detritus-feeding organisms, which populated the deep sea bottom shortly after the impact-derived material were sedimented, penetrated subbottom sediments for nearly half a meter, and passed through the impactoclastic layer. Maximum enrichment of iridium in the resulting burrow fills progressively decrease from the impactoclastic layer downwards. A dark-colored long burrow (*Planolites*) within the impactoclastic layer (sample no. 23) accumulated 287 ± 33 ppt iridium and sample no. 5 (dark burrow 15 cm below) contains 221 ± 29 ppt iridium, whereas in the *Zoophycos* burrow (sample no. 6) only 102 ± 20 ppt iridium were found. Below meter level 5.4, burrow fillings show iridium abundances of 50 to 80 ppt. A burrow at level 5.42 m (sample no. 12), containing clustered flattened Ni-rich spinel spherules, is characterized by a distinct enrichment in iridium and iron. Sample nos. 8 and 10, a small lozenge shaped burrow and a large tube with biotite flakes, show highly enriched abundances of iron, nickel, and cobalt, whereas the iridium content is indistinguishable from that of the matrix. Increased abundances of Ni can also be found in sample no. 22 (mottling spots) just below the impact-layer, possibly from the Ni-rich spinels distributed within this layer.

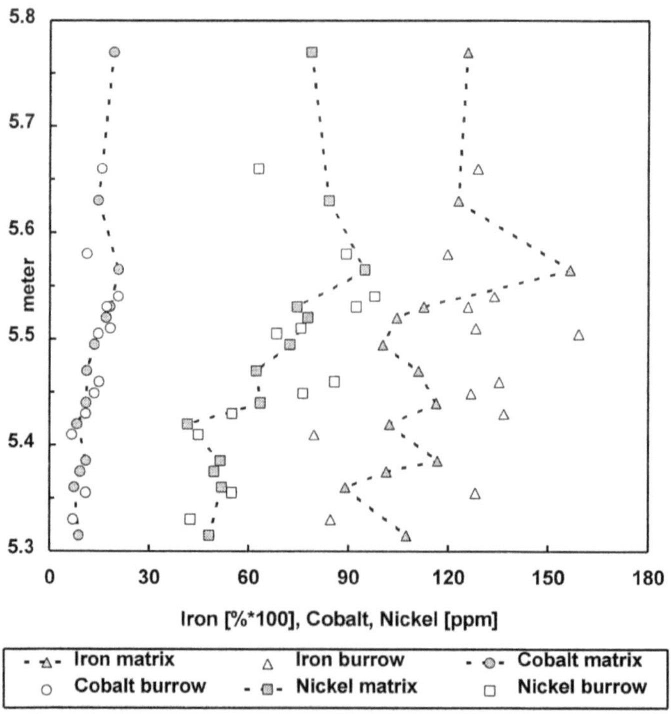

Fig. 5c: Iron, cobalt, and nickel distribution through the Late Eocene section at Massignano. Iron, Co, and Ni are enriched just below the impactoclastic layer, whereas iridium shows an abundance peak within the layer.

Obviously some burrowing organisms did not penetrate the impactoclastic layer, such as those represented by sample nos. 1, 3, 8, 10 14, and 18, or introduced only minor amounts of impact-derived material into the burrow-filling sediments. Such burrows show Ir abundances that are indistinguishable from those of the matrix.

Some burrows contain biotite flakes that were distributed downward from the volcanoclastic layer at level 5.8 m. They can be easily detected by their Fe/Sc ratio (Fig. 6). The pattern of the Fe/Sc and Ir/Sc ratios through the Massignano section shows a similar trend to that shown by Montanari et al. (1993) using Fe/Cs and Ir/Cs, except that the definitive peak at 5.61 m is missing in our study, because we did not use any samples from the 5.6 m level. Sample no. 5 (dark burrow), no. 6 (*Zoophycos*), no. 12 and 18 (both are dark pink bioturbation tubes) show higher Fe/Sc ratios than the matrix, which is characteristic for volcanic contamination (see Montanari et al. 1993). The iron, cobalt, and nickel abundance plot (Fig. 5c) shows a common maximum for these elements at meter level 5.55, whereas the curve of the iridium content is increasing up to level 5.60 m, where the impactoclastic layer begins.

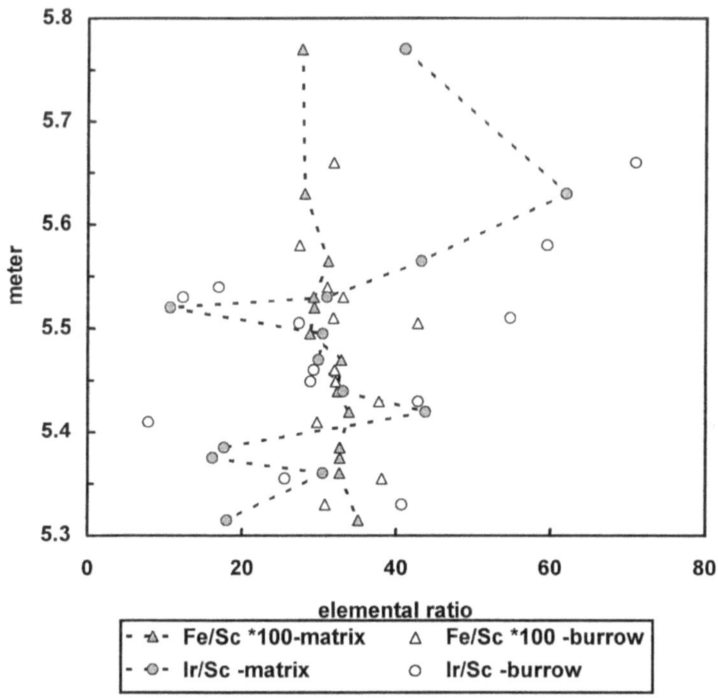

Fig. 6. Variation of ratios for Fe/Sc and Ir/Sc across the Late Eocene impactoclastic layer at Massignano. Anomalously high Fe/Sc ratios are indicative for volcanic contamination (see Montanari 1993). Therefore, six burrows cut the volcanoclastic layer 11 cm above the impactoclastic layer, such as burrow TF 23 at 5.66 m.

5.2 High-resolution Stratigraphy

The upper Eocene interval, from 5.00 m to 5.90 m, above the base of the GSSP for the E/O boundary at Massignano, contains a Late Eocene distal impactoclastic layer (5.64 to 5.72 cm), and a biotite-rich, distal volcanoclastic layer (5.83 to 5.88 cm). Both the impactoclastic and volcanoclastic layers are stratigraphic horizons (*sensu* Salvador 1994). Bioturbation involving both the impactoclastic and volcanoclastic layers has resulted in stratigraphic leakage of particles from those horizons to burrow-filling sediments below. However, some "return" burrows may have caused upward-mixing of the sediment as well, as it can be seen in many K/T boundary outcrops in this region, in the basal pink limestone of the Tertiary, which contains sparse impact spherules, and *Planolites* burrows filled with white carbonate ooze derived from the top layer of the Cretaceous (Montanari 1991; Montanari and Koeberl 2000).

The thoroughly bioturbated marly limestones of the study interval at Massignano have a relatively consistent non-carbonate component of approximately 20 percent (Mattias et al. 1992). The carbonate content of these biomicrites derives mainly from calcareous nannofossils and planktonic foraminiferal tests and, to a far lesser extent, benthic foraminifers and calcite cement. Non-carbonates in these rocks include clay, mica, silicate silt, iron oxides, sulfides, and organic material. In the distal impactoclastic layer and in burrow fillings derived from that layer, there is an impact-related component, including flattened microspherules (in the size range of 100-300 μm) mostly made of clustered, black Ni-rich spinel crystallites and more rarely of green smectitic clay, sparse Ni-rich spinel crystallites, shocked-quartz grains (in the size range of 100 μm), and Ir and other extraterrestrial siderophile elements probably dispersed into the clay fraction. In the volcanoclastic layer (5.83 to 5.88 cm), fine to medium sand-size crystals of biotite are relatively abundant. Bed thickness in the study interval ranges from 3 to 13 cm, and averages approximately 8 cm. Owing to relatively thorough bioturbation, the texture is more-or-less homogenous within each bed.

Ichnology of the study interval includes a secondary (i.e., superimposed) suite of tiered *Planolites* and *Zoophycos*, which are filled with sediment derived from overlying layers. In most instances, *Zoophycos* is the younger of the two superimposed traces. In some instances, provenance of burrow-filling sediment is suggested by burrow-fill color, which can be matched with distinctive color shades of overlying layers. In addition, volcanically derived biotite flakes (derived from the interval between 5.83 and 5.88 m) and impact-generated spherules (derived from the interval between 5.64 and 5.72 m) may be found within the secondary burrow suite (see Fig. 7). Specifically, biotite flakes comprise part of back-filling layers inside *Zoophycos* and *Planolites* as low as 5.20 m. Similarly, impact-generated spherules are found within *Planolites* as low as 5.06 m.

Stratigraphic leakage is the process whereby sediments and (or) fossils of a younger age are deposited within or under older rocks. The product of such leakage is a stratigraphic leak (*sensu* Foster 1966). Bioturbation-related stratigraphic leakage from two horizons is clearly shown by leaked biotite flakes and impact-generated spherules within burrows. However, geochemical evidence also documents stratigraphic leakage regarding iridium-enriched materials from the impactoclastic layer.

An iridium enrichment comparable to that found in the impactoclastic layer (i.e., approximately 200 ppt) is also found in *Planolites* fill material as far as 14 cm below the base of the impactoclastic layer. An iridium content nearly one half that of the impactoclastic layer (or approximately 100 ppt) is found in *Planolites* filling as far as 22 cm below the impactoclastic layer. In addition, an iridium content similar to the latter occurs in apparent laminae that may be highly compressed *Planolites* at 28 cm below the base of the impactoclastic layer. *Planolites* fillings with iridium concentrations nearly equal to the impactoclastic layer are interpreted to have been derived almost entirely from the impactoclastic layer. Burrow fillings with significantly lower iridium content are interpreted to have been derived from the impactoclastic layer, plus one or more overlying layers.

Stratigraphic leakage is apparently not the only process at work in redistributing iridium in the sediment about the impactoclastic layer. For example, the iridium content in a micro-drill hole, 5 cm above the top of the impactoclastic layer, is almost as high as in the impactoclastic layer itself. Robin et al. (1991) have shown, by comparing the distribution of Ni-rich spinels and iridium abundances at different K/T-boundary sites, how extraterrestrial iridium is postdepositionally reworked. They used the theoretical approach of Officer and Lynch (1983), resulting in the conclusive interpretation that chemical diffusion yields the symmetrical "shoulders" of the iridium distribution at impact sites, which extend beyond the Ni-rich spinel peak. This scenario can also be seen at Massignano (see matrix in Fig. 5b) from meter level 5.64 down to level 5.53, where iridium abundances for burrows and matrix are indistinguishable. Nevertheless, the burrows in the region of meter level 5.50 to 5.53 and at 5.33 show an enrichment in iridium by a factor of two or more compared to matrix values, which is obviously the result of vertical distribution by a non-chemical process, such as bioturbation.

Another possible explanation for the enrichment in iridium above the impactoclastic layer could be through a so-called return-burrow, as previously explained in the case of the K/T boundary (see Montanari and Koeberl 2000), where *Planolites* filled with white Cretaceous ooze are found in the first 10 cm of the pink basal Tertiary limestone, a process that is called remanié (Kidwell et al. 1986). This upward redistribution may be due to bioturbation or a diagenetic effect.

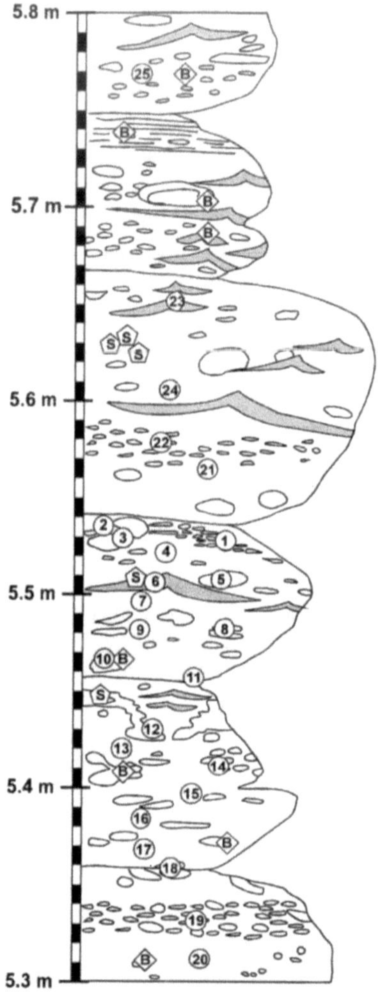

Fig. 7. Stratigraphic section of strata between 5.0 and 5.9 m above base at Massignano GSSP. The impactoclastic layer spans 5.64 to 5.72 m, whereas the volcanoclastic layer appears at 5.83 to 5.88 m. Stratigraphic leakage of biotite flakes and impact-derived spherules by bioturbation is indicated where the symbol is superimposed on a burrow. Biotite flakes and spherules outside burrows are indicated by "B" and "S" in the matrix. Sample numbers are shown at the approximate locations of micro-drill holes. Where sample numbers are superimposed on burrows, the drill hole is in the burrow fill. Details of burrow shapes and cross-cutting relations, laminae, biotite and spherule distribution, and location of micro-drill holes were obtained from rock slab surfaces.

6 Summary and Conclusions

The impactoclastic layer of Late Eocene age in Massignano has recently been interpreted as a result of one of two large-scale impacts (Chesapeake Bay, USA, and Popigai, Russia). Furthermore, a paleo-environmental change apparently occurred immediately after the impactoclastic layer was deposited (Montanari and Koeberl 2000). This dramatic paleoenviromental change is related to the extinction of seventy percent of the then-extant total radiolaria and can be seen in the change to reducing conditions on the sea-floor. Such a change could help explain the low number of fossil traces through the impactoclastic layer. However, bioturbation by organisms perturbed the sedimentary record and, thus, left detectable iridium traces in the underlying sediments, a form of stratigraphic leakage. An effective disturbance of the iridium profile in the sedimentary record can only be seen within a layer of about 20 cm although impact-generated microkrystite spherules have been found far below the impactoclastic layer. Still unsolved is the influence of Ni-rich spinel stored in some of the burrow fillings upon the Ni distribution in the section. Late-stage organic bioturbation redistributed some of the iridium that was deposited and dispersed chemically. Nevertheless the influence of bioturbation causes only minor variations within the original iridium distribution of the underlying sediments. Additionally, the distribution of Ir in the overall sediment volume (or strata column) will show up with different intensities according to the mode of sampling, and sample size. For instance, when collecting small samples, such as drill cores or even small hand samples, it is very likely to find a burrow enriched in iridium, which will result in a high concentration for that particular level. On the other hand, by pure chance, the concentration may be near the detection limit, when drilling in the matrix or collecting a sample that contains mostly matrix. A large sample, properly crushed and homogenized, will yield a value closer to the average composition, i.e., the amount of vertically-mixed contaminant material. Thus, bioturbation is also an effective factor for the redistribution of extraterrestrial iridium, although chemical diffusion is the most important process for element mixing affecting the immediate vicinity (± 15 cm) of an impactoclastic layer.

Acknowledgements

This work was supported by the Austrian Fonds zur Förderung der wissenschaftlichen Forschung, project Y58-GEO (to C.K.), the Austrian-Italian Scientific and Technical Cooperation (project #14, to C.K. and A.M.), and the Coldigioco Research Fund. We want to thank Eric Robin and Jan Smit for their constructive and helpful reviews.

References

Alvarez W, Asaro F, Michel H, Alvarez LW (1982) Iridium anomaly approximately synchronous with terminal Eocene extinctions. Science 216: 886–888

Berggren WA, Kent DV, Swisher C III (1995) A revised Cenozoic geochronology and chronostratigraphy. In: Berggren WA, Kent DV, Hardenbol J (eds) Geochronology, time scales and global stratigraphic correlations: A unified temporal framework for an historical geology. SEPM Spec Pub 54, pp 129–212

Bice DM, Montanari A (1988) Magnetic stratigraphy of the Massignano section across the Eocene-Oligocene boundary. In: Premoli Silva I, Coccioni R, Montanari A (eds) The Eocene-Oligocene Boundary in the Marche-Umbria Basin (Italy). IUGS Special Publication, F.lli Aniballi Publishers, Ancona, pp 111–117

Bohor BF, Betterton WJ, Foord EE (1988) Coesite, glass, and shocked quartz and feldspar at DSDP Site 612: Evidence for nearby impact in the late Eocene [abs]. Lunar Planet Sci 19: 114–115

Bottomley R, Grieve R, York D, Masaitis V (1997) The age of the Popigai impact event and its relation to events at the Eocene/Oligocene boundary. Nature 388: 365–368

Brinkhuis H, Coccioni R (1995) Is there a relation between dinocysts changes and iridium after all? Preliminary results from the Massignano section. In: Montanari A, Coccioni R (eds) The effects of impacts in the evolution of the atmosphere and biosphere with regard to short- and long-term changes. European Science Foundation, Ancona, 40 pp

Clymer AK (1996) Discovery and trace element geochemistry of Late Eocene shocked quartz: Insights into the Late Eocene impacts. Master's Thesis, Univ California, Berkeley, 85 pp

Clymer AK, Bice DM, Montanari A (1996) Shocked quartz from the Late Eocene: Impact evidence from Massignano, Italy. Geology 24: 483–486

Coccioni R, Monaco P, Monechi S, Nocchi M, Parisi G (1988) Biostratigraphy of the Eocene-Oligocene boundary at Massignano (Ancona, Italy). Int Subcomm Paleog Strat, E/O Meeting, Ancona, 1987, Spec Pub II, pp 59–76

Farley KA, Montanari A, Shoemaker EM, Shoemaker CS (1998) Geochemical evidence for a comet shower in the Late Eocene. Science 280: 1250–1253

Foster NH (1966) Stratigraphic leak. Am Ass Petrol Geol Bull 50: 2604–2611

Gardin S, Spezzaferri S, Basso D, Coccioni R (1999) Calcareous nannofossil and planktonic foraminiferal response to a meteorite impact in the Umbria-Marche Massignano section (Northeastern Apennines, Italy) [abs]. Abstracts, EUG-10 Strasbourg Meeting, Journal of Conference Abstracts 4(1): 271

Glass BP, Dubois DL, Ganapathy R (1982) Relationship between an iridium anomaly and the North American microtektite layer in core RC9-58 from the Caribbean Sea. Proc Lunar Planet Sci Conf 13th, J Geophys Res 87: A425–A428

Glass BP, Burns CA, Crosbie JR, DuBois DL (1985) Late Eocene North American microtektites and clinopyroxene bearing spherules. Proc Lunar Planet Sci Conf 16th, J Geophys Res 90: D175–D196

Glass BP, Hall CM, York D (1986) $^{40}Ar/^{39}Ar$ laser-probe dating of North American tektite fragments from Barbados and the age of the Eocene-Oligocene boundary. Chemical Geology 5: 181–186

Gonzalvo C, Molina E (1992) Estudio cuantitativo de los foraminiferos planctònicos en el estratotipo del limite Eoceno/Oligoceno en Massignano (Apeninos, Italia). Geogaceta Soc Geol Espana 12: 64–67

Hut P, Alvarez W, Elder W, Hansen TA, Kauffman EG, Keller G, Shoemaker EM, Weissman P (1987) Comet showers as a cause of mass extinctions. Nature 329: 118-126

Keller G, D'Hondt SL, Orth CJ, Gilmore JS, Oliver PQ, Shoemaker EM, Molina E (1987) Late Eocene impact microspherules: Stratigraphy, age, and geochemistry. Meteoritics 22: 25–60

Kidwell SM, Fursich FT, Aigner T (1986) Conceptual framework for the analysis and classification of fossil concentrations. Palaios 1: 228–238

Koeberl C (1993) Instrumental neutron activation analysis of geochemical and cosmochemical samples: a fast and reliable method for small sample analysis. J Radioanalyt Nucl Chem 168: 47–60

Koeberl C, Huber H (2000) Optimization of the multiparameter - coincidence spectrometry for the determination of iridium in geological materials. J Radioanalyt Nucl Chem 244: 655–660

Koeberl C, Poag CW, Reimold WU, Brandt D (1996) Impact origin of the Chesapeake Bay structure and the source of the North American tektites. Science 271: 1263–1266

Lanci L, Lowrie W, Montanari A (1996) Magnetostratigraphy of the Eocene/Oligocene boundary in a short drill-core. Earth Planet Sci Lett 143: 37–48

Langenhorst F (1996) Characteristics of shocked quartz in late Eocene impact ejecta from Massignano (Ancona, Italy): Clues to shock conditions and source crater. Geology 24: 487–490

Lowrie W, Lanci L (1994) Magnetostratigraphy of Eocene-Oligocene boundary sections in Italy: No evidence for short subchrons within chrons 12R and 13R. Earth Planet Sci Lett 126: 247–258

Lowrie W, Alvarez W, Napoleone G, Perch-Nielsen K, Premoli Silva I, Toumarkine M (1982) Paleogene magnetic stratigraphy in Umbrian pelagic carbonate rocks: The Contessa sections, Gubbio. Geol Soc Am Bull 93: 414–432

Lowrie W, Alvarez W, Asaro F (1990) The origin of the white beds below the Cretaceous-Tertiary boundary in the Gubbio section, Italy. Earth Planet Sci Lett 98: 303–331

Mattias P, Crocetti G, Barrese E, Montanari A, Coccioni R, Farabollini P, Parisi E (1992) Caratteristiche mineralogiche e litostratigrafiche della sezione eo-oligocenica di Massignano (Ancona, Italia) comprendente il limite Scaglia Variegata-Scaglia Cinerea. Studi Geologici Camerti 12: 93–103

Meyer G, Piccot D, Rocchia R, Toutain JP (1993) Simultaneous determination of Ir and Se in K-T boundary clays and volcanic sublimates. J Radioanalyt Nucl Chem 168: 125–131

Michel HV, Asaro F, Alvarez W, Alvarez LW (1991) Geochemical study of the Cretaceous-Tertiary Boundary region at hole 752B. Proc ODP, Sci Results 121: 415–422

Molina E, Gonzalvo C, Keller G (1993) The Eocene-Oligocene planktic foraminiferal transition: Extinctions, impacts and hiatuses. Geol Mag 130: 483–499

Montanari A (1988) Geochemical characterization of volcanic biotites from the Upper Eocene-Upper Miocene pelagic sequence of the Northern Apennines. In: Premoli Silva I, Coccioni R, Montanari A (eds) The Eocene-Oligocene Boundary in the Marche-Umbria Basin (Italy). IUGS Special Publication, F.lli Aniballi Publishers, Ancona, pp 209–228

Montanari A (1991) Authigenesis of impact spheroids in the K/T boundary clay from Italy: new constraints for high-resolution stratigraphy of terminal Cretaceous events: J Sed Petrol 61: 315–339

Montanari A, Koeberl C (2000) Impact stratigraphy – The Italian record. Lecture Notes in Earth Sciences 93, Springer Verlag, Heidelberg-Berlin, 364 pp

Montanari A, Deino A, Drake R, Turrin BD, De Paolo DJ, Odin SG, Curtis GH, Alvarez W, Bice DM (1988) Radioisotopic dating of the Eocene-Oligocene boundary in the pelagic sequence of the Northern Apennines. In: Premoli-Silva I, Coccioni R, Montanari A (eds) The Eocene-Oligocene Boundary in the Marche-Umbria basin (Italy), IUGS Special Publication, F. Aniballi Publishers, Ancona, pp 195–208

Montanari A, Asaro F, Michel H, Kennett JP (1993) Iridium anomalies of Late Eocene age at Massignano (Italy), and ODP site 689B (Maud Rise, Antarctic). Palaios 8: 420–437

Oberli F, Meier M (1991) Age of Eocene-Oligocene boundary in the Marche-Umbria basin, Italy, by high resolution U-Th-Pb dating [abs]. Terra Abstracts 3: 286

Obradovich JD, Snee LW, Izett GA (1989) Is there more than one glassy impact layer in the Late Eocene? [abs]. Geol Soc Am, Abstracts with Programs 21(6): A134

Odin GS, Montanari A (1989) Age radiométrique et stratotype de la limite Éocène Oligocène. Comptes Rendus Acad Sci Paris 309: 1939–1945

Odin GS, Montanari A, Deino A, Drake R, Guise P, Kreuzer H, Rex DC (1991) Reliability of volcano-sedimentary biotite ages around the Eocene/Oligocene boundary. Chemical Geology (Isotope Geosci) 86: 203–224

Officer CB, Lynch DR (1983) Determination of mixing parameters from tracer distributions in deep-sea sediment cores. Marine Geology 52: 59–74

Pierrard O, Robin E, Rocchia R, Montanari A (1998) Extraterrestrial Ni-rich spinel in upper Eocene sediments from Massignano, Italy. Geology 26: 307–310

Poag CW, Aubry M-P (1995) Upper Eocene impactites of the U.S. East Coast: Depositional origins, biostratigraphic framework, and correlation. Palaios 10: 16–43

Premoli Silva I, Jenkins DJ (1993) Decision on the Eocene-Oligocene boundary stratotype. Episodes 16: 379–381

Robin E, Boclet D, Bonté Ph, Froget L, Jéhanno C, and Rocchia R (1991) The stratigraphic distribution of Ni-rich spinels in Cretaceous-Tertiary boundary rocks at El Kef (Tunisia), Caravaca (Spain) and Hole 761C (Leg 122). Earth Planet Sci Lett 107: 715–721

Rock Color Chart Committee (1991) Rock color chart, 7th printing with revised text. Geological Society of America Boulder, Colorado

Salvador A (ed) (1994) International stratigraphic guide; a guide to stratigraphic classification, terminology, and procedure. Geological Society of America, 214 pp

Vonhof HB, Wijbrans J, Smit J (1995) The Popigai impact crater: $^{39}Ar/^{40}Ar$ dating and its expression in the $^{87}Sr/^{87}Sr$ record of the Massignano section: In Montanari A, Coccioni R (eds) The role of Impacts on the Evolution of Planet Earth, Abstracts and Field Trips, European Science Foundation, Ancona, pp 163–164

Vonhof HB, Smit J, Brinkhuis H, Montanari A (1998) Late Eocene impacts accelerated global cooling? In: Vonhof HB (PhD thesis) The Strontium Stratigraphic Record of Selected Geologic Events, Academisch Proefschrift, University of Utrecht, pp 77–90

Vonhof HB, Smit J (1999) Late Eocene microkrystites and microtektites at Maud Rise (Ocean Drilling Project Hole 689B; Southern Ocean) suggest a global extension of the approximately 35.5 Ma Pacific impact ejecta strewn field. Meteoritics 34: 747–755

Vonhof HB, Smit J, Brinkhuis H, Montanari A, Nederbragt AJ (2000) Global cooling accelerated by early late Eocene impacts? Geology 28: 687–690

Whitehead J, Papanastassiou DA, Spray JG, Grieve RAF, Wasserburg GJ (2000) Late Eocene impact ejecta: geochemical and isotopic connections with the Popigai impact structure. Earth Planet Sci Lett 181: 473–487

The Ries and Steinheim Meteorite Impacts and their Effect on Environmental Conditions in Time and Space

Madelaine Böhme[1], Hans-Joachim Gregor[2], and Kurt Heissig[1]

[1]Bavarian State Collection for Palaeontology and Historical Geology, Richard-Wagner-Str. 10, D-80333 Munich, Germany. (m.boehme@lrz.uni-muenchen.de; k.heissig@lrz.uni-muenchen.de)
[2]Naturmuseum, Im Thäle 3, D-86152 Augsburg, Germany. (H.-J.Gregor@t-online.de)

Abstract. The Ries and Steinheim Impact about 15 Ma ago had no palaeontologically verifiable effect on the composition of the fauna and the flora in Southern Germany. The destruction was only of local importance. The ecosystem was quickly, about 100 years later, reconstituted by the same rich subtropical vegetation and vertebrates. The fluviatile ecosystem of the „Upper Freshwater Molasse" accelerated the re-colonization. This fast reorganization of the ecotopes was particularly influenced by fluviatile drift.

1 Introduction

About 15 million years ago (14.87 ± 0.36 Ma; Storzer et al. 1995) the Ries and the Steinheim meteorites happened to hit Southern Germany with an explosion power of 250 000 atomic bombs. A disaster of incredible extension destroyed the area where nowadays two romantic villages, Nördlingen (Ries) and Steinheim, are situated in Southern Germany. At a distance of at least 10 km from the center the rocks of the basement were shocked and melted to a depth of 5000 m. Large blocks were ejected 180 km around (Fig. 1), earthquakes occurred and widespread fires burnt the forests down. This was a significant geological event which is directly recorded by the Ries- and Steinheim craters (see Gregor 1992). In contrast to the idea of a devastation over millions of years (Schleich 1984, Spitzlberger 1984), the present authors show that the event was only of regional importance with a short-term effect on the landscape.

The literature contains abundant data regarding the geological, mineralogical, geophysical and paleontological effects, problems and hypotheses associated with the Ries impact event (see Bayerisches Geologisches Landesamt 1964, 1969, 1970, 1977).

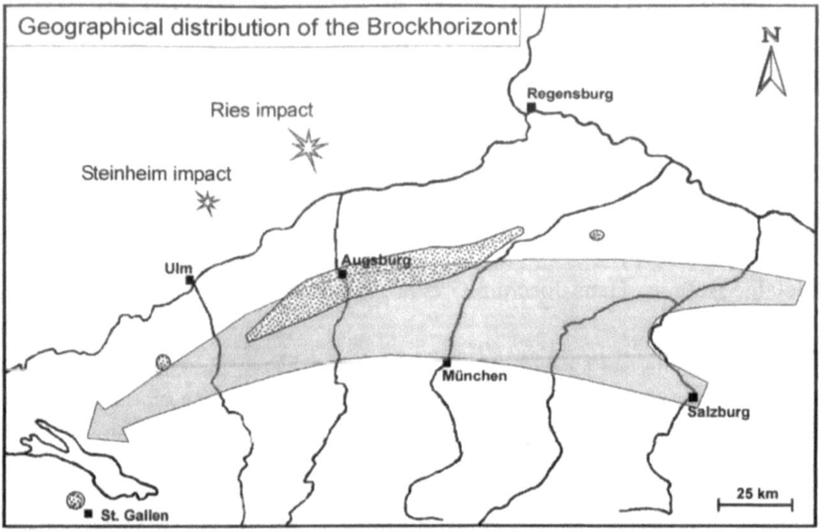

Fig. 1. Geographical location of the Ries and Steinheim impact craters and the distribution of the „Brockhorizont" (dotted areas; greyish arrow indicate the axial outlet of the Molasse basin during the Middle Miocene).

2 The Event – the Impact – the Catastrophe

From the Ries event, abundant evidence remain that help us to understand what happened at that time. Many outcrops contain suevite, a polymict impact breccia with melt rock inclusions, shocked granites, shatter cones in limestones, and brecciated upper Jurassic limestones with fossils like belemnites broken and re-cemented.

Within 200 km radius a compression wave destroyed the forests. Rock fragments, from huge blocks to microscopic fragments, were riding on the front of this wave, coming down at distances known up to 160 km west and 120 km east of the crater centre. They hit the earth with supersonic speed and killed all animals on the surface and in the air. The air was heated by the compression and set fire to the woods within at least the same distance. Burrowing animals may have been killed by the earthquake or the compression wave entering their dens.

The small quarry of Unterneul, about 60 km SSE from the Ries crater, shows some traces of the catastrophe preserved within a channel that was later covered by a small lake and its sediments. This channel was incised through 5 m of sand into a greyish marl bed. In this situation the impact event occurred. The upper 0.2 m of the marl are heavily disturbed by the falling blocks and the compression wave. The sediment, even if already consolidated, was partly reworked and formed a matrix filling the space between undissolved debris of marl. There are horizontal shearing planes with a faint glimmer of oriented mica grains on the surfaces, cutting through fossil snails. Medium-sized to small blocks intruded with sharp edges into the surface of the marl.

The whole channel was later covered by fine clay or, marginally, by the gravels of a small delta. These contain lots of flattened trunks, remains of the broken logs that escaped burning by falling into the water. On the other hand, well rounded wooden blocks were found that have a higher degree of carbonization, gagate near the surface, lignite in the centre, with folded wood structure. These pieces of wood have been compressed before being imbedded in the sediments and were, therefore, no more compressible. Probably they were thrown out from the crater or its immediate neighbourhood, compressed by the immense pressure of the explosion. During their flight into the foreland all edges and splitters burned away until they fell into the water of the channel. Where the immediate cover of the Ries debris is clay, a few centimetres above we find the first shells of the mussel *Margaritifera*, sometimes broken over the edge of a bigger block by the compression of the clay. Half a meter above that layer follows a well-preserved and rich leaf flora.

3
The Stratigraphic Context of the Destruction Horizon in the Upper Freshwater Molasse (Heissig)

During the time of the Ries impact the Alpine foreland was a sedimentary basin, where a system of large longitudinal streams accumulated the gravels, sands and silts of the Upper Freshwater Molasse. Thus, we find the traces of Ries debris and various destructions within this stratigraphic sequence, the so called „Brockhorizont“ (Figs. 1 and 2) of the local geologists.

Following Dehm (1951), the Upper Freshwater Molasse was divided into an Older, Middle and Younger Series (Ältere, Mittlere, and Jüngere Serie), defined by their faunal composition, mainly by the occurrence and size of different proboscideans. The lithological content of these series was partly studied in Lower Bavaria by Blissenbach (1957), Grimm (1965), Stiefel (1957), and other sedimentologists. A more detailed stratigraphy was elaborated by Heissig (1986, 1989, 1997) and Fiest (1986), using fluvial cyclic sedimentation in combination with the faunal composition and the size evolution of certain phylogenetic lineages. This work is still in progress. The new concept has not yet been applied to the floras and to the whole Younger Series.

The first series comprises five sedimentary cycles of the Older Series, numbered from 1 to 5, with the so called „Limnische Süßwasserschichten“ as zero. Within this range four faunal groups have been distinguished: Group A, corresponding to cycle 0, the latest fauna of MN 4 age, group B, without typical elements of MN 4, comprising the first cycle, forming the base of the real Upper

Freshwater Molasse, group C and group D both comprising two sedimentary cycles with rich faunas.

The Middle Series, also with five sedimentary cycles, has so far been divided into two faunal groups, but some additional information may be forthcoming. Two of these cycles are anterior to the Ries impact; the second one, being very incomplete, may be due to the pre-Riesian erosion. This forms a big hiatus at the northern margin of the Molasse basin with the formation of a relief of more than 100 m. Within the basin it is split into two gaps, one separating the Older and the Middle Series, the other immediately preceding the Ries impact. As the fauna of these two cycles is not exactly the same, they should be taken as subgroups E and E' (Table 1). Subgroup E has a fauna of the same composition as the reference locality of MN 5, Pont Levoy-Thenay. The three cycles after the impact, number 8 – 10, are considered as faunal group F with a fauna containing the first immigrant, characterising MN 6. The question whether E' belongs to MN 5 or MN 6 is not yet resolved.

The time of the Ries impact is, therefore, near, or at, the base of MN 6.

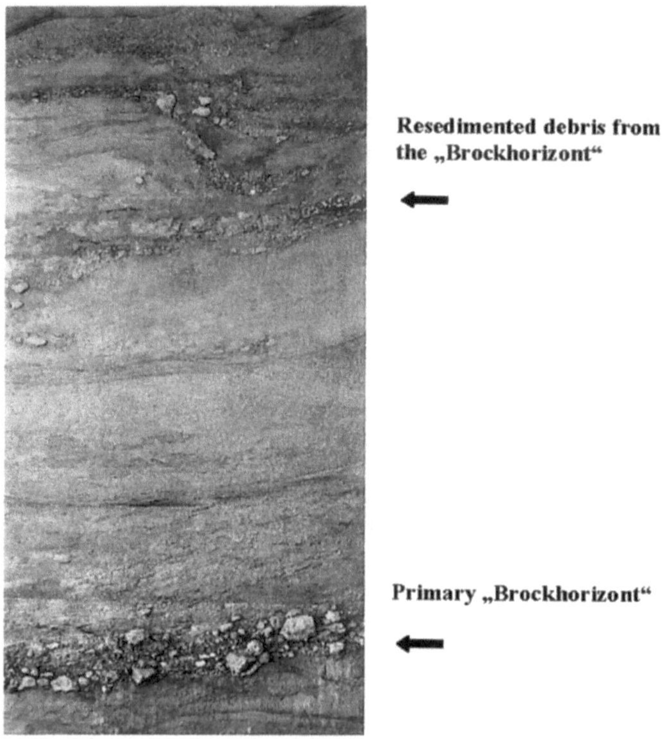

Fig. 2. Profile of the immediately post-Riesian sediments at the locality Ziemetshausen (65 km south of the Ries impact crater) showing the Ries debris in the primary "Brockhorizont" and re-sedimented debris (Section is about 2 m high)

Table 1. Sedimentary cycles and faunal groups of the Older and Middle Series of the Upper Freshwater Molasse (OSM) of Bavaria

MN-zone	OSM-units	Sedimentary cycles	Region Günz-Iller	Region Lech-Paar	Region Ilm-Isar	Eastern Bavaria
MN 6 type	none	none	Hiatus	Hiatus	Hiatus	Hiatus
	OSM F?	OSM 10		Laimering 4b, 5		
			Bentonite	Bentonite 14,6 MA	Bentonite 14,6 MA	
	OSM F	OSM 9	Thannhausen Ziemetshausen 1e	Laimering 2,3, 4a Stätzling	Sallmannsberg Göttschlag	Stürming
	OSM F	OSM 8	Ziemetshausen 1b Ries-boulders	Gallenbach 2a Unterneul 1c Ries-boulders	Unterzolling Streitdorf Ries-boulders	
			Hiatus	Hiatus	Hiatus	Hiatus
	OSM E'	OSM 7		Derching 1b		
MN 5 type	OSM E	OSM 6	Ziemetshausen 1c Ebershausen Mohrenhausen Altenstadt	Unterneul 1a Bentonite	Eberstetten	
			Hiatus	Hiatus	Hiatus	Hiatus
	OSM D	OSM 5	Oggenhof	Oberbernbach	Affalterbach Bentonite	
		OSM 4	Betlinshausen		Unterempfenbach 1b,d	
	OSM C	OSM 3	Burtenbach		Sandelzhausen Unterempfenbach 1c	Maßendorf
		OSM 2	Schönenberg Roßhaupten		Puttenhausen	
MN 5 base	OSM B	OSM 1	Bellenberg 2 Bellenberg 1	Langenmoosen	Wörth a. d. Isar	Niederaichbach
MN 4b	OSM A	OSM 0	Günzburg			Forsthart Rembach Rauscheröd

4 Flora and Fauna before and after the Event

4.1 The Megafloras (Gregor)

The western and eastern parts of the Brackish Molasse became drier and drier, huge rivers succeeded the overall muddy area, and dense riparian forests evolved, very well known from the Günzburg area. In sand- and claypits from Burtenbach, Kirrberg (Riederle and Gregor 1997), etc., we find abundant leaves and fruits from different types of wetland, bottomland, and riparian forests (Gregor 1982, Gregor et al. 1989).

The main components were small-leaved *Gleditsia* and other legumes (often misinterpreted as arid or dry floras), of *Cinnamomum*-laurel-types, of *Zelkova*, waterelm, elmtree, beech, maple, spiny oaks, and many more. The floral composition was very similar in the whole of Europe at this time, some special stands excluded (Table A-C, appendix).

Spitzlberger (1984) postulated the extinction of plant and animal life between the Alps and Northern Germany for millions of years, due to the fact that for example an aceroid *Tilia* (from Goldern, MN5, before the event) vanished from this time on in Bavaria. But fossil leaves and fruits from *Tilia* are so rare in the molasse sediments that we can hardly conclude anything from the occurrence or absence of this taxon. It is a typical accompanying type of plant with an irregular behaviour. The presence or absence of *Tilia* type plants and other uncommon ones can be easily explained by special stands and habits in the Molasse region (the Goldern site is a high plateau), not by the impact.

The case of the palms is similar. We have real, well determinable palm remains only in the Lower Miocene and none in the Middle or Upper Miocene (Gregor 1980), in contrast to wrong determinations (see Jung 1981). It is not possible to say anything about the biotopes with palms around the time of the event. Many silicified wood remains in molasse sediments come from a secondary site and are not suitable for stratigraphical interpretations.

Concerning the vegetation of the time around the impacts it may be mentioned that in sediments of the Steinheim crater we find *Gleditsia, Populus, Celtis,* etc. (Gregor 1983, in the flora before the impact (e.g., Burtenbach site). In the Ries crater we find *Cedrelospermum, Gleditsia, Spondieaemorpha* etc., as formerly in the Randecker Maar or later in the Öhningen sites (Gregor 1982a).

In the Ries crater filling we find a certain *Zanthoxylum wemdingense* (Gregor 1977) as a single element, but it also occurs in the Lower Miocene of the Mainz basin. The small pine *Pinus aurimontana* is rare in the molasse (Gregor 1982b).

Typical pioneers also occurred in the Ries crater – *Ailanthus* – the Chinese Godtree, but no Chenopodiaceae as previously postulated (Jung in Dehm et al. 1977), today also typical pioneer herbs.

Around Augsburg we find post-Riesian sites with the common *Cinnamomum, Ulmus, Platanus* and *Hemitrapa*-flora, etc., as in pre-Riesian floras (Knobloch and Gregor, in prep.). Especially in the drill cores of the Steinheim crater these plants occur immediately after the impact (squeezed limestones, tectonically destroyed), which means perhaps a return within only years or tens of years. This shows clearly a rapid return of vegetation after devastation – such as after the Krakatau eruption of the last century (Ernst and Seward 1908, Thornton 1997). There, 50 years later half of the original plant- and animal life had come back. Also if we bear in mind that this was a volcanic eruption, it must have been somewhat similar to the Ries event and it is the greatest catastrophe that we can use for such a comparison.

All the mentioned data also allow the reconstruction of climates before and after the impacts - they were of the same Cfa-type (Virginia-climate sensu Köppen), only showing the normal cooling effect observed in the whole Tertiary.

Principally, the plant elements in the molasse were of exotic character, like *Ginkgo* (China. Japan), *Ailanthus*, *Corylopsis*, *Liquidambar*, *Celtis*, *Gleditsia*, *Glyptostrobus* or *Meliosma* (China), *Acer*, spiny oaks, *Taxodium*, *Magnolia*, *Zanthoxylum* or *Chionanthus* (North America). The wetland plants belong to native genera, such as *Ulmus*, *Alnus*, *Betula* or *Fraxinus*, but to Asian and American species (see Table A-C in the appendix). The palynological data give the same indication as the megaflora.

In the accompanying diagram we try to correlate the occurrences of fossil fruits and seeds and of leaves by their coenocomplexes (Fig. 3). The laurophyllous forest in pre- and post-Riesian times was similar to that of today in SE-Asia (*Cinnamomum*) or elsewhere. All the forest types were similar to the evergreen broad-leaved forests, mixed mesophytic forests or deciduous broad-leaved forests of Asia and America, of the Indian Sholas and Littoral forests, the Japanese Oak-beech-forests, the Hardwood bottom formations of America and the Canarian Laurel forests.

The leaves from many different localities of the Molasse region are currently under study by Gregor and Knobloch, but a preliminary list can be given here (see appendix).

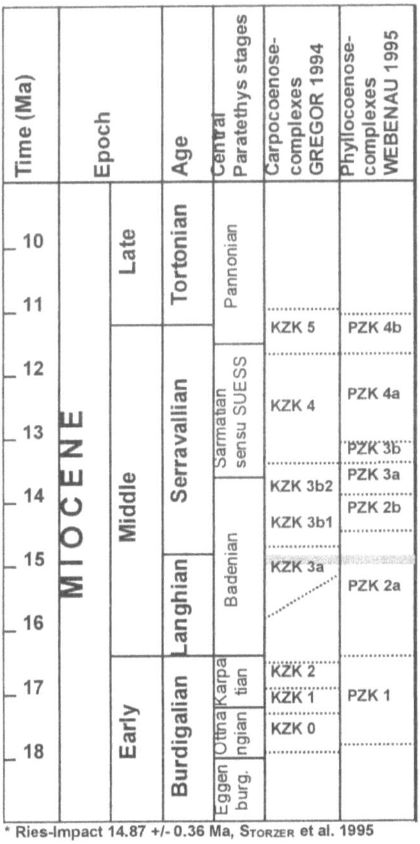

Fig. 3. Stratigraphic sequence of Carpocoeno- and Phyllocoenocomplexes. The data for this figure are given in tables A to C in the appendix. The grey line indicates the Ries impact event.

4.2
The Lower Vertebrates (Böhme)

Schleich (1984) characterised the Ries impact as a „faunal event", postulating the extinction of large reptiles, such as *Diplocynodon* and *Geochelone*. He noted (Schleich 1985): „ein deutlich unterscheidbares präriesisches und postriesisches Herpetofaunenbild" (a very distinct Herpetofauna before and after the Ries impact).

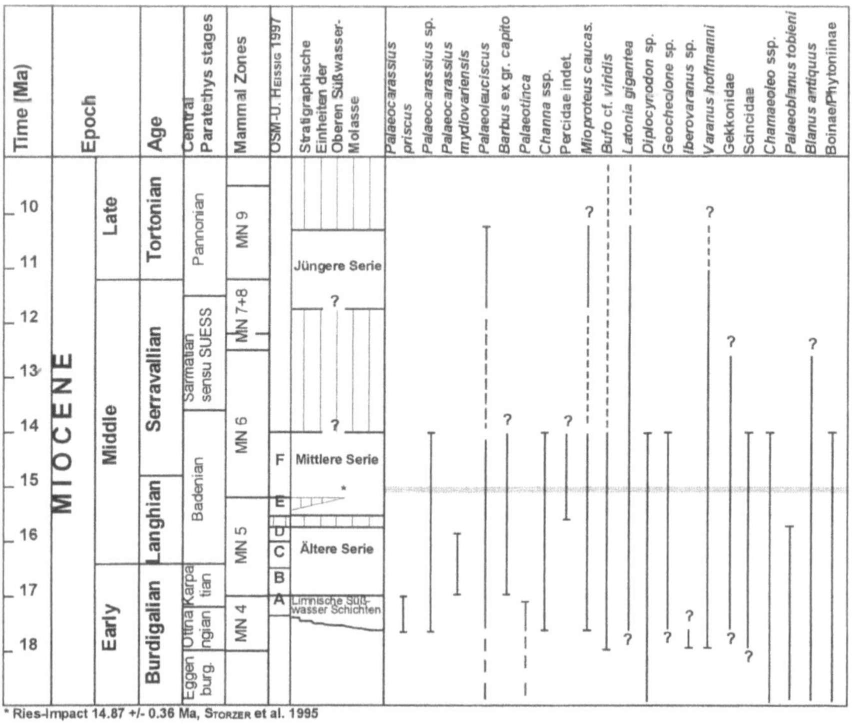

Fig. 4. Stratigraphical distribution of important fish and herpeto-taxa in Southern Germany. The grey line indicates the Ries impact

4.2.1
The Lower Vertebrates before the Ries Impact

The lower vertebrates of the Older and the Middle Series before the Ries impact come from sediments of fluviatile ecosystems (braided and meandering rivers, riparian waters). The fish fauna of the still water is dominated by species of the families Channidae und Cyprinidae, which tolerate a low oxygen content, especially some species of the air-breathing snakehead *Channa* and three species

of the genus *Palaeocarassius*. A *Channa/Palaeocarassius*-association is typical for flat, muddy, oxygen deficient, in exceptional cases temporary waters in the riparian zone (Böhme 1999a). The fish fauna of the running water systems is dominated by barbels closely related to *Barbus capito* and a species of the genus *Palaeoleuciscus*.
The amphibians reach, with 14 species, a very high diversity at this time (Böhme 1999a). The characteristic forms of the riparian waters are newts of the genus *Triturus* (*T. vulgaris*, *T.* cf. *marmoratus*) and the frogs *Rana* (*ridibunda*) sp., *Latonia gigantea* and *Pelobates* nov. sp.. The waterdog (*Mioproteus caucasicus*) and the salamandrids (*Salamandra sansaniensis*, *Chelotriton paradoxus*) are the typical elements in the running water systems.

The reptilian fauna of the fluviatile ecosystems is mostly thermophilic. Apart from the common crocodile *Diplocynodon styriacus* and the giant turtle „*Geochelone*", the Chamaeleonidae, Scincidae, Amphisbaenidae, and diverse taxa of the Lacertidae and Anguidae are present with several species (Böhme 1999a, b). Gekkonidae and two genera of the Varanidae (*Varanus*, *Iberovaranus*) appear in the drier habitats of the Franconian and Swabian Alb (fissure fillings).

These reptiles are characteristic from the Eggenburgian until the Middle Badenian, indicating a climate optimum. All taxa (especially *Chamaeleo* ssp., „*Geochelone*" and *Diplocynodon*) refer to a „thermal window" with a mean temperature during the year of at least 17-18° C, a mean temperature of the coldest month not lower than 8° C, and a mean temperature for the warmest month about 25-30° C (inferred from the recent distribution of the most closely related taxa and there hibernation, reproduction, and activity temperatures; cf. Haller-Probst 1998).

4.2.2
The Lower Vertebrates after the Ries Impact

In contrast to Schleich (1984, 1985), who postulated a „faunal event" character of the Ries impact, I could not find in my survey any hints of a palaeontologically verifiable regional extinction event. All taxa of fishes, amphibians, and reptiles in the sediments of the Older and Middle Series before the Ries impact (OSM-units A to E; Fig. 4) also occur in the sediments of the Middle Series after the Ries impact (OSM-unit F), particularly the giant turtle „*Geochelone*" (Griesbeckerzell 1a, Ziemetshausen 1b, Haberskirch) and the crocodile *Diplocynodon* (Griesbeckerzell 1a, Derching „Blauer Ton" (blue clay), Göttschlag 1b, Kirrberg, Unterneul 1b). However, these taxa, like other ecologically sensitive reptiles, such as chameleons (Laimering 2a) and skinks (Griesbeckerzell 1a), are documented after the Ries event. It is not possible to made a distinction between the herpetofauna before and after the Ries impact event. The extinction of thermophilic taxa took place between the Middle Series and the Younger Series in the Late Badenian and Sarmatian. But at this time we have a sedimentary hiatus in the Molasse Basin (Fig. 4). This extinction was caused by a climatic change during the Middle Miocene (increasing seasonality, decrease of the mean temperature).

4.3 The Mammals (Heissig)

4.3.1 The Mammal Fauna before the Ries Impact

Due to some erosion and relief formation just before the impact event we know the mammalian fauna immediately preceding the Ries impact only from one locality, "Derching 1b" (OSM-unit E', Table 1). This fauna is not yet thoroughly studied, but seems to be similar to the preceding ones, except for one stratigraphically important hamster species, which was replaced by another one (Fig. 5).

Generally, the mammal fauna corresponds to the type of the MN 5 mammal unit, characterised by the lack of the old rodent genera *Melissiodon* and *Ligerimys* and the presence of modern hamsters of the genera *Cricetodon, Democricetodon* and *Megacricetodon*. The numerous small mammal sites, covering most of the times of sedimentation in a rather dense succession, allow a more detailed subzonation than in any other region of Central Europe. It is controlled by superposition.

Within the MN 5 unit, a faunal turnover occurred throughout Western and Central Europe that can be followed in the faunal succession of the Upper Freshwater Molasse of Bavaria and Switzerland. This turnover antedates the Ries impact and is contemporaneous with a stratigraphic gap between the units OSM D and OSM E (Table 1, Fig. 5). Above this gap several immigrants are found, partly replacing older representatives of the same genus or family. The immigrants probably came not exactly at the same time, but the hiatus has condensed the first occurrence of most of them into one line of erosional discordance. Thus, we can observe the first immigrants already before the gap: the flying squirrel *Albanensia*, two dormice of the genera *Myoglis* and *Eomuscardinus* and the small ruminant *Micromeryx*. Most of the immigrations falls within the time of non-sedimentation: the primate *Pliopithecus*, the big hamster *Cricetodon,* the pig *Conohyus* and the deer *Dicrocerus* and *Stehlinoceros* have their first appearances. At the species level the hamster *Eumyarion bifidus* is replaced by the immigrating *Eumyarion medius*. There is one newcomer later than the gap: within the larger-sized lineage of the hamster genus *Megacricetodon* there appears a medium-sized species, *Megacricetodon* aff. *gersi* just a few meters below the „Brockhorizont“ in Derching 1b. It is related to the slower increasing lineage of Western Europe and replaces the very large *Megacricetodon lappi*, the final species of the rapidly increasing *M. bavaricus* lineage of Central Europe, which had survived the main turnover for maybe a hundred thousand years.

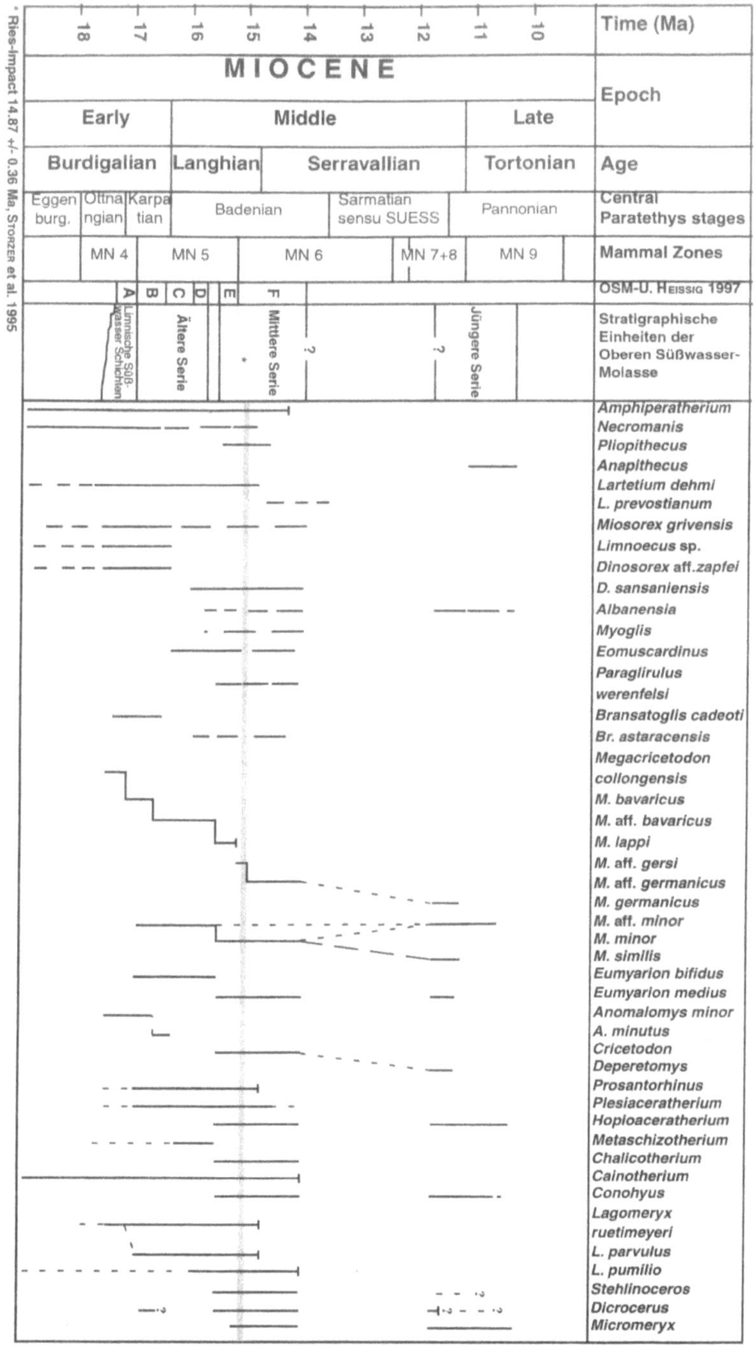

Fig. 5. Stratigraphic distribution of mammalian taxa in the Upper Freshwater Molasse (OSM) of Southern Germany. The grey line indicates the Ries impact event.

4.3.2 *The Mammal Fauna after the Ries Impact*

The first mammal faunas after the impact are found within the first reworked horizon above the autochthonous „Brockhorizont“ (Figs. 1 and 2). This horizon still contains lots of angular blocks and fragments coming from the Ries area, as well as single silicified Jurassic microfossils, which have probably been transported within blocks of residual clay, dissolved in the first process of reworking. This means that the fauna came back into the devastated landscape within a time-span that was too short for a visible weathering of small calcareous debris.

The faunas after the impact are generally put into the mammal unit MN 6, because of the immigration of the rhinocerotid *Hoploaceratherium* at that time, but not because of a real turnover.

There is no faunal element that was wiped out by the catastrophe. Thus, we can conclude that the radius of destruction was less than the area of distribution of all the documented species. This result is corroborated by the fact that even those survivors of old genera, which became extinct rather soon after the impact, had survived over the event for several hundred thousand years. The European opossum *Amphiperatherium frequens* had its last occurrence at the end of MN 6. The smaller short legged rhino, *Prosantorhinus germanicus* still occurs in the first sedimentary cycle after the event, but becomes extinct before the second one. Even *Cainotherium*, a survivor of an old-fashioned praeruminant group of the Oligocene, survived the event for at least two sedimentary cycles. It is recorded also from the calcareous crater sediments of the Ries. *Pliopithecus*, a monkey that lived almost at the northernmost boundary of its range, came just before the event and disappeared from the molasse basin within the second sedimentary cycle. Any climatic deterioration on a larger scale would have wiped out this fastidious animal. Most of the fauna established by the faunal change before the Ries event remains constant during the Middle Series of the Upper Freshwater Molasse, i.e., within MN 6. The more important changes fall in the time of the next stratigraphic gap, spanning the upper part of MN 6, MN 7, and the lower part of MN 8. These units are partly recorded in Switzerland, where the sedimentation was more continuous.

5 Conclusions

The Ries impact event had no long-time effect on the composition of the fauna and the flora in Southern Germany. The radius of destruction was less than the area of distribution of all the documented species. The ecosystem was quickly reconstituted after the catastrophic impact. The probably fast reorganization of the fluviatile ecosystems of the Upper Freshwater Molasse is explained by their dynamic properties. The immigration of the faunal and floral taxa was accelerated by fluviatile drift from refugia. Species with a high colonization ability (plankton, benthos, riparian vegetation, fishes, and some amphibians) were the founders of a new succession.

This model of the Ries and Steinheim impact should be borne in mind when postulating an extinction of organisms by other impact events (e.g., at the Cretaceous/Tertiary boundary). We should be more critical when separating biological from geological events.

Acknowledgement

We want to express our gratitude to August Ilg, Düsseldorf, for important help.

References

Bayerisches Geologisches Landesamt (1964) Erläuterungen zur Geologischen Karte von Bayern 1: 500 000

Bayerisches Geologisches Landesamt (1969) Das Ries. Geologie, Geophysik und Genese eines Kraters. (Eds: Preuss E, Schmidt-Kaler H), Geologica Bavarica 61, 478 pp

Bayerisches Geologisches Landesamt (1970) Exkursionsführer zur Geologischen Karte des Rieses 1:100 000 (by Schmidt-Kaler H, Treibs W)

Bayerisches Geologisches Landesamt (eds) (1977) Ergebnisse der Ries – Forschungsbohrung 1973: Struktur des Kraters und Entwicklung des Kratersees. Geologica Bavarica 77, 470 pp

Blissenbach E (1957) Die jungtertiäre Grobschotterschüttung im Osten des bayerischen Molassetroges. Beihefte Geologisches Jahrbuch H 26: 9-48

Böhme M (1999a) The Miocene Fossillagerstätte Sandelzhausen. 16. The Fish- and Herpetofauna – preliminary results. Neues Jahrbuch Geologie Paläontologie Abhandlungen 214 (3): 487-495

Böhme M (1999b) Doppelschleichen (Sauria, Amphisbaenidae) aus dem Untermiozän von Stubersheim 3 (Süddeutschland). Mitteilungen der Bayerische Staatssammlung für Paläontologie und historische Geologie 39: 85-90

Dehm R (1951) Zur Gliederung der jungtertiären Molasse in Süddeutschland nach Säugetieren. Neues Jahrbuch Geologie Paläontologie Monatshefte 1951: 140-152

Dehm R, Gall H, Höfling H, Jung W, Malz H (1977) Die Tier- und Pflanzenwelt aus den obermiozänen Riessee - Ablagerungen in der Forschungsbohrung Nördlingen 1973. Geologica Bavarica 75: 91-109

Ernst A, Seward AC (1908) The new Flora of the Volcanic Island of Krakatau. Cambridge Univ. Press, 74 pp

Fiest W (1986) Lithostratigraphie und Schwermineralgehalt der Oberen Süßwassermolasse im Bereich um die Gallenbacher Mülldeponien zwischen Aichach und Dasing. Unpublished diploma thesis, University of Munich

Gregor HJ (1977) *Zanthoxylum wemdingense* nov. spec. aus untersarmatischen Riessee-Ablagerungen. Mitteilungen der Bayerische Staatssammlung für Paläontologie und historische Geologie 17: 249-256

Gregor HJ (1980) Zum Vorkommen fossiler Palmenreste im Jungtertiär Europas unter besonderer Berücksichtigung der Ablagerungen der Oberen Süßwasser-Molasse Süddeutschlands. Berichte der Bayerische Botanischen Gesellschaft 51: 135-144

Gregor HJ (1982a) Die jungtertiären Floren Süddeutschlands - Paläokarpologie, Phytostratigraphie, Paläoökologie, Paläoklimatologie. Enke Verlag Stuttgart, 263pp

Gregor HJ (1982b) *Pinus aurimontana* n. sp. - eine neue Kiefernart aus dem Jungtertiär des Goldbergs (Ries). Stuttgarter Beiträge zur Naturkunde B 83: 1-11

Gregor HJ (1983) Die miozäne Blatt- und Fruchtflora von Steinheim am Albuch (Schwäbische Alb). Documenta Naturae 10: 1-45

Gregor HJ (1992) Die Ries- und Steinheimer Meteoriten-Einschläge und ihre Folgen auf die Umgebung in Zeit und Raum. Berichte des Naturwissenschaftlichen Vereins für Schwaben e.V. 96 4: 66-73

Gregor HJ, Hottenrott M, Knobloch E, Planderova E (1989) Neue mega- und mikrofloristische Untersuchungen in der jungtertiären Molasse Bayerns. Geologica Bavarica 94: 281-369

Grimm WD (1965) Schwermineralgesellschaften in Sandschüttungen, erläutert am Beispiel der süddeutschen Molasse. Abhandlungen der Bayerischen Akademie der Wissenschaften, mathematisch-naturwissenschaftliche Klasse NF 121: 1-135

Haller-Probst MS (1998) Die Verbreitung der Reptilia in den Klimazonen der Erde. Courier Forschungsinstitut Senckenberg 203: 1-67

Heissig K (1986) No effect of the Ries impact event on the local mammal fauna. Modern Geology 10: 171-179

Heissig K (1989) Neue Ergebnisse zur Stratigraphie der Mittleren Serie der Oberen Süßwassermolasse Bayerns. Geologica Bavarica 93: 239-257

Heissig K (1997) Mammal faunas intermediate between the reference faunas of MN 4 and MN 6 from the Upper Freshwater Molasse of Bavaria. in: Aguilar JP, Legendre ES, Michaux J (eds) Actes du congrès BiochroM'97. Mémoires travaux EPHE Institute Montpellier 21: 537-546

Jung W (1981) Sind die fossilen Palmenhölzer aus der Oberen Süßwassermolasse Bayerns umgelagert? Berichte der Bayerischen Botanischen Gesellschaft 52: 109-116

Knobloch E, Gregor H-J (2000) Flora Tertiaria Bavariae (The leaf- and fruit-floras from Southern Germany) in preparation

Riederle R, Gregor H-J (1997) Die Tongrube Kirrberg bei Balzhausen - eine neue Fundstelle aus der Oberen Süßwassermolasse Bayerisch-Schwabens - Flora, Fauna, Stratigraphie. Documenta naturae 110: 1-53

Schleich HH (1984) Neue Reptilienfunde aus dem Tertiär Deutschlands. 1. Schildkröten aus dem Jungtertiär Süddeutschlands. Naturwissenschaftliche Zeitschrift für Niederbayern 30: 63-93

Schleich HH (1985) Zur Verbreitung tertiärer und quartärer Reptilien und Amphibien. 1. Süddeutschland. Münchner Geowissenschaftliche Abhandlungen (A) 4: 67-149

Spitzlberger G (1984) Die Rieskatastrophe in ihrer Auswirkung auf die Florengeschichte Mitteleuropas. Naturwissenschaftliche Zeitschrift für Niederbayern 30:173-177

Stiefel J (1957) Ein Beitrag zur Gliederung der Oberen Süßwassermolasse in Niederbayern. Beihefte Geol Jahrbuch H 26: 201-259

Storzer D, Jessberger EK, Kunz J, Lange, JM (1995) Synopsis von Spaltspuren- und Kalium-Argon-Datierungen an Ries-Impaktgläsern und Moldaviten. 4. Jahrestagung Ges. Geowiss., Nördlingen, GGW 195: 79-80

Thornton I (1996) Krakatau – The Destruction and Reassembly of an Island Ecosystem. Harvard Univ. Press Cambridge, MA, United States. 346 pp

Webenau B v (1995) Die jungtertiären Blattfloren der westlichen Oberen Süßwassermolasse Süddeutschlands. Documenta naturae 98:1-147

Appendix

Table A. Distribution of carpofossils in the molasse sediments in Neogene times; the Carpocoenocomplexes (after Gregor 1982 and Gregor et al. 1989), (KZK= Karpozoenosenkomplex *sensu* Webenau 1995)

Fossil taxon	KZK-0	KZK-1	KZK-2	KZK-3	KZK-4	KZK-5
Acanthopanax solutus	----------			---------		
Acer giganteum				---------		
Ailanthus confucii			--------------			--------------
Alnus kefersteini				---------	---	
Asimina brownii						--------------
Betula sp.					---------	
Brasenia victoria	----------		----------	---------	---------	
Calamus daemonorhops	----------		----			
Caldesia cylindrica					---------	
Carex flagellata					---------	
Carpinus grandis					---------	--------------
Carpinus kisseri					---------	--------------
Cedrelospermum aquense			--------------	---------		
Celtis lacunosa		----------	----------	---------	---------	
Cephalanthus kireevskianus		----------	----------		---------	
Chionanthus kornii	----------		----------	---------		
Chionanthus rühlii		----------				
Cladiocarya trebovensis		----------		---------		
Cladium oligovasculare	----------		----------	---------		
Cladium palaeomariscus	----------		----------			
Cleome probstii		----------				
Cordia mettenii	----------	----------	----------	---------		
Coriaria collinsonae		----------				
Cornus brachysepala				---------		--------------
Corylopsis urselensis	----------	----------		---------	---------	
Cyclocarya cyclocarpa				---------		
Decodon globosus	----------		----------	---------	---------	
Eoeuryale moldavica		----------				
Eomastixia sp.	----------	----------				
Epipremnum cristatum			----------			
Epiprmnum ornatum				---------	---	
Eurya stigmosa	----------		----------			
Fagus sp.						--------------
Gleditsia knorrii				---------		
Glyptostrobus europaeus	----------	----------	----------	---------	---------	--------------
Hartziella rosenkjaeri		----------		---------		

Table A. Continued.

Hartziella vindobonensis					---------	
Hemitrapa heissigii				---------		
Koelreuteria macroptera				---------		
Leguminocarpum sp.				---------		
Limnocarpus eseri				---------		
Limnocarpus major				---------		
Liquidambar europaea	-----------			---------	---------	-------------
Liquidambar magniloculata	-----------			---------	---------	
Ludwigia ungeri						-------------
Microdiptera parva		-----------				
Mneme menzelii					---------	
Myrica ceriferiformis			-----------	---------	---------	
Myrica stoppii	-----------	-----------	-----------			
Nymphaea alba foss					---------	
Nyssa ornithobroma	-----------	-----------	-----------	---------	----	
Olea moldavica		-----------		---------		
Ostrya scholzii		-----------		---------	---------	-------------
Paliurus thurmannii					---------	
Passiflora heizmannii		-----------				
Pinus aurimontana				--		
Pinus tomasiana	-----------	-----------	-----------	---------	---------	
Polygonum leporimontanum		-----------	-----------			
Populus sp.				---------		
Potamogeton piestanensis					---------	-------------
Potamogeton schenkii		-----------	-----------			
Proserpinaca reticulata				---------	---------	
Pterocarya sp.				---------	---------	
Quercus sapperi					---------	-------------
Rhus cf. toxicodenron			-----------			
Ruppia maritima-miocenica				---------		
Ruppia palaeomaritima				---------		
Salix sp.				---------		
Sambucus pulchella					---------	
Sambucus pusilla		-----------	-----------			
Sapindoidea margaritifera	-----------	-----------	-----------	------		
Sapium germanicum	-----------		-----------	---------		
Saururus bilobatus			-----------			
Schizandra moravica		-----------		---------		
Spirematospermum wetzleri	-----------	-----------	-----------	---------	---------	
Spondieaemorpha dehmii	-----------	-----------		--		

Table A. Continued.

Swida gorbunovii					----------	----------
Symplocos lignitarum	----------				----------	
Symplocos pseudogregaria	----------	----------	----------			
Taxodium hantkei					----------	
Tilia praeplatyphyllos		------				
Toddalia latisiliquata	----------			----------		
Toddalia thieleae	----------	----------	----------			
Toddalia maii	----------	----------	----------			
Ulmus sp.				----------		----------
Umbelliferopsis molassicus			----------			
Vitis teutonica	----------					----------
Zanthoxylum ailanthiforme	----------	----------				
Zanthoxylum wemdingense				----------		
Zanthoxylum ailantiforme	----------					
Zanthoxylum giganteum	----------					
Zanthoxylum müller-stolli		----------				

Table B. Phyllocoenocomplexes of the molasse region in time, connected with the impacts of the Ries- and Steinheim meteorites (PZK=Phyllozoenosenkomplex after Webenau 1995 and Gregor et al. 1989)

Fossil taxon	PZK 1	PZK 2a	PZK 2b	PZK 3a	PZK 3b	PZK 4a	PZK 4b
Acer palaeocacharinum						----------	
Acer tricuspidatum		----------	----------	----------	----------	----------	----------
Alnus ducalis							----------
Alnus julianaeforms		----------	----------				
Berchemia multinervis		----------	----------	----------	----------	----------	
Betula subpubescens							----------
Carpinus grandis						----------	----------
Celtis begonioides					----------	----------	
Daphnogene bilinica	----------	----------	----------	----------	----------		
Daphnogene polymorpha	----------	----------	----------	----------	----------		
Fagus attenuata							----------
Ginkgo adiantoides					----------	----------	
Gleditsia lyelliana		----------	----------		----------		
Liquidambar europaea				----------	----------	----------	----------
Monocotyledoneae	----------	----------	----------				

Table B. Continued.

Myrica sp.		-----------	-----------	----------			
Parrotia pristina		-----------			-----------		
Persea princeps		-----------				-----------	
Platanus leucophylla			-----------	----------	-----------	-----------	-----------
Populus balsamoides	-----------	-----------	-----------	----------	-----------	-----------	-----------
Populus mutabilis			-----------				
Populus populina		-----------	-----------	----------	-----------		-----------
Quercus cruciata	-----------						
Quercus ex gr. Kubinyi					-----------	-----------	-----------
Quercus gregori							-----------
Quercus pseudocastanea					-----------	-----------	-----------
Salix angusta	-----------	-----------	-----------	----------	-----------	-----------	-----------
Salix lavateri	-----------	-----------	-----------	----------	-----------	-----------	
Sapindus falcifolius		-----------	-----------	----------	-----------	-----------	
Smilax sagittifera		-----------				-----------	
Tilia atavia	---						
Tilia sp.				-------			
Ulmus pyramidalis		-----------	-----------	----------	-----------	-----------	-----------
Ulmus minuta		-----------		----------	-----------	-----------	-----------
Ulmus ruszovensis					-----------		-----------
Viscum morlottii	-----------					-----------	
Zelkova ungeri					-----------	-----------	
Zelkova zelkovaefolia		-----------		----------	-----------	-----------	-----------
"Palms"	---						

Table C. Palynomorphs in their time distribution in the molasse sediments (after Planderova in Gregor et al. 1989)

Taxa	Ottnang	Karpat	Baden	Sarmat	Pannon
Alnipollenites verus	-------------	-----------	-----------	-----------	-----------
Arecipites wiesenensis	-------------	-----------			
Caryapollenites ssimplex	-------------	-----------	-----------	-----------	
Engelhardtioidites microcorypheus	-------------	-----------	-----------	-----------	
Ephedripites treplinensis	-------------	-----------	-------		
Ginkgoretectina neogenica	-------------	-----------	-----------	-----------	-----------
Graminidites media	-------------	-----------	-----------	-----------	-----------
Liquidambarpollenites styraciflua eformis	-------------	-----------	-----------	-----------	-----------
Magnolipollis neogenicus			-----------		
Myricipites rurensis	-------------		-----------		
Pinus haploxylon typus	-------------	-----------	-----------	-----------	-----------
Pityosporites cedrisacciformis	-------------	-----------	-----------	-----------	-----------
Porocolpopollenites vestibulum	-------------	-----------	-----------		
Pterocaryapollenites stellatus	-------------	-----------	-----------	-----------	-----------

Table C. Continued.

Quercoidites petraea			------------	------------	------------
Rhoipites pseudocingulum	------------		------------		
Salixipollenites div. spec.			------------	------------	------------
Sapotaceoidaepollenites microrhombus	------------	------------	------------		
Sciadopityspollenites serratus	------------		------------	------------	------------
Sequoiapollenites polymorphosus	------------	------------	------------	------------	------------
Trapa sp.			------------	------------	------------
Tricolpopollenites liblarensis	------------	------------	------------		
Tricolporopollenites cingulum	------------	------------	------------	------------	------------

Radiation Effects of the Chicxulub Impact Event

Valery V. Shuvalov

Institute for Dynamics of Geospheres, Leninsky pr. 38, bld. 6, Moscow 117979, Russia. (valery@vshuvalov.mccme.ru)

Abstract. Numerical simulations of all stages of the Chicxulub impact event (including flight through the atmosphere, cratering and plume rising) are used to calculate the radiation impulse on the Earth's surface and to estimate the area of possible wildfires. The results show that direct plume radiation could be responsible for ignition of fires in an area of about 3-10 % of the total Earth's surface. Some other possible mechanisms of global wildfires origin are also discussed.

1 Introduction

A link between large impacts and global mass-extinctions seems to be highly probable now. At least a large body of evidence suggests the possibility of a link between the Chicxulub impact and the biota mass-mortality at the Cretaceous-Tertiary (KT) boundary (Alvarez et al. 1980; Smit 1996). Nevertheless, the mechanisms of the impact-linked mass extinctions are still poorly known. The light impulse and the following global wildfires resulting from radiation have been considered as a probable reason for global changes in the biosphere (Toon et al. 1997). This radiation is emitted both during the entering flight of the impactor and by a hot air-vapor cloud, or plume, expanding through the Earth's atmosphere just after the impact. Moreover, Melosh et al. (1990) proposed that global wildfires would be caused even by the radiation generated due to ejecta re-entry.

The first estimates of the radiation produced in impact events were based on an analogy between impacts and nuclear explosions (Nemtchinov and Svetsov 1991). However, recent investigations (Boslough and Crawford 1997; Shuvalov 1999a) showed that there is a great difference between an atmospheric "meteor" and nuclear-like explosions. The cratering process adds yet another distinction. In this study I used numerical simulations of the main stages of the Chicxulub impact (including the flight through the atmosphere, cratering and plume expansion) to calculate the radiation impulse on the Earth's surface and to estimate the area of wildfire ignition.

2
Statement of the Problem and Numerical Techniques

A vertical impact of a stony asteroid 5 km in radius impacting at a velocity of 20 km/s against a granite target was modelled with the use of SOVA, a multi-dimensional multi-material hydrocode (Shuvalov 1999b). In some sense this code is similar to the hydrocode CTH (McGlaun et al. 1990), which is widely used in the USA.

SOVA was developed to study multi-dimensional, large deformation gasdynamic flows, with accurate description of the boundaries between different materials (e.g., vapor, air, soil, etc.). A two-step Eulerian solution scheme is exploited to integrate the problem through time. The two-step procedure is more suitable for multi-material problems. In the first step the Lagrangian forms of hydrodynamic equations are solved, then the distorted cells are remapped back to initial mesh (or some new mesh changing in time). At the beginning of the remap step a boundary (or boundaries) between different materials are constructed. The process of multi-dimensional equation integration is simplified by using operator splitting techniques which replaces the multi-dimensional equations with a set of one-dimensional equations. This techniques is used both in the Lagrangian and remap steps.

Non-uniform computational mesh changing in time consists of 300×1000 cells with 20 cells across initial impactor radius.

Detailed tables of thermodynamical and optical properties of air (Kuznetsov 1965; Avilova et al. 1970) and H-chondrite vapor (Kosarev et al. 1996) were used to compute the temperature and density distributions, as well as the radiation fluxes on the Earth's surface.

The simulations do not include material strength; therefore, the stress tensor is considered to be spherical. This, however, does not affect the results of the simulations since the strength of the target becomes important only at the late stage of crater formation (Melosh and Ivanov 1999). During that stage a non-evaporated cold target substance is ejected at a rather low (less than 1 km/s) velocity, and does not contribute to the plume radiation.

The radiation efficiency (i.e., a ratio of the emitted radiation energy to the initial kinetic energy) of large impacts in the presence of an atmosphere is of the same order of magnitude as for impacts in vacuum, because the atmospheric mass involved in the process is negligible in comparison to the impactor mass. The value of radiation efficiency in vacuum was found to be very small, 10^{-5}-10^{-3} (Nemtchinov et al. 1998). As a result, the radiation can not considerably influence the impact hydrodynamics and can be ignored in the numerical simulations of the large impact events. At high altitudes, however, the radiative cooling of the hot, rarefied air and vapor can become significant. To account for this cooling a volumetric approximation of radiation transfer was used:

$$\frac{\partial e}{\partial t} = -4k\,B \cdot \exp\left(-\frac{k}{\rho H}\right) \tag{1}$$

where e is the specific thermal energy, k is average mass absorption coefficient, B is the Plank function, ρ is the gas density, and H is the atmospheric scale. The term $\exp(-k/\rho H)$ determines the portion of radiation emitted from the plume.

The temperature and density distributions obtained in the course of numerical simulations are used to calculate radiation intensities I_ε and radiation fluxes q on the ground surface at different distances from the impact site at different moments of time. The value of q is determined from the equation

$$q = \int\int I_\varepsilon(\Omega)\vec{i}\,\vec{n}\,d\Omega\,d\varepsilon, \tag{2}$$

where Ω is the space angle, $\vec{n}$ is the unit vector normal to the radiated surface, and i is the unit vector directed along the ray (Zel'dovich and Raizer 1967). To calculate the integral (2) at some point **X** of the Earth's surface, the equation of radiation transfer

$$\frac{\partial I_\varepsilon}{\partial s} = \rho \quad k_\varepsilon(B_\varepsilon - I_\varepsilon) \tag{3}$$

was integrated along a large number (10^3–10^4) of rays (crossing radiating volume and passing through the point **X**) and for a large number (about 10^3) of photon energies ε. Here s is the distance along the ray, k_ε is the spectral mass absorption coefficient, and B_ε is the Planck function.

3
Results of Numerical Simulations

The numerical simulations of impactor penetration into the Earth's atmosphere start at an altitude of 400 km. Figure 1 shows density and temperature distributions just before the collision with the ground surface. A rarefied wake forms behind the falling body, where the density is 1–2 orders of magnitude lower than the ambient air density. The temperature of the shock compressed (and heated) air reaches values of 30–40 kK close to the impactor. The shock wave radiates intensively and at the surface it can cause ignition of woods within an area of tens of kilometers. However, the calculations show that this radiation is considerably less than thermal radiation emitted by the vapor cloud (plume) at the late stage of the impact.

Figure 2 shows temperature and density distributions 5 seconds after the impactor reaches the ground surface. There is a clear difference between a contact nuclear-like explosion and an impact event. Two main reasons for this difference can be identified. The first one is connected to energy release. At the moment of the collision shock waves are generated both in the impactor and in the target. The shock compressed and heated mass initially moves mainly downwards. As a result the effective energy source is below the ground surface. After being decelerated, the hot substance begins to expand upwards, but this expansion is restricted by the cavity walls and the conic curtain formed by the dense ejecta from the cavity boundaries. As a result an upward directed jet (plume) is formed, instead of an isotropic flow, which is common result of point-like explosions.

In the case of a dense atmosphere and a moderate-size impactor this jet is decelerated quickly and it transfers its kinetic energy to the ambient atmospheric gas (Nemtchinov and Shuvalov 1992). This makes the flow isotropic again. However, in the case of a large impactor, like the Chicxulub case considered here, the atmospheric drag is small, and a great portion of vapor and condensed material is ejected to high altitudes.

The second distinction between a cosmic body impact and nuclear-like explosion results from the influence of the meteoroid wake (Artem'eva and Shuvalov 1994). This effect was found to be crucial for "meteoroid explosions" caused by small cosmic bodies burning in planetary atmospheres (Boslough and Crawford 1997; Shuvalov 1999a). In the case under consideration the mass ejected in the impact (equalled to tens of impactor masses) is much greater than the air mass within the cone of ejection. Therefore, the main portion of ejecta does not sense the presence of the atmosphere, and much less the presence of the wake in the atmosphere. However, the leading, very hot, and rarefied, part of the plume is affected by the atmosphere and it experiences a considerable atmospheric drag, expanding preferentially within the rarefied wake. In some sense the wake is like an empty tube through which the hot rarefied vapor moves to high altitudes.

The vapor temperature 5 s after the impact equals 5–7 kK. The vapor expansion generates a blast wave in the atmosphere. This wave propagates preferentially along the wake as well. The air temperature behind the upward directed shock reaches a value of about 20–30 kK. It is this part of the plume which emits the main portion of the radiation at this time. The upper (not shock compressed) part of the wake is rather cold (4–5 kK).

The next phase of the plume evolution is shown in Fig. 3 which represents the plume 15 s after impact. The blast wave accelerates in the low density upper atmosphere and goes far from the vapor cloud. The acceleration leads to an increase of the shock compressed air temperature. At high altitudes (above 100 km) the atmospheric density becomes very small, and the vapor and air lifted from below begin to expand also horizontally, and a large cap of compressed air arises above the impact site. However, the radiation emitted by the air decreases because of the high transparency of air at these low densities. It is well known that at high altitudes even the equilibrium (undisturbed) air temperature is around 1.4–1.5 kK, but because of low density and high transparency of the air there is no detectable radiative flux to the Earth's surface. As a result the main source of radiation at the late stage of the impact is the expansion vapor plume, whose size increases considerably with time. The average plume temperature remains almost constant (close to the temperature of phase transition) for a long time due to the energy release resulting from a continuing condensation. The top boundary of the radiating volume is controlled by adiabatic and radiative cooling of the vapor and the decrease of its density and optical thickness. Figure 3d shows a view of the impact plume from the side by displaying the intensity shading of the brightness, expressed in terms of effective temperature T_{eff}. The value of T_{eff} is defined from the relation:

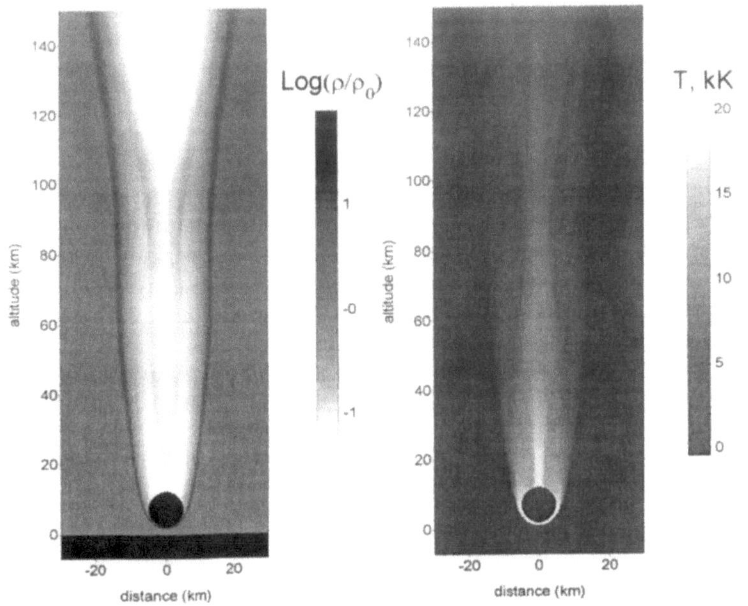

Fig. 1. Relative density (ρ/ρ_0) and temperature (T) shading just before the collision of a stony body 5 km in radius with the ground surface. $\rho_0(h)$ is the equilibrium atmospheric density at an altitude h.

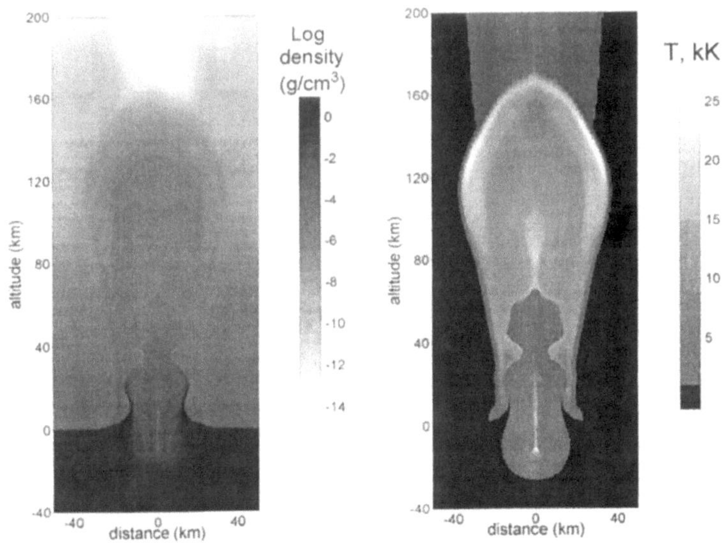

Fig. 2. Density and temperature shading 5 s after the impact of a stony body 5 km in radius.

$$\sigma T_{eff}^4 = I, \tag{4}$$

where σ is the Stefan-Boltzmann constant and

$$I = \int_0^\infty I_\varepsilon d\varepsilon$$

is the total radiation intensity along the horizontal line crossing the plume. The source image is obtained without taking into account the absorption of light by the cold undisturbed atmosphere. In other words, it does not depend on distance of observation. The effective temperature is low at low altitudes (below 30 km), because this part of the plume is screened by a dense cold material ejected from the boundary of the growing cavity (dust curtain).

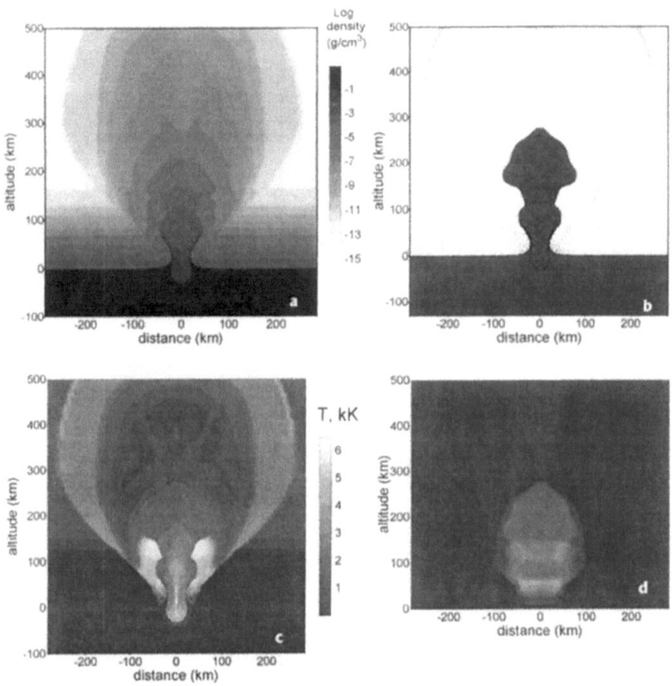

Fig. 3. Density (**a**), material (**b**), temperature (**c**) and effective (radiative) temperature (**d**) distributions 15 s after the impact of a stony body 5 km in radius. In (**b**) the black area designates the impactor and surface material (both solid and vapor).

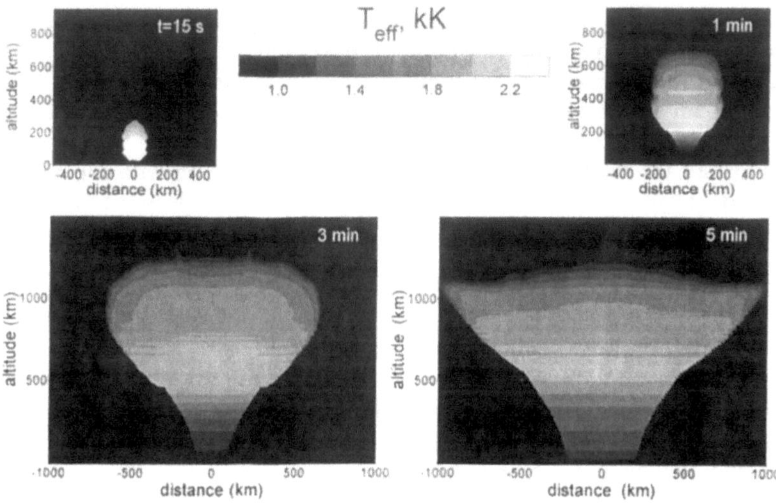

Fig. 4. Radiation source images (effective temperature shading) at various stages of the plume evolution.

The value of T_{eff} varies in a rather narrow range from about 1.5 kK to 2.5 kK. This temperature is close to the temperature of condensation. Within several seconds after the collision and during the impactor flight through the atmosphere a fireball is much brighter (about 20–30 kK), but it is very small. In addition the duration of this bright phase is very short compared to the total emission time. Therefore, most of the radiation is emitted at a later stage, 1–5 min after the collision. The source images (i.e., effective temperatures) at that time are shown in Fig. 4. Almost all the radiation is emitted by the impactor and ground vapor and/or two-phase vapor-solid mixture. The radiation of the shock compressed air is negligible compared to the total radiation emitted, although the brightest initial peak results mainly from the air emission.

It follows from the numerical simulations that a rough estimate of the radiation impulse can be obtained by assuming that the plume temperature equals about 2 kK, its characteristic scale is around 1 km, and the height of the top boundary is 1.2 km. The time scale for light emission is 5–10 min. The results of direct calculations are shown in Fig. 5.

The radiation intensity at the Earth's surface depends on several factors, such as geometrical divergence, curvature of the Earth's surface and its topography, absorption by the ambient (cold) atmospheric gas, absorption and scattering by aerosol particles suspended in the air, etc. Among these factors the most undefined one is aerosol absorption and scattering. Typical estimates of the radiation free path L_r in the atmosphere are around 40 km for very good sunny days and 10 km for rainy weather conditions (Glastone and Dolan 1977). Figure 5 shows the time-integrated incident radiation energy per unit surface versus distance from the impact site. In this calculation it is assumed that the surface being radiated is optimally oriented (from the viewpoint of maximum radiation). A critical value of the incident radiation energy for ignition is estimated to be around 20–100 J/cm^2 depending on forest type, moisture and other conditions (Glastone and Dolan 1977).

Almost all the radiation is emitted in the infra-red region with exception of the short initial peak. The maximum of the radiation spectra falls to 0.5–0.7 eV.

4 Discussion

The results presented above show that wildfires could arise at distances of 2 000–3 000 km from the impact cite. Twofold increase or decrease of the impactor size can extend this range to 1 500–4 000 km.

Some peculiarities can be induced by the angle of impact. Recent investigations (Pierazzo and Melosh 1998, 2000) have shown that the impact angle can influence considerably the fate of the projectile and the degassing of the sedimentary layer. However, numerical simulations of light flashes generated by meteoroids impacting the Moon (Nemtchinov et al. 1998) have shown that the difference in the total radiation impulse between a vertical and an inclined impact does not exceed a factor of two. The radiation is larger for the vertical impact because the larger part of impactor kinetic energy is transformed into the heat. Such variations can not change an area of forest ignition significantly.

On the other hand, it follows from the results of numerical simulations, that the size R of the radiation source is controlled by the vapor rarefaction and can be roughly estimated as

$$R = \left(\frac{M_{ejecta}}{\rho_*} \right)^{1/3}, \tag{5}$$

where M_{ejecta} is the mass of hot ejecta, and ρ_* is the density crucial for radiation (around 10^{-9} g/cm^3). The relation (5) also suggests a weak dependence of radiation on the impact angle. Nevertheless, these speculations should be verified by direct 3D numerical simulations.

It should be noticed that in the vicinity of the crater (at a distance of several crater radii) all the ground surface will be covered by a thick layer of ejected soil. Most likely, this could prevent the forests (and other biomass) from total burning.

One more factor which could influence the expansion plume evolution is the target lithology. The presence of a volatile-rich sedimentary layer can significantly affect the outcome of the impact and environmental perturbations (Toon et al. 1997). A large amount of gases resulting from the outgassing of the sedimentary layer can produce a stronger initial expansion of the plume. However, at the late stage (several minutes after the impact, when the most part of radiation is emitted) the size and temperature of the light source are defined mainly by impactor and soil vapor (partially condensed) because of low mass concentration of other constituents.

To summarise, this work suggests that the direct radiation induced by the Chicxulub impact expansion plume could be responsible for the ignition of wildfires over about 3–10% of the Earth's surface, close to the impact point. This is an upper estimate because this area was partially covered by water.

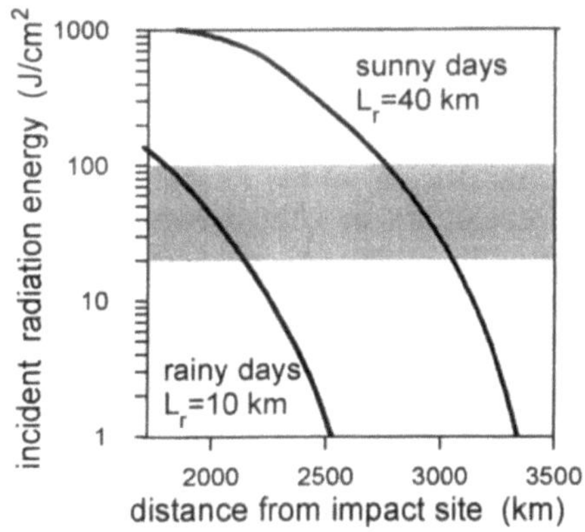

Fig. 5. Incident radiation energy versus distance from the impact site, using the values of Glastone and Dolan (1977) values for the radiation free path L_r in the atmosphere (see text).

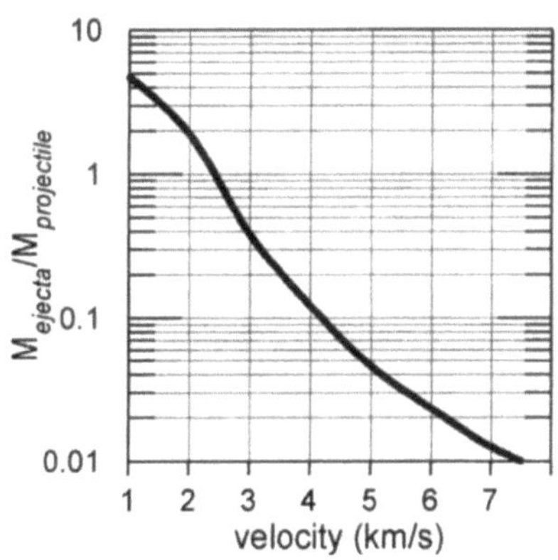

Fig. 6. Cumulative velocity distribution of ejected mass normalised to impactor mass.

Melosh et al. (1990) assumed that global wildfires could be caused by ejecta re-entry. In their calculations the fallback of the impact created debris was treated like a large number of small meteoroid impacts. However, these speculations were based on rather arbitrary assumptions of the ejecta velocity distribution. In the present study this distribution was a direct result of numerical simulations of the impact. A simple consideration shows that all the mass ejected with velocities less than 5 km/s falls within the area affected by direct plume radiation. The proportion of high velocity ejecta (velocity exceeding 5 km/s) appears to be rather small — around a few percent of the impactor mass (see Fig. 6). Melosh et al. (1990) assumed that the mass of ejecta with velocities 5 to 10 km/s was about $5 \cdot 10^{15}$ kg (i.e., 3–5 times larger than the mass of 10 km stony impactor). Under this assumption the average irradiance from the re-entering ejecta was near the lower limit of that required for ignition of solid wood. It follows from the present investigation that the mass of high velocity ejecta (about 10^{14} kg) is almost by two orders of magnitude less.

Therefore, in my opinion the process of ejecta re-entry is probably not responsible for global wildfires. However, more detailed investigation is needed. In particular, the interaction of the re-entering ejecta aerosol cloud with the Earth's atmosphere probably can not be described as a sum of independent microimpacts (airbursts). Correspondingly, the irradience from the re-entering ejecta can not be estimated to be a sum of independent micrometeors (shooting stars).

Another scenario may be considered to explain a great amount of soot which was produced after the KT impact (Wolbach et al. 1990). The local wildfires and other mechanisms detailed in (Toon et al. 1997) could lead to a global mortality of forests and other plants. Dead forests are known to be much more subject to ignition than living forests. Moreover, in dry forests the fraction of totally burned mass of wood increases considerably (several times) leading to an increase of the total mass of soot produced. The ignition of dead forests could result from the natural thunderstorm activity enhanced by strong atmospheric perturbations due to the impact and its consequences. There is no evidence of instant global fires and extinction. The period (or periods) of wildfires could continue for several years (or even more). In other words, a great amount of soot could be produced by a large number of local wildfires (maybe not instantaneous) in the dead forests. In this case the global mortality of forests was not the result of global wildfires, but on the contrary, wildfires themselves and KT soot could result from the forest mortality.

Acknowledgements

The author would like to thank C. Koeberl, E. Pierazzo and M. Boslough for valuable remarks and I. Trubetskaya for the assistance with preparing this paper.

References

Alvarez LW, Alvarez W, Asaro F, Michel HV (1980) Extraterrestrial cause for the Cretaceous-Tertiary extinction. Science 208: 1095–1108

Artem'eva NA, Shuvalov VV (1994) Oblique impact: Atmospheric effects [abs]. Lunar and Planetary Science XXV: 39-40

Avilova IV, Biberman LM, Vorob'ev VS, Zamalin VM, Kobsev GA, Lagar'kov AN, Mnatsakanian AK, Norman GE (1970) Optical properties of the hot air. Nauka, Moscow (in Russian)

Boslough MB, Crawford DA (1997) Shoemaker-Levy 9 and plume-forming collisions on Earth. In: Remo JL (ed) Near-Earth Objects. New York Academy of Sciences, New York, pp 236-282

Glastone S, Dolan P (1977) The effects of nuclear explosions. 3rd ed. US Government Printing Office, Washington DC, 653 pp

Kosarev IB, Loseva TV, Nemtchinov IV (1996) Optical properties of vapor and ablation of large chondritic and icy bodies in the Earth's atmosphere. Solar System Research 30, 4: 265-278

Kuznetsov NM (1965) Thermodynamic Functions and Shock Adiabats for air at High Temperatures. Mashinostroyenie, Moscow (in Russian)

McGlaun JM, Thompson SL, Elrick MG (1990) CTH: a three-dimensional shock wave physics code. International Journal of Impact Engineering 10: 351-360

Melosh HJ, Ivanov BA (1999) Impact crater collapse. Annu Rev Earth Planet Sci 27: 385-425

Melosh HJ, Schneider NM, Zahnle KJ, Lathan D (1990) Ignition of global wildfires at the Cretaceous/Tertiary boundary. Nature 343: 251-254

Nemtchinov IV, Svetsov VV (1991) Global consequences of radiation impulse caused by comet impact. Adv Space Rev 11: 95-97

Nemtchinov IV, Shuvalov VV (1992) Explosion dynamics in meteor impacts on the surface of Mars. Planet Space Sci 26, 4: 19-31

Nemtchinov IV, Shuvalov VV, Artem'eva NA, Ivanov BA, Kosarev IB, Trubetskaya IA (1998) Light flashes caused by meteoroid impacts on the lunar surface. Solar System Research 32, 2: 99-114

Pierazzo E, Melosh HJ (1999) Hydrocode modeling of Chicxulub as an oblique impact event. Earth Planet Sci Lett 165: 163-176

Pierazzo E, Melosh HJ (2000) Hydrocode modeling of oblique impacts: The fate of the projectile. Meteoritics Planet Sci 35:117-130

Pittock AB, Ackerman TP, Crutzen PJ, MacCracken MC, Shapiro CS, Turco RP (1986) Environmental Consequences of Nuclear War 1: 320 pp

Shuvalov VV (1999a) Atmospheric plumes created by meteoroids impacting the Earth. J Geophys Res 104: 5877-5890

Shuvalov VV (1999b) 3D hydrodynamic code SOVA for interfacial flows, application to thermal layer effect. Shock Waves 9: 381-390

Smit J (1996) The K/T boundary Chicxulub impact event: a review [abs]. Abstracts of International workshop: The role of impact processes in the geological and biological evolution of planet Earth. Postojna, Slovenia: p 83

Toon OB, Turco RP, Covey C (1997) Environmental perturbations caused by the impacts of asteroids and comets. Reviews of Geophysics 35: 41-78

Wolbach WS, Gilmour I, Anders E (1990) Major wildfires at the Cretaceous/Tertiary boundary, In: Sharpton VL, Ward P (eds) Global Catastrophes in Earth history. Geological Society of America Special Paper 247: 391-400

Zel'dovitch YaB, Raizer YuP (1966) Physics of Shock Waves and High-temperature Hydrodynamic Phenomena. Academic Press, New York, 678 pp

Extraterrestrial Material Deposition after Impacts into Continental and Oceanic Sites

Natalia N. Artemieva and Valery V. Shuvalov

Institute for Dynamics of Geospheres, Leninsky pr., 38, bld.6, Moscow 117979, Russia. (nata_art@mtu-net.ru)

The search for meteoritic material at large, ancient craters is only slightly more rewarding than the search for the Loch Ness Monster (Palme et al. 1978)

Abstract. The idea that the bulk of the projectile is ejected from the crater with high velocity at the initial stage of impact and then is deposited outside the crater, is tested by direct numerical modeling. Simulations are performed for vertical and oblique impacts both on land and into the ocean. The results show that an appreciable amount of the melted projectile is conserved in the crater only in the case of a low impact velocity (~10 km/s) impact. The possible role of target lithology (e.g., the presence of volatiles), as well as special characteristics of the projectiles (iron), are also discussed.

1 Introduction

Accretion of extraterrestrial material onto the Earth ranges from everyday falls of 10^5 kg of cosmic dust and small meteoroids to a catastrophic impact of a 10^{15} kg asteroid once in about 10^8 years. Impact signatures on the Earth are summarized in Table 1 according to Grieve (1982).

Identification of the projectile material is based on the enrichment in siderophiles (Ni and Ir are the most popular elements). They are sensitive indicators for meteorite contamination on the Earth, because crustal rocks are depleted in these elements relative to average solar composition. For large impact structures geological facts are as follows: ballistically emplaced unmelted ejecta that dominate near-crater ejecta deposits are poor in traces of projectile (Hörz 1982; Grieve 1982); tektites, which are associated with high-temperature melted ejecta and form strewn fields, also contain very little of projectile material (<0.1 wt%), and their compositions strongly resemble that of crustal target rocks (Koeberl 1990); the relative abundance of meteoritic material in impact melt rocks with well-characterized siderophile signatures is about 1% or less, with the East Clearwater crater being an exception with 7.8% (Palme et al. 1978). Detectable amounts of meteoritic material allow, in some cases, to define the projectile type, whereas an absence of meteoritic signatures may be due to a specific type of meteorite (eucrites) or high impact velocity (Palme et al. 1978).

Laboratory experiments (Schultz and Gault 1990; Evans et al. 1994; Schultz 1996) are very important, but are strongly constrained by the low value of impact velocity (less than 7 km/s). Therefore, they do not reproduce valid ratios of vapor/melt/solid ejecta and usually demonstrate an extremely high content of impactor material in the ejecta. Numerical experiments, based on the real lithology of impact sites, may be an important instrument in modern geophysics to achieve a better understanding of impact cratering.

Table 1. Flux of cosmic material onto the Earth

	Flux	Records
Cosmic dust and micrometeorites	~10^5 kg/day	Stratosphere Deep sea sediments
Meteorites	~100 kg/year	Impact pits and impact craters D <100 m
Large meteorites	~10^8 kg/100 kyr	Simple craters D = 100–1000 m
Asteroids	~10^{12} kg/Myr	Complex craters and impact structures D >1 km

2 Numerical Simulations of the Impact

The hypothesis that the bulk of the projectile material is presumably contained in the early-ejected high-speed part of ejecta, widely dispersed, and apparently lost from the crater area, is tested here through direct numerical simulations. Several probable scenarios for the impact of a 1-km-diameter stony asteroid with a velocity of 10–40 km/s are considered. Some estimates for the case of an iron projectile are also included.

2.1 Dimensionless Parameters of the Cratering Flow

Some dimensionless ratios are commonly used in hydrodynamics to evaluate the importance of different physical processes. For impact cratering modelling, two ratios are important (Melosh 1989). The first one is the Cauchy number – the ratio between inertial stress and material strength: $Ca = \rho U^2/Y$. Here ρ and Y are the target material density and strength; respectively, U is the impact velocity. Material strength effects play a significant role in the cratering flow when this number is of the order of 1 or less. The importance of gravity is gauged by the

Frude number, which is defined as the ratio of inertial and gravitational stresses: $Fr = U^2/gD$. Here g is the gravity acceleration and D is the crater diameter. Gravity modifies the cratering flow when $Fr < 1$. The ratio of these two numbers, that is $\rho gD/Y$, marks the boundary between two different cratering regimes. When this ratio is less than unity — strength dominates excavation, otherwise the gravity is a more significant factor. For the purposes of the present investigations, one more criterion is of great importance, namely, the melt (or vapor) number: $H = U^2/H_m$, where H_m equals melt (or vapor) enthalpy of a rock. This value of H_m varies from 15 to 35 kJ/g for various types of naturally occurring rocks. The magnitudes of the Ca, Fr, and H_m numbers are measures of the dominant mechanisms controlling the cratering process.

Thus, for the impacts under consideration, gravity is dominant, although rock strength also plays an important role. Complete melting and considerable vaporization occurs for all impact velocities in the range from 10 to 40 km/s.

2.2 Hydrocode and Thermodynamics in Use

We used the SOVA hydrocode (Shuvalov 1999), which allows to model high-velocity vertical and oblique impacts with accurate reconstruction of the boundaries between distinct substances (impactor, target, atmosphere) and proper conservation of mass, momentum and energy.

The effect of the computational mesh resolution in the determination of the amount of melt/vapor was studied by Pierazzo et al. (1997). The lowest resolutions, 5 and 10 cells per projectile radius (cppr), result in artificially small, by a factor of 2, estimates of vapor/melt production in the case of low-velocity (10–20 km/s) impact. High resolution (40 cppr) produces runs with unreasonably long time for calculations. Accordingly, 20 cppr are usually used for 2D numerical simulations and 10 cppr for 3D simulations.

The material strength is considered via a simplest approximation using rigid-plastic model (Hill 1950). In the framework of this model, each component of the stress tensor is considered to be proportional to the strain rate. Assuming that the total deformation obeys the plastic criteria, the stress tensor components may be expressed through the velocity gradients and plastic limit of the material. The plastic limit Y depends on the pressure P, temperature T and the velocity of deformation: $Y = (Y_0 + kP)(1 - T/T_m)$. This relationship reflects the growth of the strength with increasing pressure (for natural rocks Y_0 is about 0.1 GPa and $k = 1$), and the decrease of the strength with temperature increase. Obviously, the strength approaches zero when the melting point is approximated ($T = T_m$).

To describe material thermodynamics (i.e., pressure and temperature dependencies on density and internal energy) analytical equation of state ANEOS (Thompson and Lauson 1972) is used both for the target and the projectile. In this equation of state (EOS) pressures, temperatures, and densities are derived from the Helmholtz free energy and are thermodynamically self-consistent. The main disadvantage of ANEOS for granite (Pierazzo et al. 1997) is that the heat of complete vaporization (about 20 kJ/g) is much higher than the real one (about 7 kJ/g). This EOS also does not describe melting and vapor chemistry. Therefore, we can only estimate the amount of melt within the crater through material temperature and density or through the value of maximum pressure in each lagrangian volume. To follow a motion of the lagrangian volumes we use tracers, that is, passive mass-less particles that move through the mesh with local velocity at the current tracer position. Up to 20 000 tracer particles are regularly distributed in the projectile. The tracers have no effect on the flow field, but allow to follow the thermodynamic history of some given material point in time. The decompression of matter from the shocked state can be approximated as an isentropic (thermodynamically reversible) process and shock pressures, required for melting of any material, are obtained by the intersection of the Hugoniot with the corresponding isentropes for melting. Values of shock pressures for incipient and complete melting at 1 bar are extracted from the literature (Chase et al. 1985; Ahrens and O'Keefe 1972; Pierazzo et al. 1997). They are about 60 GPa for stone melting and about 390 GPa for iron melting.

3 Results of Numerical Simulations

3.1 Impacts into Continental Sites

Peak pressures inside stony and iron projectiles resulting from a 10 km/s impact are shown in Fig. 1. It is clear that the main portion of a stony asteroid projectile is evaporated (light-gray area) and only a few percent of the total mass remain solid (black area). In contrast, the bulk of an iron projectile is conserved in the solid (but disrupted) state.

Figure 2 illustrates initial stages of impacts with velocities of 10 and 40 km/s. The compression stage is very short in each case: its duration t may be estimated from the equation: $t = d/U = 0.02–0.1$ s, where d is the projectile diameter, U– projectile velocity. The shock wave quickly reaches the rear end of the projectile and is reflected back as a rarefaction wave, which releases the compressed material to low pressure and density. At the velocity of 40 km/s projectile and target vapors fill the growing crater and form a narrow jet, which transports extraterrestrial material back into the atmosphere. The high-temperature vapor jet is surrounded by more dense and colder melted ejecta from the rim of the crater.

The amount of projectile material within the crater quickly diminishes with time, and 10 s after the impact the main portion of projectile substance has left the crater. The dependence of projectile mass within the crater with time for different velocities is shown in Fig. 3. This mass changes only slightly during later stages of the cratering process. Thes results support the idea that deposition of meteoritic material decreases as the impact velocity increases. A high velocity (30–40 km/s) projectile is completely vaporized after the impact and escapes from the growing crater during the first 3–10 s. The higher the initial velocity, the shorter the escape time. In the case of an impact velocity of 20 km/s, the process continues much longer, but it seems certain, that all the impactor material will have left the crater 40–50 s after the impact. However, in this case the rocks in the crater may have some evidence of extraterrestrial material. Only for the lowest velocity under consideration (10 km/s) appreciable amounts of the projectile material are not vaporized and are conserved in the crater. Our calculations demonstrate that the the thickness of projectile melt layer is smaller than the size of one computational cell. In reality, the projectile material is fragmented and mixed with melted and unmelted disrupted rocks.

In Fig. 4 the dynamics of crater growth is shown for two computing runs. The gray curves correspond to the run without strength consideration, black ones — to a run where strength was included. In this latter case, the crater reaches its maximum depth of about 3 km and stops to grow in depth, although the crater radius continues to increase. In the run without strength, the crater depth increases to a larger value (of about 4.5 km) and than sharply decreases to an unreliably small value. This is the result of the strengthless model calculations.

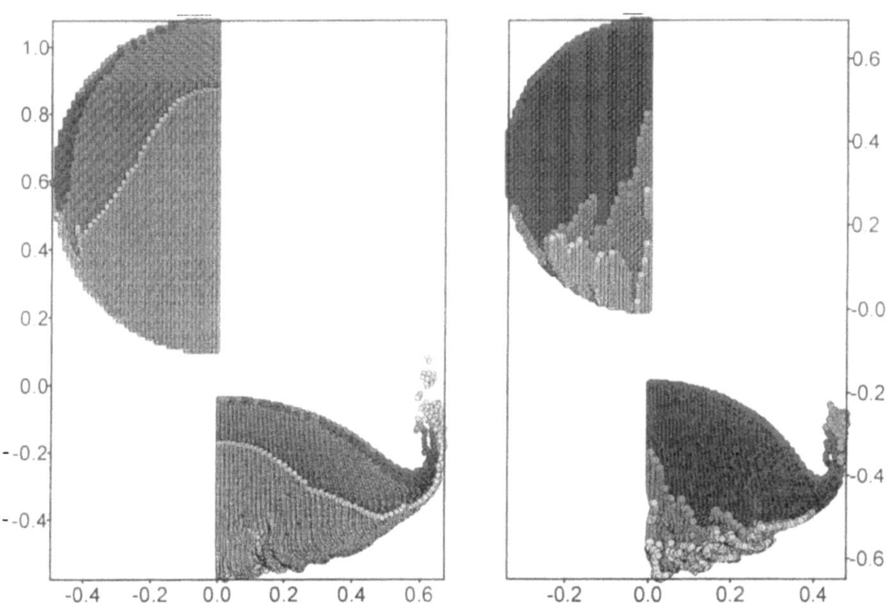

Fig. 1. Peak pressures inside the 1 km–diameter stony (on the left plate) and iron (on the right plate) projectile for the impact with velocity of 10 km/s. Material colored with light-gray is evaporated, with gray — melted, black — remains solid. Left side of each plate corresponds to the initial position of the projectile material, right — to the projectile material position after compression stage. All distances are in km.

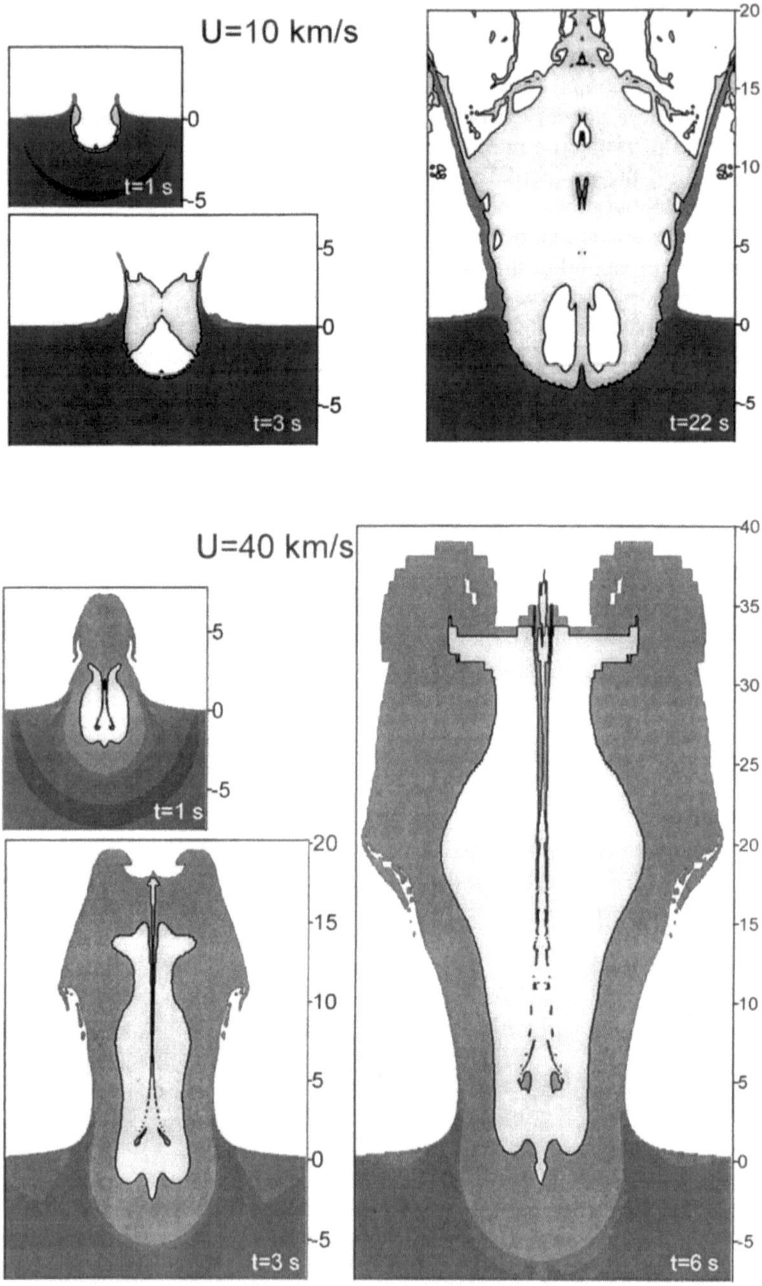

Fig. 2. Projectile (light-gray) and target (gray-black) material distribution after the vertical impact of a 1 km–diameter stony asteroid into a continental site. Impact velocity equals 10 km/s (upper group of plates) and 40 km/s (bottom group). All distances are in km.

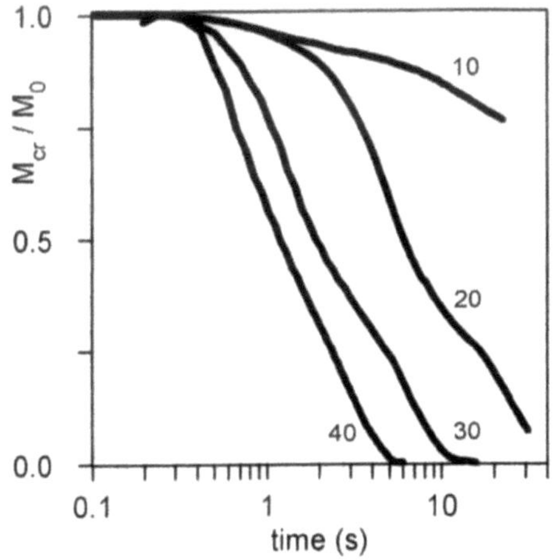

Fig. 3. Ratio of projectile mass within the crater M_{cr} to the total projectile mass M_0 versus time. Figures near the curves indicate the initial impact velocity.

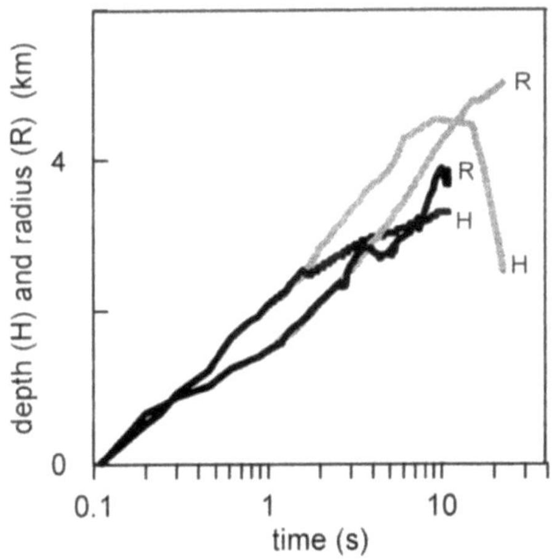

Fig. 4. Crater depth H and radius R versus time for the impact velocity of 10 km/s. Gray lines correspond to the run without material strength.

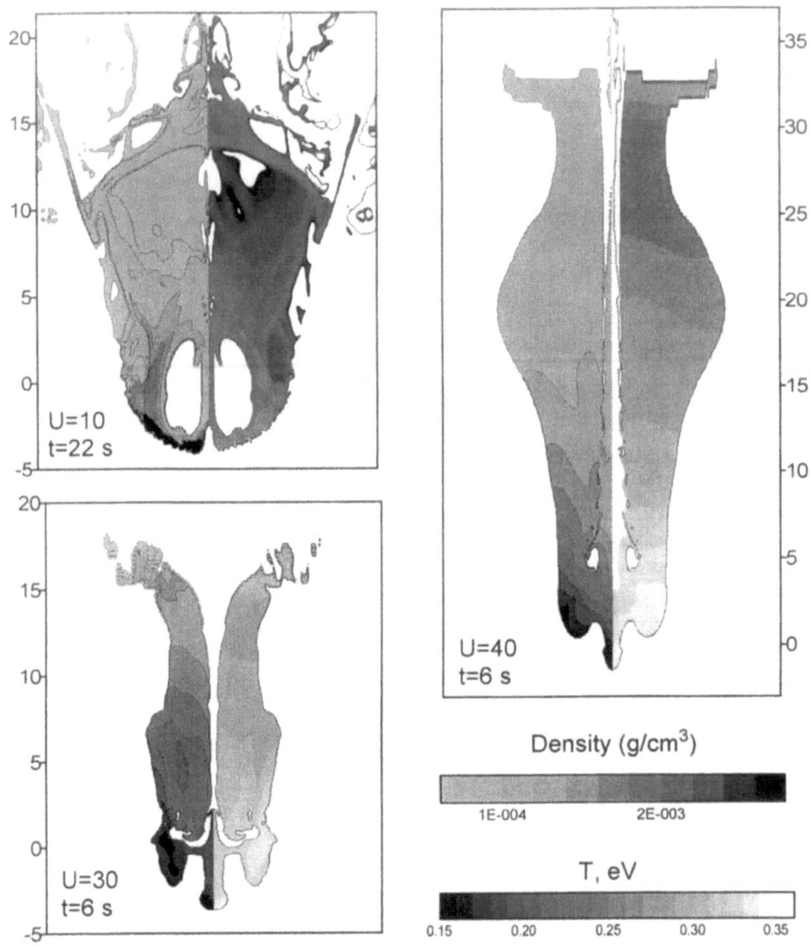

Fig. 5. Density (left) and temperature (right) distribution in projectile plume after the impact of a 1-km-diameter stony asteroid. Impact velocities range from 10 to 40 km/s. All distances are in km. Temperature is measured in electron-volts: T[K] = 11 600 T[eV].

The evolution of an atmospheric plume consisting of projectile-target vapor, is shown in Fig. 5. Six to twenty seconds after the impact, the density of the plume varies from 10^{-2} g/cm^3 near the ground surface to 10^{-4} g/cm^3 at the top (15–30 km above the surface). Vapor is rather hot at the bottom and cools to the liquid-vapor equilibrium temperature (about 2000 K) at an altitude of several kilometers. Condensation of the vapor starts and, in theory, at this moment the homogeneous vapor cloud should be replaced by a mixture of gas and solid (liquid) particles. Satisfactory ways to define particles size and ratios of solid/gas fractions have not been found yet. Still later these particles move ballistically and can be dispersed over an area of about hundred kilometers width by the process of sedimentation of fallout. Identification of this ejecta type is difficult due to its wide dispersion, mixing with cold ejecta, and erosion due to long-term, post-impact geological processes. The influence of the material strength on the plume formation was found to be negligible.

3.2 Impacts into Oceanic Targets

Earlier, some results for impacts of kilometer-sized asteroid into ocean were presented by Artemieva and Shuvalov (1999). We estimated the peak pressure values at the oceanic floor rocks for various impact velocities (in the range from 15 to 40 km/s) and water-depth/projectile-diameter ratios (from 1 to 4) on the basis of accurate numerical simulations similar to those described above. Vertical and oblique impacts were considered. If an oceanic depth exceeds two impactor diameters, the influence of the water layer is important. This layer strongly smoothes down all impact effects. Underwater crater looks like a shallow bowl – the deeper the ocean, the more shallow the bowl. For an oceanic depth of 4 km the underwater crater is practically invisible. Tracer particles were used to define the amount of shock modified and melted material in the floor.

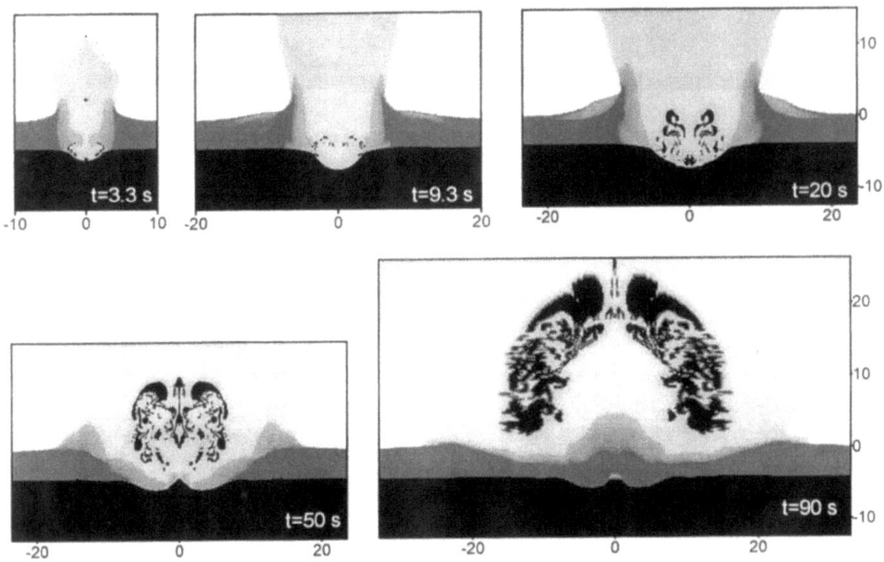

Fig. 6. Projectile and target material distribution for different times after the vertical impact of a 1-km-diameter stony asteroid into deep water. Impact velocity equals 20 km/s, the ratio of oceanic depth to projectile diameter equals 4. All distances are in km. Colors – black corresponds to granite (both – the bottom and the projectile), gray – water (shading intensity is proportional to water density), white – air.

In the present study we have continued our simulations to a later stage of the process. Figure 6 demonstrates the results for a vertical impact of a 1-km-diameter stony asteroid at a velocity of 20 km/s into an ocean basin of 4 km depth. This case is comparable to that of the well-known Eltanin impact (Gersonde et al. 1997). At such a depth, the main part of the impact energy is released in the water layer. Only a small transient cavity with a small depth to diameter ratio is formed in the ocean floor. The impactor is totally fragmented while passing through the water. Furthermore the fragments (partially evaporated) are entrained by hot vapor and ejected into the atmosphere. About 90 s after the impact the projectile material is dispersed within an air-vapor–aerosol cloud extending to an altitude of 20 km and having a width in excess of 30 km. At this moment the transient water cavity has totally closed up.

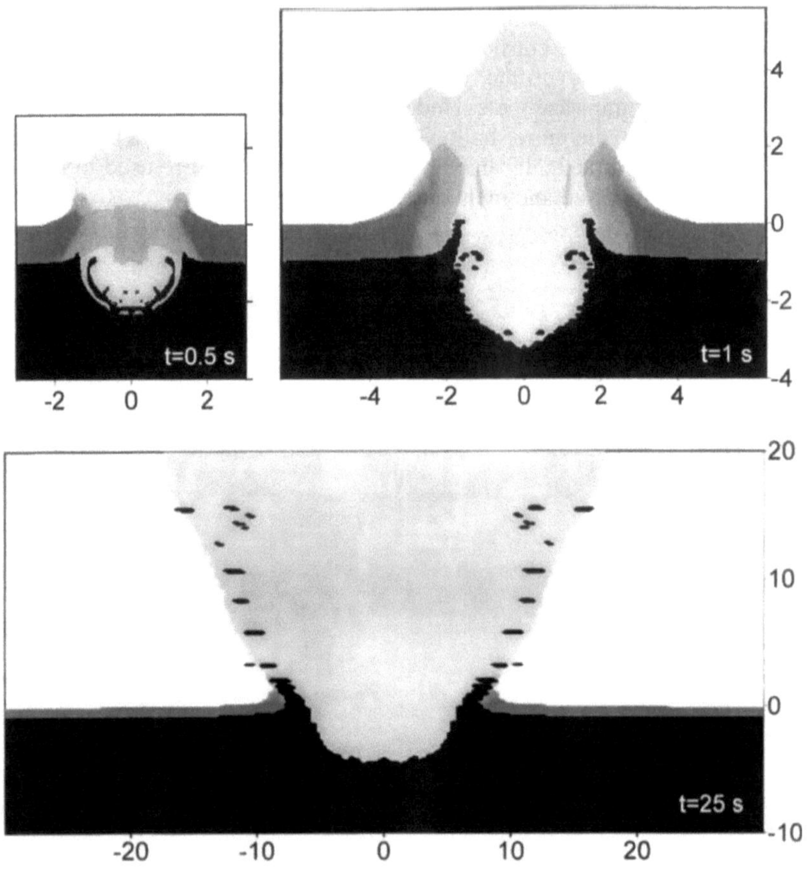

Fig. 7. Projectile and target material distribution for different times after the vertical impact of a 1-km–diameter stony asteroid into shallow water. Impact velocity equals 20 km/s, the ratio of oceanic depth to projectile diameter is 1. All distances are in km. Colors are the same as in Fig. 6.

Another scenario occurs in the case of reduced ocean depth. Figure 7 shows the flow patterns for a water depth of 1 km, but otherwise the same conditions. The projectile is highly deformed during the passage through the water, but is not fragmented and is not totally decelerated. A typical crater grows in the oceanic floor. The envelope-like impactor strikes the surface of the transient cavity (the cavity has been produced by water column, not by impactor itself) and projectile material is mixed with the target substance. The hydrocode doesn't allow to distinguish these materials properly because the scale of mixing is less then the size of computational cell. The mixture is gradually ejected from the crater rim.

Twenty five seconds after the impact the altitude of the rising ejecta plume and spatial distribution of ejecta differ from the previous (deep impact) case. At this stage of ejecta evolution the problem of condensation becomes significant (Nemtchinov et al. 1998). The further evolution of extraterrestrial material depends largely on particle size distribution, defined by initial fragmentation and condensation, and should be described by means of two-phase hydrodynamics. Moreover, the evolution of the cloud material depends on the local atmospheric conditions, because impact induced disturbances become comparable with natural atmospheric motion.

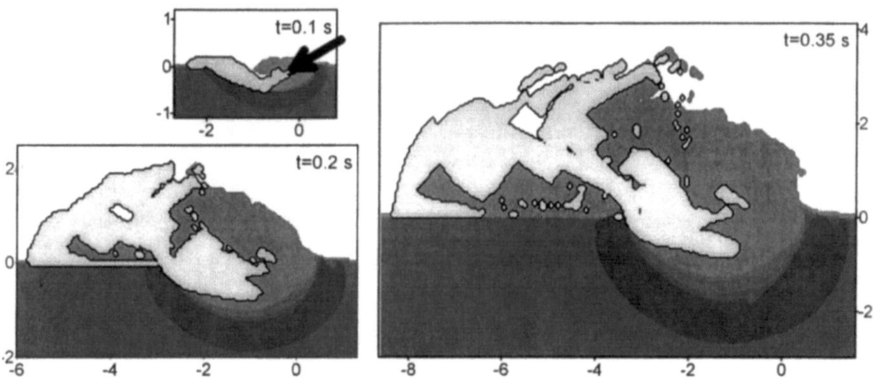

Fig. 8. Projectile (light-gray) and target (gray-black) density in the central cross-section of the flow for different times after the oblique impact of a 1 km–diameter stony asteroid into granite. Impact velocity equals 20 km/s, impact angle is 30° to horizon. All distances are in km.

3.3 Oblique Impacts

The results, described above, concern vertical impacts. But it is a well-known fact that all impacts are oblique to some degree and that the most likely angle of impact is 45° (Shoemaker 1962). The effects of an oblique impact event on the target are known, producing craters that appear circular even for low impact angles (Bottke et al. 2000). However, the fate of the projectile in oblique impact event is poorly investigated with the exception of early 2.5D simulations by Brown (1981), O'Keefe and Ahrens (1986), and a recently published paper by Pierazzo and Melosh (2000).

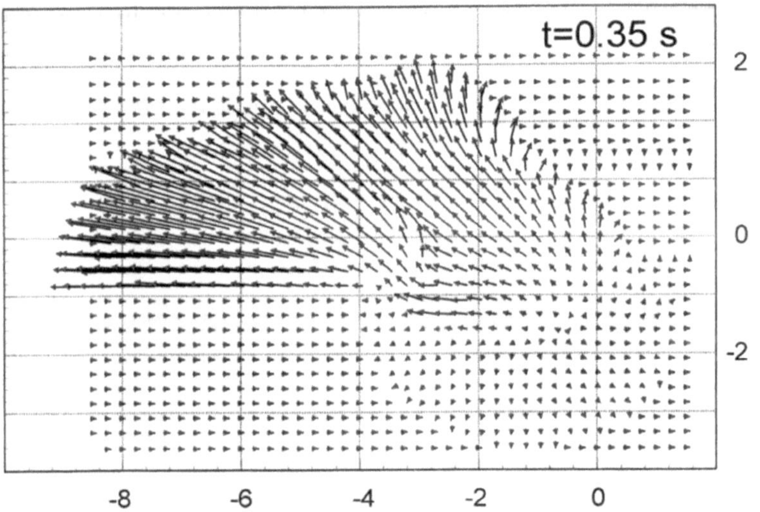

Fig. 9. Velocity field for the same impact as shown in Fig. 8.

We have obtained some preliminary results the impact at an angle of 30° and a velocity of 20 km/s (Figs. 8 and 9). During the compression stage, only the vertical component of the projectile velocity diminishes substantially, whereas the horizontal one changes only slightly. The projectile continues to move along the surface, forming a high-velocity jet, consisting of a projectile-target material mixture. The rear edge of the impactor emerges from the high pressure regime even before the compression phase has been totally completed. The growing crater is shallow, with its depth not exceeding 1 km. About 1 s after the impact, all of the projectile material has been melted and has escaped from the crater. Our results coincide qualitatively with those obtained by Pierazzo and Melosh (2000) for the case of low impact angles. They used the Sandia National Laboratories' three-dimensional hydrocode CTH in a series of high-resolution simulations for the impact at 15-90° from the horizontal, keeping projectile size (10 km in diameter) and impact velocity (20 km/s) constant. Our computations, as well as the simulations by Pierazzo and Melosh (2000), have been made in poor hydrodynamic approximation, i.e., we assume that neither the projectile or target material possess any intrinsic strength. This is in contrast with the observations by Schultz (1996) that shear heating plays an important role in his laboratory-scale experiments. We plan to include strength in our 3D modelling and make new computing runs. Another purpose is to prolong three-dimensional simulations up to 10–100 seconds to follow meteoritic material deposition in the frame of two phase hydrodynamics (as mentioned above for the case of vertical impacts).

4 Discussion

The simulations, presented above, illuminate some aspects of impact cratering and projectile fate after the impact. As the granite EOS was used, the results apply only to impacts of stony asteroids into crystalline basement. In the cases when the amount of sediments in the target is not negligible, the presence of volatile materials (for example, CO_2 and H_2O, which are produced from degassing of calcite) greatly changes the process of ejecta formation, melt and vapor production. These gases may operate as a giant piston and force out additional projectile material.

In the case of an iron impactor of the same mass, the projectile is 1.5 times smaller in diameter, than a stony one with the same velocity and energy, but penetrates much deeper into the target. Moreover, the melt enthalpy for iron is twice as large as that for granite. Thus, the total amount of projectile vapor and melt is smaller. Schnabel et al. (1999) demonstrated that after the impact of a 15 m-radius iron projectile with the velocity of 20 km/s (this impact corresponds to the one that might have formed the well-known Meteor Crater) about 35–40% of the total impactor mass remain solid. Unmelted pieces produce the so called Canyon Diablo meteorites, whereas some part of melted material yield the Canyon Diablo spheroids. These parts of the projectile differ substantially in their ^{59}Ni enrichment due to a different thermodynamic history.

The results of numerical simulations for a 1-km-diameter asteroid may be easily extrapolated to smaller impactors. The minimum diameter of a meteoroid, for which the influence of atmospheric disruption and deceleration is negligible, depends on the projectile composition and ranges from 100 to 300 m. Thus, we should obtain practically the same relative mass and velocity distribution of ejecta over a smaller area and altitude. Extrapolation to large impacts (up to 10 km) is not so clear-cut. The Earth's atmosphere does not influence the passage of an asteroid, but decrease of atmospheric density with altitude leads to a sharp increase of plume velocity. Some part of the extraterrestrial material may reach near-escape velocity and be dispersed globally as in the case of the impact at the Cretaceous-Tertiary boundary geological periods (e.g., Alvarez et al. 1980).

Acknowledgements

This study was partially supported by the Russian Fund of Basic Research (Grant # 00–05–81152 Bel2000_a)

References

Ahrens TJ, O'Keefe JD (1972) Shock melting and vaporization of lunar rocks and minerals. Moon 4: 214–249

Alvarez LW, Alvarez W, Asaro F, Michel HV (1980) Extraterrestrial cause for the Cretaceous-Tertiary extinction. Science 208: 1095–1105

Artemieva NA, Shuvalov VV (1999) Shock metamorphism on the ocean floor (numerical simulations). Deep-Sea Research, submitted

Bottke WF, Love SG, Tytell D, Glotch T (2000) Interpreting the elliptical crater populations on Mars, Venus, and the Moon. Icarus 145: 108–221

Brown WT (1981) Numerical modeling of oblique hypervelocity impact using two-dimensional plain strain models. In: Nellis WJ, Seaman L, Graham RA (eds) Shock waves in condensed matter. Am Inst Physics, Washington DC, pp 529–533

Chase MW, Davies CA, Downey JR, Frurip DJ, McDonald RA, Syverud AN (1985) JANAF thermochemical tables, 3rd ed. J Phys Chem Ref Data 14

Evans NJ, Shahinpoor M, Ahrens T (1994) Hypervelocity impact: Ejecta velocity, angle, and composition. In: Dressler BO, Grieve RAF, Sharpton VL (eds) Large meteorite impacts and planetary evolution. Geol Soc Amer Spec Pap 190: 93–101

Gersonde R, Kyte FT, Bleil U, Diekmann B, Flores JA, Gohl K, Grahl G, Hagen R, Kuhn K, Sierro FJ, Vulker D, Abelmann A, Bostwick JA (1997) Geological record and reconstruction of the late Pliocene impact of the Eltanin asteroid in the Southern Ocean. Nature 390: 357–363

Grieve RAF (1982) The record of impact on Earth: Implications for a major Cretaceous/Tertiary impact event. In: Silver L, Schultz P (eds) Geological implications of impact of large asteroids and comets on the Earth. Geol Soc Amer Spec Pap 190: 25–37

Hill R (1950) The mathematical theory of plasticity. Oxford Univ Press, London, 368 pp

Hörz F (1982) Ejecta of the Ries Crater, Germany. In: Silver L, Schultz P (eds) Geological implications of impact of large asteroids and comets on the Earth. Geol Soc Amer Spec Pap 190: 39–55

Koeberl C (1990) The geochemistry of tektites: An overview. Tectonophys 171: 405–422

Melosh HJ (1989) Impact cratering: A Geologic Process. Oxford University Press, Oxford 246 pp

Nemtchinov IV, Shuvalov VV, Artemieva NA, Ivanov BA, Kosarev IB, Trubetskaya IA (1998) Light flashes caused by meteoroid impacts on the lunar surface. Solar System Res 32, 2: 99–114

O'Keefe JD, Ahrens TJ (1986) Oblique impact: a process for obtaining meteorite samples from other planets. Science 234: 346–349

Palme H, Janssens M-J, Takahashi H, Anders E, Hertogen J (1978) Meteoritic material at five large impact craters. Geochim Cosmochim Acta 42: 313–323

Pierazzo E, Vickery AM, Melosh HJ (1997) A reevaluation of impact melt production. Icarus, 127: 408–423

Pierazzo E, Melosh HJ (2000) Hydrocode modeling of oblique impacts: the fate of the projectile. Meteoritics Planet Sci 35: 117–130

Schnabel C, Pierazzo E, Xue S, Herzog GF, Masarik J, Cresswell RG, di Tada ML, Liu K, Fifield LK (1999) Shock melting of the Canyon Diablo impactor: constraints from Nickel-59 contents and numerical modeling. Science 285: 85–88

Schultz PH, Gault DE (1990) Prolongated global catastrophes from oblique impacts. In: Sharpton VL, Ward PD (eds) Global catastrophes in Earth history; an interdisciplinary conference on impacts, volcanism, and mass mortality. Geol Soc Amer Spec Pap 247: 239–261

Schultz PH (1996) Effects of impact angle on vaporization. J Geophys Res 101: 21,117– 21,136

Shoemaker EM (1962) Interpretation of lunar craters. In: Kopal Z (ed) Physics and Astronomy of the Moon. Academic Press, San Diego, pp 283–359

Shuvalov VV (1999) 3D hydrodynamic code SOVA for interfacial flows, application to thermal layer effect. Shock Waves 9, 6: 381–390

Thompson SL, Lauson HS (1972) Improvements in the Chart D radiation-hydrodynamic CODE III: Revised analytical equations of state. Sandia National Laboratory Report SC-RR-71 0714.

Petrophysics Hints at Unexplored Impact Physics

Vladimir V. Svetsov

Institute for Dynamics of Geospheres, Leninsky pr. 38, bld.6, Moscow 117979, Russia.
(svetsov@idg1.chph.ras.ru)

Abstract. This paper deals with thermoluminescence and magnetization anomalies discovered around the epicentre of the 1908 Tunguska explosion of a large bolide in Central Siberia. Several mechanisms that can be caused by the event and influence the magnetization of soil and the level of thermoluminescent emission in minerals are considered. Simple estimates show that strong currents and magnetic fields can be generated through magnetohydrodynamic effects when a disintegrated meteoroid is severely decelerated by aerodynamic forces at the terminal point of the trajectory in the geomagnetic field. Available experimental results on electrification of projectiles are considered in relation to bolides. Debris particles moving in the atmosphere collect electric charges and their precipitation can cause lightning. Short-circuiting of the ionosphere and the Earth's surface through the ionized wake and the discharge channels can happen in this case, triggering a discharge of the all-Earth capacitor.

1 Introduction

It is generally accepted that on June 30, 1908, a 50 to 100 m diameter bolide entered Earth's atmosphere and fully ablated (Vasilyev 1998; Chyba et al. 1993; Hills and Goda 1993; Svetsov 1996). It released $4 \cdot 10^{16}$ to $8 \cdot 10^{16}$ J of energy (Hunt et al. 1960; Ben-Menahem 1975; Korobeinikov et al. 1976; Vasilyev 1998) mostly at altitudes between 5 and 10 km (Ben-Menahem 1975; Korobeinikov et al. 1976) devastating a forest area of 2150 km^2 (Fast 1967; Fast et al. 1983; Vasilyev 1998). Burnt vegetation was found over a 200 km^2 area around ground zero (Lvov and Vasilyev 1976; Vorobyov and Demin 1976; Korobeinikov et al. 1982). The ground zero determined from the pattern of felled trees is located at 101°53′40″E, 60°53′09″N (Fast 1967). The bolide moved from the east-south-east to the west-north-west so that its trajectory probably was within a sector with azimuth from 99° (as measured from north to east) to 115° (Zotkin 1966; Vasilyev 1998; Bronshten 1999). The inclination angle (measured from the horizontal) of the

Tunguska meteoroid's trajectory is still a question of debates. Its values were estimated from very low, in the range 5°–15° (Sekanina 1983; Bronshten 1998), to more typical 30°–45° (Zotkin and Tsikulin 1966; Korobeinikov et al. 1976, 1998).

The aftermath of the explosion was felt in one way or another practically all over the globe but little or nothing could be discovered that could have shed light on the nature of this phenomenon. The recent surge of interest in it resulted in various conjectures many of which cannot be substantiated because since then time has obliterated much of the evidence of the catastrophe.

Among other studies geological investigations have been carried out during the last decades searching for long lasting marks of the Tunguska impact (Florensky 1962; Plekhanov 1964; Krinov 1966; Sapronov 1986; Vasilyev 1988; Kolesnikov et al. 1998, 1999; Rasmussen et al. 1999). They led to the discovery of anomalous variations in thermoluminescent (TL) emission of some minerals (Vasilenko et al. 1967; Bidyukov et al. 1990) and magnetization anomalies in the soil (Boyarkina and Sidoras 1974) around the explosion epicentre. The fact that hundred meters in size impactors, that are fairly frequent in terms of geological time scale, might produce geological marks on Earth other than craters is very interesting and deserves further studies.

The general concept of the Tunguska explosion is based on estimates and numerical modelling of the main impact processes and mechanisms, including meteoroid break-up and fragmentation (Korobeinikov et al. 1998), movement and flattening of the fragmented mass with rapid deceleration and energy release at the final point of the trajectory (Chyba et al. 1993; Hills and Goda 1993), ablation (Lyne et al. 1996) and radiation transfer (Putyatin 1980; Svetsov 1996). Theoretical and field studies show that there was no hard radiation or high-energy particles that might explain the TL increase in minerals (Kirichenko 1975; Kolesnikov et al. 1975; Vasilyev 1998). Electric and magnetic phenomena that probably accompany impacts and could cause the variations in the soil magnetization remain unstudied both for small meteors and the Tunguska sized impactors. Thus, the TL and magnetic anomalies remain something of an enigma, similar to magnetic storm that began approximately 6 minutes after the Tunguska explosion and lasted for about 5 hours (Vasilyev 1998). There is also no full clarity in understanding of unusual atmospheric phenomena, such as light dusks, colourful dawns and sunsets, abnormal amount of noctilucent clouds, which were observed over Eurasia for several days after the event. It has been suggested that a significant amount of dust, aerosols, nitrogen oxides and water spread from the site by easterly winds (Turco 1982; Chyba et al. 1993). But the fact that the unusual atmospheric phenomena started essentially simultaneously with the explosion (Vasilyev et al. 1965) is puzzling. A strike of the putative comet's tail (an explanation supported by Bronshten in 1991b) seems improbable in the light of recent work (Chyba et al. 1993; Lyne et al. 1996) that indicates that the body was probably an asteroid.

One of the purposes of this paper is to draw attention to, and discuss the results of, geological investigations of the Tunguska thermoluminescent and magnetic anomalies that until now have only been reported in Russian scientific literature with restricted circulation (see below). The second objective is to consider probable mechanisms responsible for these phenomena and suggest explanations for the origin of these anomalies. And, lastly, to suggest the direction for future investigations.

It is likely that the deviations in soil magnetization vectors could be due to a strong magnetic field, which, in turn, had to be generated by electric currents. There is evidence that meteoroid impacts and bolides produce electromagnetic phenomena (Crawford and Schultz 1988, 1991, 1993, 1999; Keay 1980, 1992; Bronshten 1983; Keay and Ceplecha 1994; Beech and Foschini 1999), but the problem with large bolides has received little attention and remains unexplored. Eyewitness reports show that the largest of the recent bolides, the 5 kiloton Sikhote-Alin iron meteoroid that struck the Earth on 12 February 1947, and the 10 kiloton Chulym bolide fallen in Central Siberia on 26 February 1984, acted on local electric circuits (Astapovich 1958; Anfinogenov and Fast 1985; Ovchinnikov and Pasechnik 1988).

Eyewitnesses of the Tunguska event, who happened to stay 30 kilometres from ground zero, were interviewed in 1926 (Souslov 1967). They saw several very bright lightning flashes accompanied by thunder immediately after the explosion when the trees had fallen and the fire had started. Time intervals between the flashes could not be exactly determined, but might have been from some seconds to some minutes. Many of the Tunguska event eyewitnesses that lived some hundred kilometres from the explosion told that they had heard several bangs similar to thunder or artillery fire with intervals of some minutes (Vasilyev et al. 1981). Though this phenomenon was attributed to a complicated mechanism of sound propagation in the atmosphere (Pasechnik 1976), we cannot exclude that those were consequences of powerful electric discharges.

Electric and magnetic phenomena receive primary emphasis in this paper. It examines possible mechanisms of generation of electric currents in ionized air during the impacts, with particular emphasis to magnetohydrodynamic effects. The magnetic anomaly could have also been caused by multiple electric discharges between the meteoroid and the ground. It seems quite probable that powerful electric discharges, similar to lightning, could also cause the TL anomaly due to high temperatures in discharges and hard UV radiation in the vicinity of the discharges at the ground. The discharges could result from electrification of the Tunguska meteoroid in flight or electrification of the debris after the explosion. Another possibility is the "grounding" of the ionosphere or electrosphere by the ionized wake formed behind the body that resulted in a kind of a short circuit.

2 The Tunguska Thermoluminescent Anomaly

As is well established, some minerals exhibit thermoluminescence (TL), that is, the property to emit light when heated to some hundred degrees centigrade (McKeever 1985). Natural TL of a mineral is the result of its previous exposure to ionizing radiation or high energy particle bombardment that occur during the mineral geological life time. In these processes, electrons are knocked out of the crystal lattice, thrown into the conduction band and then drop into trapping centres existing due to lattice defects. The electrons can remain in these traps infinitely at normal temperatures. Heat frees electrons from the traps and they produce light in a wavelength range from 250 to 800 nm. TL glow curves (temperature dependence of light intensity) usually have several peaks which depend on impurities, typical TL peak temperatures of quartz and feldspar lie in three intervals 200–220 °C, 280–300 °C and 350–380 °C (Marfunin 1975; McKeever 1985; Aitken 1985).

During the last three decades of records at the Tunguska explosion site, investigations of TL have been carried out in rock samples consisting mainly of plagioclase, calcite, and quartz (Vasilenko et al. 1967; Vasilyev et al. 1976), in soil mineral fraction involving mainly feldspar and quartz (Bidyukov 1988; Bidyukov et al. 1990), and in pure transparent quartz extracted from soil (Bidyukov 1997). Soil samples were taken from subsurface layers at 3–5 cm depth below the vegetation layer and the typical size of the extracted mineral grains was 0.25 to 0.5 mm. Sampling was made over the territory around the epicentre in certain directions with a step of about 0.5 km in the close vicinity to the zero point, 1–2 km to the periphery, and 4–5 km beyond the area of felled trees, but only part of the area has been covered. Background values of TL emission of the samples from the subsurface layers, both at the impact site and far beyond, vary in a relatively wide range. Samples with low TL intensity alternate with high TL ones, probably due to natural selective action of sunlight, fires, or lightning. For this reason, a relatively long time is necessary to collect a sufficiently great number of specimens for reliable statistics.

Statistical processing of data on more than 400 soil samples and about 200 rock samples led to the discovery of an anomalous area in which samples are found with TL emission that is much lower or much higher than the limits of the background range. Locations of soil samples with anomalously low TL emission (either total TL output below 2, or the third emission peak below 0.3, in relative units respectively) and anomalously high TL emission (total TL output above 147, or the third peak above 9.5, in relative units respectively) are shown in Fig 1, which follows research by Bidyukov et al. (1990) and Bidyukov (1997). The average background value of TL output is about 20–30 in relative units. The data of Vasilyev et al. (1976), who examined rock samples, are also shown in Fig 1. The anomalously high level of TL for rocks was chosen above 1000 relative units and the low level was below 10 units. As a function of the distance from the zero point, the TL intensity level in minerals is lower on average within 6 km diameter and higher within 15 km diameter (Bidyukov et al. 1990). A distinguishing feature of the TL emission field is its nonuniform angular distribution – the highest values of TL emission are found in samples taken around the line through ground zero to the east relatively close to the range of projection of the probable Tunguska meteoroid trajectories. The Tunguska mineral samples with the highest TL net output showed no first TL peak, in contrast to other samples (Bidyukov et al. 1990), though minerals showing the first peak are common for the whole region.

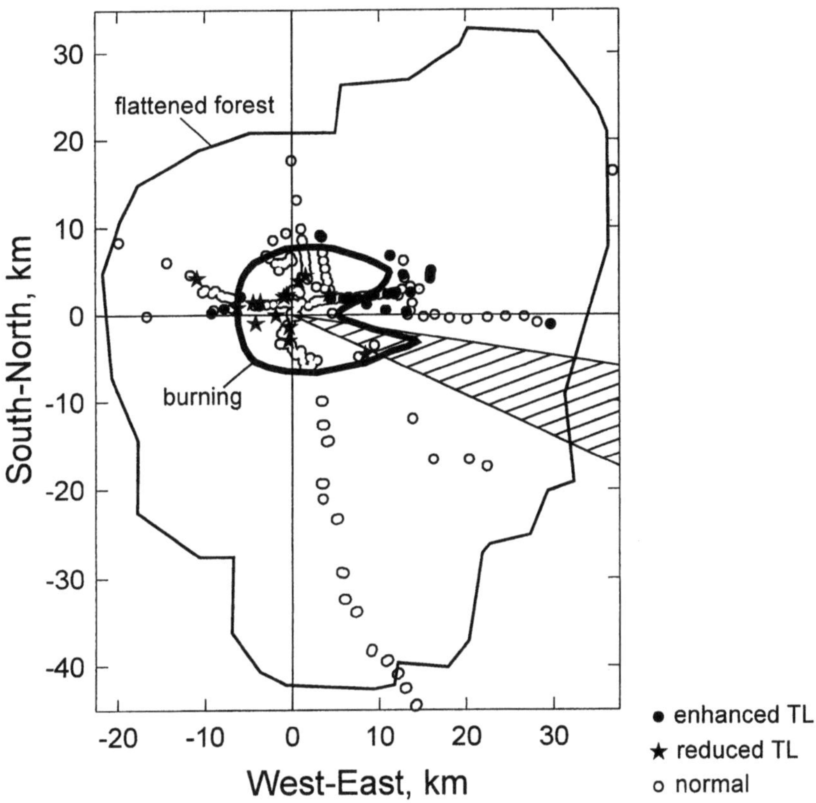

Fig. 1. Distribution of samples studied for TL properties in the Tunguska catastrophe area. The solid circles and stars show the sites of the samples with enhanced and reduced TL output, respectively. The "butterfly" shaped contour is the conventional boundary of the flattened forest (Fast et al. 1967), and the inner contour is a boundary of the area where trees bore marks of burning (Lvov and Vasilyev 1976). The origin of coordinates is placed at the ground zero defined from the pattern of fallen trees. A sector of probable trajectories is hatched.

Tunguska researchers generally believe that lower TL values could have been caused by annealing action of thermal radiation from the fireball (Bidyukov et al. 1990; Bidyukov 1997; Vasilyev 1998). This agrees with the fact that the area of lower TL samples correlates with the area where trees bear evidence of radiation burns from the explosion (Fig. 1). On average, TL is only half of the background level there. Some of the samples were completely annealed, which usually requires an energy of about 300 J/cm^2 in laboratory experiments (Vlasov et al. 1979; Shakhovets and Shlyukov 1987). This energy estimate is in reasonable agreement with the Tunguska explosion radiation flux at the ground, according to computations made by Putyatin (1980) and Svetsov (1996). It should be mentioned that minerals may also lose a significant portion of their TL energy when stored at moderate, close to room, temperatures (Fragolis and Readhead 1991), and a low temperature 220° peak in irradiated quartz may have relatively short life time, about 100 years (Godfrey-Smith 1985). Thus, the interpretation of results involves some difficulties.

Vasilyev et al. (1976) and Bidyukov et al. (1990) argued that some factor enhancing TL also acted on minerals, but the origin of this factor and the cause of the specific distribution of samples with high TL emission remains unclear. Attempts to explain the increased TL by ionizing (X-ray) radiation or neutron flux conflicted with the absence of appropriate marks of such factors (Vasilyev 1998). Laboratory experiments and geological investigations of impact structures show that high pressures (Angino 1964) and strong shock waves with kilobar pressures (Ramseyer et al. 1992; Ivliev et al. 1995, 1996) can influence TL, probably due to deformation in crystals and creation of new defects, although little is known about the mechanisms responsible for these effects. Shock waves can result in both an increase and decrease of natural TL for various minerals and shock pressures (Ivliev et al. 1995, 1996). But the pressure behind the aerial blast wave from the Tunguska explosion seems to have been too small, not higher than some bars on the ground, to affect the TL level.

Hard UV radiation could be another reason for an increase in the TL. The energy gap between the valence band and conductivity band in the crystals is about 10 eV (Antonov-Romanovskiy 1966). Thus, UV radiation with photon energies higher than 10 eV can increase the amount of trapped electrons just as X-rays do. The necessary photon energies can be even lower if UV acts simultaneously with thermal radiation which cause redistribution of electrons among defect centres (Marfunin 1975). However, hard UV emitted by the fireball from altitudes of some kilometres cannot reach the ground due to very strong absorption – photon free-paths in the atmosphere are not more than 10 cm (Lee 1965; Churchill et al. 1966).

Softer UV radiation from the fireball, with photon energies below 6.5 eV, could reach the Earth's surface. Solar radiation, heating the minerals, releases electrons from the traps and thereby reduces TL light output (Vlasov et al. 1979). However, it is necessary to stress that the solar radiation at the ground level involves only soft UV below 4 eV, while higher energies are taken up by the ozone layer at an altitude of about 20 km. A major energy release of the Tunguska explosion happened at lower altitudes, from 5 to 10 km. Therefore, the radiation spectrum involved photons with energies ranging from 4 to 6.5 eV, which could influence the trap population in different ways. Experiments of Vaz et al. (1968), in which natural calcite samples were illuminated with monochromatic UV light of different photon energies of up to 5.2 eV, show that charges released by the light from one trap can be redistributed to other traps. The TL peak, which occurred in the calcite samples at 220 °C, was diminished after UV irradiation, but began to grow rapidly when photon energies exceeded 4.9 eV. The TL peak at 320 °C was only slightly diminished by UV light and began also to grow when the samples were irradiated by photons with energies above 4.9 eV.

In experiments by Medlin (1963), UV radiation from two sources with 6.7 eV (mercury line) and 4.9 eV photon energies led to a trap-filling process in quartz, probably due to stepwise transitions to the conductivity band through some intermediate levels, but in this case the UV excitation was ineffective, unless the quartz samples had been preliminary annealed at 900 °C. These experimental results show that UV light can enhance TL peaks at some conditions. The conditions of irradiation and heating of mineral grains have not been adequately explored as far as the Tunguska event is concerned. The radiation spectrum of the source is uncertain, although it is very likely that it included multitude of lines in visible and UV range emitted by hot meteoroid vapour, which complicates the problem. The light impulse from the fireball lasted probably about 10 s (Svetsov 1996). Mineral grains could be heated by thermal radiation, including infrared radiation, either during or after the major UV impulse, or it could happen both during and after it. Another source of heating is fire. In all likelihood, thermal radiation from the Tunguska fireball set flammable materials on the ground on fire, but the blast wave then extinguished it (Hills and Goda 1993). If the fire continued at some distance from the zero point, small mineral particles lifted by the blast wave and wind could have been heated and annealed in the fire flames prior to the end of the UV radiation impulse. (This process can be also important for magnetic aspects.) It is not improbable that the UV radiation could fill the traps in the annealed particles, as in the experiments mentioned above, and release electrons from the traps in minerals not heated to sufficient temperatures in the epicentre.

Laboratory experiments with a light source similar to that probably involved in the Tunguska event should be done to see the effect of soft UV radiation emitted by the fireball on the minerals. It is also necessary to look more closely at the probable scenarios of the motion and heating of mineral grains, which could be exposed to the UV light from the fireball. Below an attempt is made to substantiate probable existence of another source of harder UV radiation at the ground – electric discharges triggered by the impact.

3
Tunguska Soil Magnetic Anomaly

Between 1969 and 1976 the soil magnetization in the area of the Tunguska catastrophe was studied by Boyarkina and Sidoras (1974), Sidoras and Boyarkina (1976), and Boyarkina et al. (1980). About one thousand soil samples were taken from upper soil layers and examined for their magnetic properties over an area of 15 000 km^2. The samples were cubic in shape, several centimetres in size, and were oriented along the geomagnetic field at the site. Magnetization vectors, that is, their absolute values, declination and inclination, and magnetic susceptibility have been measured for every sample.

The magnetization vector field of individual samples varied in a wide range. For this reason, the measurements were statistically processed in the following way. The area was covered by a rectangular grid with a step of 10 km. The mean values of magnetization at the grid points were determined using a routine averaging procedure (Hramov and Sholpo 1967). The averaging was made within 10 km circles, each circle included from 10 to 100 soil samples.

Measurements of geomagnetic field declination over an area of 1000 km^2 around the epicentre were made by Kardash (1984) several years later. Values of declination varied from –2° 30′ to 2° 30′, an average declination being about 0°. This differs from the normal declination value of 2° given on geographic maps of the global magnetic field for the same time (see, e.g., Fabiano 1983). The values of geomagnetic field declination slightly grew from the west to the east within the area of flattened trees. The variations in the geomagnetic field over this area are associated with a buried complex of paleovolcanoes which leads to variations in the field intensity up to 2600 nT (Sapronov 1986). The normal value of geomagnetic inclination is 78°. The maps of the geomagnetic field (Neumayer 1891; Teterin 1939) show that the normal inclination in 1908 was probably about 76° while normal declination was 6°. The real magnetic field at the site in 1908 could be different from the normal values and is properly unknown.

Figure 2 illustrates the results of magnetic measurements in soil following to Boyarkina and Sidoras (1974) and Boyarkina et al. (1980). Some parts of the territory have not been studied far from ground zero. The following peculiarities of the area have been revealed. First, a pronounced anomaly is seen in soil magnetic declination. The averaged magnetic declination measured in the samples around the epicentre differs from the normal value by 10–30°, becoming close to the normal value at distances of 20 to 40 km from the zero point. Second, the magnetic inclination of the magnetization vectors goes down by several degrees as compared to its normal value in the area of 10 to 15 km in radius. Third, the absolute values of magnetization vectors increase less than 50% at ground zero. The area, where the absolute values of magnetization are 10% higher than the value averaged over the whole investigated territory, extends to the north-west form the zero point. And, at last, averaged values of soil magnetic susceptibility are approximately doubled in the central part of the anomalous area. The area where susceptibility is 30% higher than average stretches to the north-west and north, as shown in Fig. 2.

Boyarkina and Sidoras (1974) explained the increase in magnetic susceptibility and absolute values of magnetization by precipitation and distribution of ferromagnetic material of the Tunguska meteoroid in the atmosphere, assuming that the bulk susceptibility had to be proportional to amount of ferromagnetic minerals in the soil. However, geochemical analysis of soil samples has not been carried out and it remains unclear what ferromagnetic particles provide the magnetic effect. On the other hand, winds at altitudes from 2 to 10 km, where condensed meteoroid microparticles could fell, are fairly stable and usually blow from the west to the east (Guterman 1979). The maximum speeds of these winds in June and July reach 50 to 60 m/s, the average speed being 20 to 30 m/s. Winds at the ground are weaker, of the order of 1 m/s, and can vary from day to day or during a day. According to observations in Vanavara, 70 km from the epicentre, the averaged direction of morning winds at the surface, as measured from north to east, varies from –106° in June to 18° in July (Guterman 1986). Thus, it seems more probable that the anomaly in magnetic susceptibility and absolute values of magnetization was produced by ferromagnetic particles of soil lifted to the air by the shock wave that initiated strong air motion at the surface, uprooting trees and stirring surface soil layers. The surface wind could transfer microparticles to the north-west and north.

The fall of lifted ferromagnetic particles could also be responsible for the anomalous declination, if the geomagnetic field at the site in 1908 had western declination, different from the normal field. Then small particles, settling in the air, re-oriented, on the average, along the magnetic lines. Samples beyond this area are likely to reflect the value of the geomagnetic field averaged over a relatively long time. The correlation of the anomalous area with the area of flattened trees counts in favour of this hypothesis.

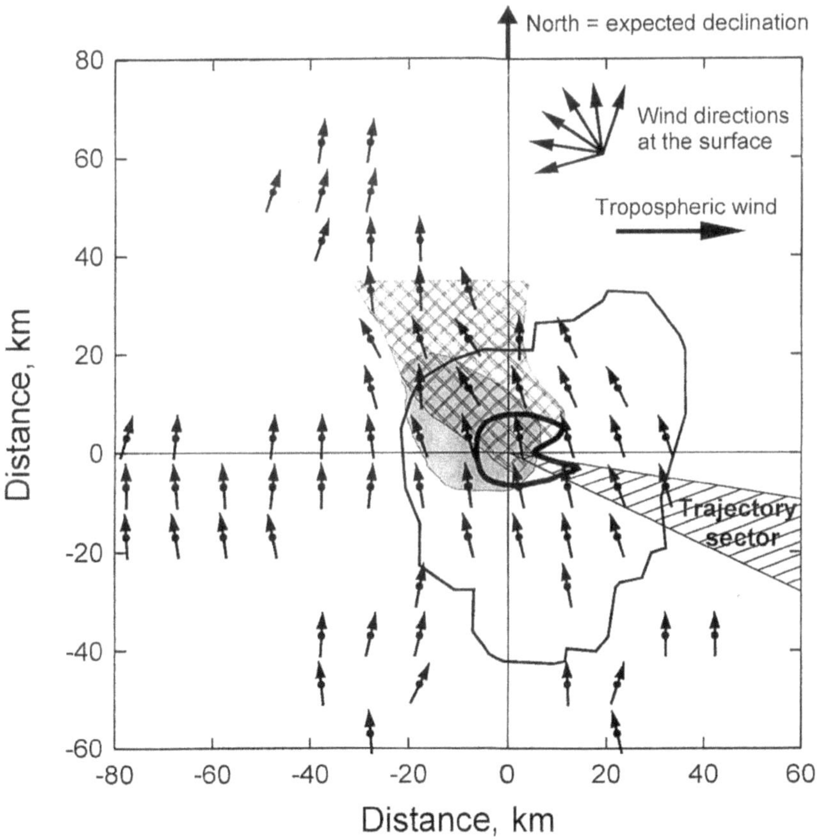

Fig. 2. The arrows show averaged magnetic declination in soil samples investigated at the site of the Tunguska explosion. The areas where the absolute values of magnetization and magnetic susceptibility exceed the average level are shaded and cross-hatched, respectively.

Boyarkina et al. (1980), however, suggested that the anomalous declination in the epicentre, as well as the geomagnetic storm registered in Irkutsk, was caused by a magnetic field generated by the Tunguska explosion. But the mechanisms responsible for the origin of this enigmatic field and the field configuration are still unclear. Laboratory demagnetization of the Tunguska soil samples in alternating magnetic field shows that the abnormal magnetization is unstable and disappears at a field intensity of about 2.5–3 mT (Boyarkina and Sidoras 1974). These values could serve as a rough estimate for the magnetic field that could induce the secondary magnetization (Petrova 1961). However, this magnetic field could be lower if it acted only on ferromagnetic microparticles on the ground. It remains unclear at what depth the effect exists at the Tunguska site, because of the natural difficulties with the exact determination of which the ground layer was the surface in 1908. Note that remanent magnetization induced by the magnetic field of a lightning is usually destroyed in alternating magnetic fields from 2 to 35 mT (Hramov and Sholpo 1967).

The electric currents, if they were strong enough and provided the soil magnetization along the trajectory projection, had to flow perpendicular to the trajectory. These currents could be generated in plasma surrounding the bolide. It is not improbable that the discharges similar to lightning could also change the soil magnetization. A problem with such an explanation is that a single cylindrical discharge would create a circular pattern of magnetization. For example, peak currents in return stroke of cloud-to-ground lightning in thunderstorm can reach 250 kA (Ogawa 95) and a circular magnetic field of 2 mT can be created by such a lightning at a distance of 25 m from it. In the case of the Tunguska magnetic anomaly, it should then be suggested that multiple lightnings happened almost simultaneously in a plane parallel to the trajectory and inclined to the ground. Then, however, the magnetization must also bear marks of the discharges – the induced magnetic field had to have the opposite sign in different parts of the area.

4 Magnetic Storm

Perturbations of the geomagnetic field on 30 June, 1908, have been registered by the observatory working in Irkutsk, 900 km to the south from the epicentre, but the results of registration have been recognised only in 1959 (Plekhanov et al. 1961; Ivanov 1961). The variations of horizontal and vertical components of the geomagnetic field are shown in Fig. 3 following to Ivanov (1964). Known average values of S_q-variations of magnetic field components are subtracted from the registered quantities in this figure. Another observatory, nearest to the Tunguska, was in action in Ekaterinburg, roughly 2400 km to the west from the epicentre. Magnetograms show that the magnetic field was quiet there, as well at other remoter observatories.

The universal time of the Tunguska explosion determined from seismic records is from 13.5 to 14.5 min (Ben-Menahem 1975, Pasechnik 1976, 1986). Though the meteoroid did not explode in the usual sense, the time of its deceleration at the terminal point of the trajectory was so short (less than 1 s) that the process of energy release can be treated as an explosion of a linear charge (Zotkin and Tsikulin 1966). The perturbations of the geomagnetic field began about 6 to 6.5 min after the explosion (Pasechnik 1986). Perturbations in the horizontal component *H* started with a 3.5 nT jump, then this component was constant during about 2 min, and then grew further by about 30 nT during 18 min. The field returned to its undisturbed state in about 5 hours. The difference between the maximum and minimum of perturbed values is approximately 53 nT (Ivanov 1964). The perturbations in the vertical component *Z* were smaller (Fig. 3), and only a very short and small perturbation of 3 nT during about 1 min has been registered for the west-eastern component of the geomagnetic field.

The effect of the geomagnetic storm probably stems from disturbances in the normal current system of the ionosphere (Ivanov 1964, 1967). The disturbances in ionization and conductivity of the ionospheric E-layer at a 100 km altitude could be caused by an ionized trail behind the meteoroid or a shock wave propagating through the ionosphere. However, the explanation of the 6 min span between the explosion and the beginning of magnetic perturbations presents a problem. No adequate theory has been put forward. Only recently, Nemtchinov et al. (1999a) proposed a promising idea that a ballistic plume, which develops along the wake after the major energy release (Boslough and Crawford 1997), could strike the ionosphere and cause the heating of the E-layer. Estimates show that the time of ballistic motion of ejected vapour and air above the E-layer is likely to be of the order of several minutes. However the theory needs further development.

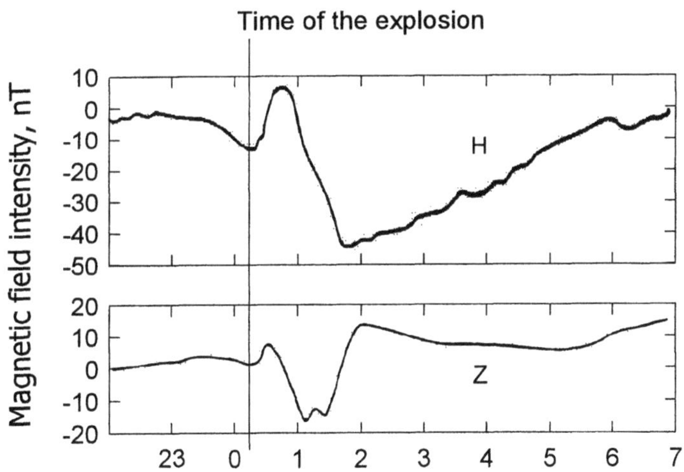

Fig. 3. The time dependence of horizontal (*H*) and vertical (*Z*) components of geomagnetic field (relative to S_q-variations) registered by the Irkutsk observatory.

Evidently, magnetic and electric fields generated during the fall and explosion-like deceleration of the Tunguska meteoroid in the atmosphere could not cause the geomagnetic storm.

5 Meteoroids as a Magnetohydrodynamic Generator

Some mechanisms of magnetic field generation have been considered recently in application to meteoroid airbursts (Crawford and Schultz 1993; Beech and Foschini 1999). All of them, including field amplification due to formation of a cavity in the Earth's magnetic field, field generation from non-aligned electron density and temperature gradients, field generation from charge separation and transport, are insufficient for the magnetization at the ground. Scaling relationships given by Crawford and Schultz (1999) show that, even if a 60 m meteoroid impacts the ground at 20 km/s, the magnitude of generated magnetic field would be only 0.1 mT at 30 km from the impact point. This suggests to consider other possible mechanisms that have not been yet reported.

The air heated in a bow shock wave in front of the meteoroid has a high temperature and electric conductivity. The flows of this ionized air both around the meteoroid and in its wake can generate electric currents. Rotation of meteoroid can involve the air plasma in rotational motion, and then the rotational acceleration experienced by the conductive air will generate circular currents. Though it seems unlikely that sufficiently strong magnetic field can be generated through this mechanism, let us estimate the currents.

The meteoroid, during time Δt, intercepts a mass of air equal to $\pi\rho r^2 V\Delta t$, where ρ is the air density in front of the bow shock wave, r is the meteoroid radius, and V is its velocity. This mass, being compressed in the shock wave to a density ρ_s and accelerated to linear rotational velocity V_r, takes on the shape of a torus around the body, the radius of the torus being close to that of the body. The mass balance can be written as

$$2\pi r \Delta A\, \rho_s = \pi r^2 \rho V \Delta t \tag{5.1}$$

where ΔA is the torus cross section by a plane through the axis of symmetry.

As is known from classical physics, a charge flows through a conductive accelerated ring due to inertial forces acting on electrons, and the total amount of the charge is

$$\Delta q = \frac{m V_r \lambda\, \Delta A}{e} \tag{5.2}$$

where m is the electron mass, e is the elementary charge, and λ is the electric conductivity. The induced electric current then is

$$I = \frac{m\lambda V_r V r \rho}{2e\rho_s} \tag{5.3}$$

For typical values $\lambda = 100$ S/cm (air plasma at 20 000 K), $V = 30$ km/s, $r = 100$ m, $\rho_s / \rho = 10$, we obtain that even for extremely high rotational velocity V_r of 0.3 km/s the electric current would be only about 3 A. This is absolutely insufficient for creation of a substantial magnetic field at the ground. As estimates show, inertial forces associated with meteoroid deceleration along the trajectory could not also create noticeable magnetic fields.

Another mechanism that can act during meteoroid flight stems from the fact that the plasma intersects the geomagnetic field lines and, moving ahead and behind the meteoroid at different velocities, can function as a magnetohydrodynamic electrical generator. The geomagnetic field at the site of the Tunguska event is directed approximately vertically down. If the trajectory inclination to the horizontal is 30° (Zotkin 1966; Zotkin and Chigorin 1991), the magnetic field component perpendicular to the trajectory B_p is about 0.05 mT. The plasma in the cap ahead of the body moves with the velocity of the body V, and the plasma in the wake decelerates due to development of hydrodynamic instabilities and turbulence. The velocity in the wake is about $0.1V$ at a distance of 100 body's diameters downstream (Heckman et al. 1968). The plasma cap can be considered, in a rough approximation, as a conductor moving perpendicular to the external magnetic field $B_{p.}$ The circuit will be closed in the wake where the velocities drop and the Earth magnetic field only slightly acts on electrons (Fig 4a). Let us make the simplest estimates. The electromotive force created by a moving conductor of length $d = 2r$ is

$$E = \frac{VB_p d}{c} \tag{5.4}$$

where c is the velocity of light. All the formulas are written here in the absolute Gaussian system of units (in which electric and magnetic units originate from mechanical ones).

Magnetic fields created by all elements of the circuit can lessen the geomagnetic field acting on the moving conductor, but we will neglect this influence because our purpose is to make a tentative evaluation of the maximum current. Then the electric current is

$$I = \frac{VB_p d}{cR} \tag{5.5}$$

where R is the full resistance of the circuit.

Let us assume for a rough estimate that the resistance of the circuit is determined by the narrow cap of shock-compressed air ahead of the body – the temperature and the conductivity are somewhat lower behind the body, but the cross-section is much larger. As the bow shock wave typically stands at a distance of $0.1d$ from the body surface, the resistance R is about $10\, d/\lambda$.

For $\lambda = 100$ S/cm we get $R = 10^{-3}$ ohm and then $I \approx 100$ kA for $V = 15$ km/s. The current flowing in the wake perpendicular to the meteoroid motion would create a magnetic field at the ground

$$B = \frac{I d_w}{c h^2} \tag{5.6}$$

where d_w is the mean diameter of that part of the wake where the circuit is closed (it is assumed that $d_w >> d$), and h is the altitude of this part of the wake. The field is directed along the trajectory but it would be too small for the magnetization of the soil – for $h = 10$ km and $d = 300$ m we get $B \approx 30$ nT. It is hardly probable that more accurate estimates could give a significantly higher value of the magnetic field.

A more effective magnetohydrodynamic generator can work at the end point of the trajectory when severely fragmented meteoroid is flattened due to aerodynamic forces, almost fully ablates and dramatically decelerates. The pancake models (Chyba et al. 1993; Hills and Goda 1993) assume infinite enlargement of the cross section, but we will restrict the diameter of the debris swarm to a value equal to 5 initial meteoroid diameters, that is, 300 m, if we assume, following to Chyba et al. (1993), that the Tunguska object was a 60 m asteroid. When the cloud of debris is abruptly aerobraked at an altitude of about 5 km, the velocity of the plasma in the wake is still significant – of the order of 10 km/s at altitudes of about 7 km or higher. In this case we have a situation opposite to the one considered above – a conductor moving in the magnetic field towards another motionless conductor which closes the circuit (Fig 4b). The geomagnetic field induces an electromotive force in the wake and the current will flow through the circuit, but the currents in this case are so arranged that they do not reduce the magnetic field at the site of the moving conductor but, vice versa, enhance it. Let us use the simplest estimates again.

Let us set a contour along which the maximum current presumably flows in the plane perpendicular to the magnetic field as shown in Fig 4b. The law of electromagnetic induction for this contour can be written as

$$IR = -\frac{1}{c}\frac{\partial \phi}{\partial t} \tag{5.7}$$

where I is the current, R is the resistance of the contour, and ϕ is the magnetic flux through the contour, t is the time. Taking the magnetic flux in the form

$$\phi = B_p A + \frac{L\,I}{c} \tag{5.8}$$

where A is the area embraced by the contour and L is the inductance, we can write

$$IR = \frac{B_p d_w V}{c} - \frac{I}{c^2}\frac{\partial L}{\partial t} - \frac{L}{c^2}\frac{\partial I}{\partial t} \tag{5.9}$$

where d_w is the diameter of the oncoming gas stream. If the gas moves inside the contour and the contour contracts, the time derivative of L is negative and, correct to the first order, is about

$$\frac{\partial L}{\partial t} \approx -V \tag{5.10}$$

Then (5.9) can be rewritten as

$$\frac{L}{c^2}\frac{\partial I}{\partial t}+I\left(R-\frac{V}{c^2}\right)=\frac{B_p d_w V}{c} \tag{5.11}$$

The current rapidly grows if the bracketed expression in (5.11) is coming to zero, while other parameters can be quite common, e.g., $V = 10$ km/s and $R = 10^{-3}$ ohm. If $R \ll V/c^2$, the current grows exponentially with characteristic time L/V. However, the time of atmospheric deceleration is not longer than 1 s, and the current cannot increase more than by three orders of magnitude in comparison to the value given by expression (5.6). Then the field at the ground will be smaller than 1 mT. Thus, it is unlikely that this mechanism could provide soil magnetization at 30 km distances on the ground.

Of course, our estimates are too simplified – the velocity varies continuously from the wake to the end point of the trajectory and the electromotive forces and currents are distributed over the whole plasma region. Numerical hydromagnetic computations based on Maxwell equations and, possibly, laboratory experiments are needed to determine whether the full current of a great magnitude can be really achieved or not.

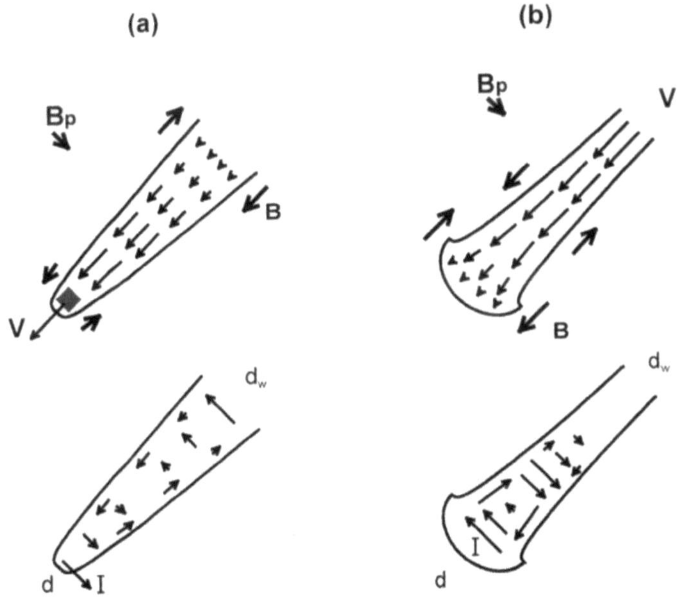

Fig. 4. An outline sketch showing how electric currents can be generated by a meteoroid moving in the geomagnetic field **(a)** and at the final point of the trajectory **(b)**. Upper panels show schematically the velocity fields in a vertical plane going through trajectory. B_p is the component of the geomagnetic field acting perpendicular to the trajectory. The arrows around the flow field show the direction of generated magnetic field B. Lower panels show the directions of electric currents I in a plane going through the trajectory perpendicular to the vertical plane (that is, the upper pictures are rotated through 90°).

6 Meteoroid Electrification

Moving meteoroids can gain electric charge. Ballistic experiments with projectiles moving in the air produced controversial results. Experiments made by ter Haseborg and Trinks (1980) with metallic projectiles of 2 to 155 mm in diameter at modest velocities between 50 and 1200 m/s showed that a projectile in the free flight acquires a positive charge of 10^{-13}–10^{-7} C, probably due to triboelectric effect by stripping material particles from the surface. In contrast, experiments of Arseniev et al. (1989) with spherical projectiles 20 mm in diameter at velocities of 2–2.5 km/s revealed that the projectiles were charged negatively in flight regardless of the body material (both conductive and nonconductive) and the trail was charged positively. The temperature behind the shock wave was above 2500 K at V = 2 km/s and the air was partially ionized.

Electrification of projectiles 4.8 mm in diameter moving at higher velocities of 2 to 5.7 km/s was studied by Pilyugin and Baulin (1993). Various materials including chondrite were used. The negative charge from $2\cdot10^{-12}$ to $2\cdot10^{-11}$ C was detected on the body and the nearest (1000 body's calibres) part of the conductive trail. The charge of the opposite sign probably spread at the far wake in the colder gas. The negative charge on the body existed until the ionized trail had disappeared. The electrification of the chondritic projectiles seems to have been caused by removal of a thin layer of melted material. The charge in these experiments was associated with an ionized trail; electrification began only at Mach numbers higher than 9.

If the electrification process means stripping of small particles carrying a definite charge, a charge of the opposite sign on the body at the saturation point will be defined by the balance between the mechanical forces acting on the particle and the force of electric attraction. If the particle is close to the body's surface, the electric force depends only on the charge surface density, which is inversely proportional to the surface area. If hydrodynamic forces acting on the particles are the same for the bodies of different sizes, the body charge is proportionate to the body's radius squared. If we assume that the body is dispersed into a cloud of small projectiles, which have the same size and charge as those in the experiments, the total charge of all the particles created by the body will be proportional to the cube of the body's radius. The maximum charge of 10^{-7} C had been detected in the experiments of ter Haseborg and Trinks (1980) on a 76 mm projectile. The cubic scaling will result in the total charge of the debris cloud of the order of 100 C for a Tunguska-size body, but the accuracy of this estimate is obviously very low.

Theoretical attempts to consider other mechanisms that can act at high meteoroid velocities are scarce. Nevskiy (1978) assumed that the meteoroid heated surface emitted electrons similar to heated metals in vacuum and argued that the potential difference U between the body and the surrounding ionized gas could be up to several volts. For an extremely idealised case, when positive and negative charges are separated at a distance of Debye's radius $d \sim 10^{-5}$ cm, he considered the body as a capacitor and derived that a charge of the body was

$Q = Ur^2/d$, that is about 100 C for $r = 100$ m. Note that both assumptions of Nevskiy about a high level of thermoelectronic emission and an extremely small distance of charge separation seem very improbable. The surface of the ablating meteoroid is a mixture of solid particles, molten and vaporised material and air, so that there is no real boundary.

Another mechanism that can charge the body is fragmentation (Bronshten 1991a). When a crystal lattice breaks up in vacuum, the powerful electric field at the crack accelerates ions, which, in turn, knock out electrons with energies of some keV (Molotskiy 1977). It is generally accepted that the Tunguska meteoroid was heavily fragmented before the explosion, but it remains unclear how to estimate the possible electron flux emitted from the surface via this mechanism. These models, however, cannot explain how the ionized and highly conductive trail can bear charge opposite in sign to that of the body. All isolating layers at the surface would be destroyed by instabilities and turbulence. The experiments of Pilyugin and Baulin (1993) mentioned above show that the body and the ionized wake have a charge of the same sign.

It is likely that meteoroid debris can gain a charge at the late stage of meteoroid disintegration when velocities drop below 2 km/s, temperatures are under 3000 K, and the wake is only slightly ionized. In the case of the Tunguska event, a portion of the cloud of vapour and dust could come very close to the ground, down to about 1 km (Svetsov 1998). Then, particles of various sizes fall under gravitational forces from the cloud stretched along the trajectory, and their speeds are determined by aerodynamic drag. Only small particles of submillimetre size have been found at the site of the Tunguska explosion (Longo et al. 1994; Serra et al. 1994). The particles had various compositions, including metals and dielectrics, low-Z and high-Z elements, but magnetic properties of these particles have not been studied. Such particles could be charged due to electrification processes during condensation, interaction with air molecules, collisions between the particles and shedding of microdrops. The physics of these processes is insufficiently studied, and it is very difficult to predict the net charge.

Experiments have been made with small dust particles dispersed by air flows at about 10 m/s (Kamra 1973; Rulenko et al. 1986). Smaller and larger particles in these experiments gained charges of opposite sign and on average dust particles 2–32 μm in size carried a charge equivalent to 100 elementary charges. Assuming that the 60 m body is fragmented into a swarm of 30 μm particles dispersed in the atmosphere by air flow, we obtain a total charge (that is, a sum of the charges of all $(60\ \text{m}/30\ \mu\text{m})^3$ particles) of about 100 C, which is too small even in comparison with a typical thunderstorm cloud. After the electrification the separation of charges in the atmosphere could happen due to gravitational forces. Millimetre-sized particles (let them be the largest among all and let us assume that a small amount of them survive) fall at about 10 m/s in the atmosphere. When they come closer to the ground, covering in some minutes a few kilometres from the end point of the trajectory to the ground, electric discharges can happen. Note that the charge of 100 C is close to a typical charge transferred by continuing current in a thunderstorm lightning (Ogawa 1965). The situation with the meteoroid, however, seems to be closer to lightning produced during volcanic eruptions. It was estimated by Anderson et al. (1965) that 0.1 to 0.5 C is neutralized in volcanic lightning, and that lightning occurs after a volcanic eruption with 10 s interval between separate discharges. In this case the 100 C cloud of meteoroid debris could result in 200 to 1000 lightning discharges of lower energy.

The mechanisms of particle formation and electrification after the complete disintegration of large bolides are far from clear. Important mechanisms may still be unknown or poorly understood. For example, microparticles probably could gain a charge when a blast wave reflected from the ground moves upward through the debris cloud. We obtain relatively small charges but it is not improbable that the net charge gained by the meteoroid microparticles can be much higher than our rough estimates may predict. The whole problem of meteoroid electrification is inherently imprecise and difficult. Note that when a charged meteoroid moves forward or rotates it can only produce small electric currents, and the electric discharges seem to be the most important consequence of the electrification. Despite their relatively low energy these discharges could trigger other phenomena.

7 Grounding of the Ionosphere

The globe may be considered as a 1 farad capacitor with a total charge Q of about $5 \cdot 10^5$ C. Soil is negatively charged and the lower layers of the atmosphere, at heights of 5 to 10 km, have a positive charge (Bering et al. 1998; MacGorman and Rust 1998). The leakage of the positive charge through the atmosphere (the total atmospheric resistance is equivalent to about 200 ohm) is constantly compensated by thunderstorm activity – cloud tops supply positive charge to the ionosphere and negative charge goes mainly from bottoms of the clouds to the ground through lightning. The ionosphere has almost the same potential of 300 kV as the 10 km layer over the surface, but the conductivity of the ionosphere at altitudes above 90 km is more than 10 orders of magnitude higher. When a large meteoroid reaches the ground at a high velocity, the ionosphere is connected to the earth's surface by an ionized wake. The ionized wake of the Tunguska meteoroid dissipated high in the air where the body was decelerated by aerodynamic forces, but lightning could connect it to the ground at some instant of time. Let us estimate the magnitude of the electric current from the ionosphere to the earth caused by the short circuiting through the wake and the discharge channels (Fig 5).

The maximum current through the wake could be defined as the ionospheric potential U of 300 kV divided by a resistance R of the circuit. This resistance is the sum of resistances of the wake, discharge channels, the ionosphere and the ground. However, there is another restriction for the maximum current – the time of the global short circuiting discharge cannot be shorter than the time necessary for light to propagate to the opposite point on the Earth surface, which time is $t = \pi R_E / c \approx 0.07$ s, where R_E is the Earth's radius. The ground-ionosphere capacitance is R_E^2/H where H is the height of the ionosphere bottom and, hence, the maximum current is

$$I_m = \frac{U R_E c}{\pi H} \tag{7.1}$$

that is about 200 kA. On the other hand, the time constant of the discharge for the ground-ionosphere capacitor is $\tau = RR_E^2/H$. If $\tau > t = 0.07$, the current will be lower than I_m. Therefore, the short circuit resistance must be lower than $\pi H/cR_E \approx 1.5$ ohm to provide the maximum current.

The inviscid part of the wake behind a flying meteoroid generated by the bow shock wave is a column of ionized and highly conductive gas. After heating in the shock wave, the air gradually cools down to about 5000 K and then emits little light because the absorption coefficients of air and vapour drop significantly. When the wake expands and the pressure in it drops to the value of the ambient pressure, the diameter of the wake d_w reaches some constant value independent of the atmospheric density. Simple estimates based both on an analogy with the cylindrical explosion and adiabatic expansion of the gas heated in the bow shock wave show that d_w is proportional to the body diameter d and the square root of its velocity V. The value of d_w is close to $10d$ for $V = 10$ km/s. At 5000 K, the electric conductivity is about 100 S/m for the air (Bazelyan and Raizer 1997) and about1000 S/m for H-chondrite vapour (Nemtchinov 1999b). (There is no information about other types of chondrites). The temperature behind the bow shock wave is over 5000 K until the meteoroid is decelerated to about 4 km/s. The Tunguska body, according to the model of Chyba et al. (1993), was probably aerobraked to this velocity, having released the major part of its energy, at about 5 km altitude. Using these data, we see that the resistance of the wake of a length of 100 km is very low, not higher than 10^{-3} ohm for $d = 60$ m and $V = 15$ km/s.

The wake resistance presents, however, a complicated problem because the hot air at the walls of the wake mixes up with cold air by a turbulence mechanism. The turbulence develops at the boundaries of the plasma wake at a distance of about 200 body diameters (Wilson 1967). For the Tunguska body this distance is 12 km, and, therefore, the turbulence can significantly reduce the wake conductivity at high altitudes. This problem, however, has not yet been quantitatively investigated. According to experimental results (Wilson 1967), the wake diameter grows due to the turbulence at the wake's boundary by a factor of 1.2 at a distance of $400d$ (24 km in our case) and doubles at distances of about $3000d$ to $4000d$. The air density in the wake at 5000 K is about 50 times lower than the density of the ambient air at the same pressure and, therefore, even a small enlargement of the wake breadth implies arrival of a relatively great amount of the cold surrounding gas to the wake. Assuming fine mechanical stirring of cold air entering the wake, which contains hot gas, we can conclude that the wake gas temperature would be about 1000 K at $400d$ and would approximately be that of the ambient air at a distance of $3000d$ to $4000d$. The electric conductivity of air drops by about three orders of magnitude at T = 1000 K in comparison to 5000 K and this means that the total resistance of the wake will grow significantly, well above 1.5 ohm, and the conductive wake will be separated at the top from the ionosphere by a span of rather cold air. Nevertheless, it remains unclear what is the degree of homogeneity in the wake behind the flying body at distances of some thousands of *d*. If some amount of hot gas, which is the main charge conductor, survives in the wake for some time, it is not impossible that the overall resistance remains relatively low. Further studies are necessary to solve the problem of the turbulence influence on the wake conductivity.

However, the problem of the wake resistance is not exhausted by the mechanism of turbulent mixing. The wake of the Tunguska meteoroid differs from those of the ballistic laboratory experiments in that a mixture of the hot vaporised meteoroid and air begin to move upward along the wake due to the vertical pressure gradient in the atmosphere – the Tunguska explosion is a plume-forming event (Boslough and Crawford 1997). The top of the plume involving the hot gas reaches the ionosphere about half a minute after the impact. Estimates show that the total resistance of the hot column inside the plume is below 1.5 ohm at this instant of time and the current through this column can reach its possible maximum.

Let us consider how the grounding of the ionosphere could happen in the Tunguska event when the meteoroid did not reach the ground losing most of its kinetic energy at about 5 km altitude. A possible way of short circuiting is a spark ground-wake discharge through the 5 km span of cold but slightly ionized air perturbed by the debris. This discharge could probably happen via the mechanism of leader propagation between the electrodes. The leader is a term for a head of a channel of high ionization which propagates through the air (between electrodes or between clouds and ground in lightning) by virtue of the electric breakdown at its front (Allibone and Schonland 1934). The ionospheric potential varies in the range 145–608 kV (Mühleisen 1977) with a mean value of 278 kV. This value coincides with the minimum potential necessary for existence and propagation of a leader from the anode at the initial stage of a spark discharge (Bazelyan and Raizer 1997). Laboratory experiments show that the leader usually cannot exist in air if the potential difference between electrodes is below 300–400 kV. There is an additional potential difference between the leader and the electrode for a 5 km channel and, therefore, it seems improbable that spark discharges spontaneously occur between the highly ionized wake area at the 5 km altitude and the ground.

The presence of charged particles in the atmosphere can change the situation and trigger discharges that would shorten the circuit. As considered above, relatively heavy particles of debris that carry the charge, probably positive, fall to the ground. The opposite charge, carried by smaller particles and/or negative ions of air, is distributed at higher altitudes, beneath the bottom of the ionized wake. This arrangement, along with fires at the ground that can start under the radiation impulse from the fireball (Svetsov 1993) or due to the fall of heated particles (Melosh et al. 1990), significantly enhances the electric field and probably results in lightning discharges between the charged zones located in the atmosphere, on the ground and in the wake, similar to fairly well-studied processes of lightning in thunderstorms. The charge carriers, however, are closer in nature to those of volcanic origin and act at smaller altitudes than in charged clouds in a typical thunderstorm.

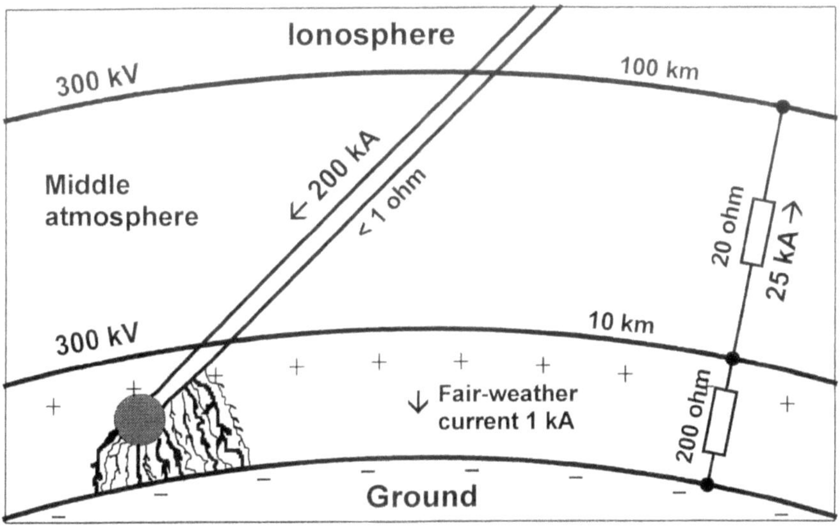

Fig. 5. Sketch explaining the global discharge caused by ionosphere-ground short circuiting by the wake and spark discharges.

If a discharge channel connects the wake to the ground, the process coming next to the lightning discharge can change the whole picture in comparison to ordinary lightning, which dies out when the charge accumulated in some localised zone of the cloud is exhausted. The closing of the circuit by a single discharge will cause a continuing current fed by a positive charge source with a constant potential in the ionosphere. When the shock wave caused by the lightning discharge decays and the channel pressure reaches that of the atmosphere, the radius of the channel can be about 10 cm depending on the energy input per unit length of the channel (MacGorman and Rust 1998). The temperature in the channel will be determined by the energy balance between the heating of the air by the electric current and its cooling by radiation. This balance seems to be quite similar to that for an ordinary electric arc, where typical temperatures are in the range 6000–12 000 K at the sea-level pressure. Note that the maximum temperatures in thunderstorm lightning are higher, from 20 000 to 30 000 K, at the stage of the return stroke, which usually propagates at a speed of about 50 000 km/s (Ogawa 1995). Conductivity of the air heated up to 12 000 K is about 50 S/cm, so that the resistance of the channel 10 cm in diameter and 5 km in length will be about 100 ohm, and the electric current through the channel at 300 kV potential difference will be only 3 kA.

Another possible consequence of the circuit-closing by a discharge channel is that the first continuing discharge between the wake and the ground can trigger multiple ground-wake discharges within the area of debris precipitation. This can happen because a spark discharge through the leader mechanism goes at lower potentials when the leader propagates along a conductor or in the vicinity of a conductor when the capacity of the channel behind the leader is enhanced. In both cases more energy is supplied in the channel behind the leader from the anode (Bazelyan and Raizer 1997). Leaders propagate at a high speed, of the order of 1000 km/s in a thunderstorm ground discharge, and, thus, in our case multiple arc-like discharges could develop within milliseconds.

Hundreds of cylindrical channels 10 cm in diameter would have a total resistance of 1 ohm, which allows the 200 kA current to pass through the circuit. The resistance of the circuit also depends on the resistance of the ionosphere and the ground. The resistance of the ionosphere in our case is mainly determined by a zone around the plume and will be well below 1 ohm. The resistance of the ground part of the circuit will depend on the ground conductivity in the area where the discharge enters the surface. Note that if the ground is used as a wire for electric signals, the resistance of the whole circuit is $1/(2\pi\lambda a)$ where a is the radius of the grounding electrode and λ is the ground conductivity at this electrode.

No measurements of the rock resistance at the site of the Tunguska explosion are available. Typical conductivity of rocks measured in various other sites is from 0.1 to 0.01 S/m at depths up to hundreds of meters (Keller 1982). The conductivity of the near surface layers is strongly dependent on the moisture of rocks and soil. The area of the Tunguska explosion is rich in swamps and ponds and the permafrost layer at a depth of about 1 m keeps water at the surface. Typical conductivity of natural waters is 1 S/m (Keller 1982) and we may use this value as a rough guess for conductivity of near surface layers at the site. The total cross-section of a hundred channels is about 1 m^2, which seems sufficient to provide an electrode less than 1 ohm for $\lambda = 1$ S/m. However, the entire resistance will also depend on the channel-ground contacts, but the physical properties of such contacts are not known. It is likely that if the amount of the discharge channels is insufficient to pass the maximum current, new channels and discharges can form. If the number of channels is too big, the currents through some of them and the temperatures drop and the channels disappear. There must be some optimum number of discharge channels to support the maximum current.

If the process goes as described, after about 70 ms of the global discharge the ionosphere acquires a negative charge equal to $UR_E^2/H \approx 1.5\cdot10^4$ C and its potential drops to that of the earth. The positive charge of about $5\cdot10^5$ C will still be dispersed over the atmosphere below 10 km altitude. As the global resistance of the atmosphere between the ionosphere and the height of 10 km is about 20 ohm, an order of magnitude lower than the global fair weather resistance between the ionosphere and the earth's surface, an electric current will go through the atmosphere from the layers 10 km high, where the global charge resides, to the ionosphere and then to the ground through the wake and the discharge channels. The time of this secondary global discharge will be about 20 ohm x 1 F = 20 s and the average current at this stage of the global discharge will be $5\cdot10^5$ C / 20 s = 25 kA, an order of magnitude higher than the usual everyday constant worldwide current. When the charge reservoir in the atmosphere is exhausted in 20 s, smaller currents can flow through the circuit until the wake or discharge channels disappear as a result of cooling, lifting due to buoyant forces, or simply are blown away by the wind. From this moment the atmosphere will begin to gain its normal charge again. Note that the energy of the global discharge is relatively low, only about 10^{11} J, six orders of magnitude lower than the Tunguska bolide kinetic energy, but, nevertheless, this discharge could lead to some detectable consequences.

Arc-like discharges could be a source of radiation at the surface. At a typical discharge temperature 12 000 K, 5 per cent of black-body radiation belongs to photons with energies above 10 eV. Photon free-paths at this temperature are about 10 cm for these energies and about 10 m for softer radiation. Therefore, a major portion of emission spectrum from a cylindrical channel 10 cm in diameter consists of hard UV radiation. But this radiation is absorbed by cold air at a distance less than 10 cm from the light source (Romanov et al. 1993). Thus, only mineral particles that were in close contact with discharge channels could be exposed to hard UV and acquire enhanced TL. At a first glance, the surface area covered by the discharge channels is very small. But note that, first, discharges are likely not motionless, and they could change their position along the surface, disappearing and arising again, horizontal spark discharges along the ground surface could also happen. Second, the discharges most probably began after the blast wave reached the ground and raised mineral dust which, being lifted into the air, could be subjected to lightning. Multiple discharges could take 20 s, until the whole atmospheric charge was exhausted. This could lead to significant enlargement of the amount of mineral grains exposed to the radiation emitted by the lightnings. At the same time, the recovered samples with the high TL output were scarce – only about 25 of 600 were identified as having enhanced TL level. And, at the very last, the hypothesis of the discharge-related TL is in agreement with the fact that the samples with high TL output are concentrated along the direction of probable trajectories, where discharges could occur with a higher probability.

The magnitude of the magnetic field caused by the 200 kA current passing through the 200 km wake or plume (inclined to the surface) would be about 4 nT in Irkutsk. An abrupt rise in the horizontal magnetic field of such an amplitude was registered in Irkutsk in the very beginning of the geomagnetic perturbations (Plekhanov et al. 1961; Ivanov et al. 1961), and it cannot be excluded that the global discharge was the reason for this increase in the magnetic field.

It is unclear what type of environmental consequences could have arisen if the atmosphere had quickly lost its charge. If ordinary thunderstorms are independent of the atmospheric electricity, the global charge would be restored in some tens of minutes. It seems unlikely that the grounding of the ionosphere can influence the weather conditions, although it may not be accidental that the thunderstorm activity and the amount of rain in Europe increased during the next 2–3 weeks after the Tunguska event (Fast and Fast 1976; Vasilyev 1998).

8 Conclusions

1) The soil magnetic anomaly revealed in the area of the Tunguska catastrophe has several distinctive characteristics: a) Soil magnetic susceptibility at ground zero is 100% higher than far from the epicentre. The area where susceptibility is 30% higher than the average extends to the north-west and north more than 30 km from the zero point. b) The magnetic inclination in soil samples is several degrees less than the normal value over an area of 10–15 km in radius. c) The magnetic declination of soil differs from the average geomagnetic field declination by –10 to –30 degrees over an area which correlates with the area of flattened trees (2000 km^2). d) The absolute values of soil magnetization vectors are 50% higher at ground zero. The area with enhanced magnetization extends for about 25 km from the zero point to the north-west.

The soil magnetic anomalies are likely to result from the shock wave which lifted dust particles in the air. The dust re-oriented on settling, and the shift in declination and inclination possibly reflects the magnetic field of 1908. This field has not been measured and remains unknown. The increase in magnetic susceptibility and absolute values of magnetization vector to the north-west and north could be caused by surface winds which transferred dust particles having ferromagnetic properties.

2) Tunguska thermoluminescent anomaly shows itself as enhanced and reduced values of TL output in the soil minerals. The area of the TL anomaly, about 15 km in radius, correlates fairly well with the area of burnt trees. The samples with reduced TL level are concentrated towards the zero point, whereas the samples with enhanced TL are found mainly closer to the periphery of the anomalous area.

The reduction in TL was probably caused by annealing action of thermal radiation from the fireball. The enhanced TL in the minerals could arise from ultraviolet radiation with photon energies up to 6.5 eV emitted by the fireball and not absorbed by the cold air. It seems plausible that UV radiation activated minerals over the whole anomalous area, but the heat flux of softer radiation was insufficient to anneal the minerals at the periphery.

3) The electromagnetic impulse, probably generated during the meteoroid deceleration in the atmosphere, could reach a noticeable magnitude which is still insufficient for soil magnetization at the distances of 30 km.

4a) The 200 kA current of the global discharge seems to be too weak to influence soil magnetic properties over the vast area of tens of kilometres, although it could result in re-orientation of soil magnetic particles due to radial currents along the ground in a narrow zone just below the trajectory, may be about 1 km wide. b) Arc-like discharges produced in the process of the global grounding could be an additional source of ultraviolet radiation at the surface and enhance stored TL energy in the minerals over a narrow area stretched along the trajectory. c) The magnetic field caused by the global discharge could be registered in Irkutsk as a small jump in field intensity. This discharge could disturb the system of local ionospheric currents.

5) The following sequence of processes is likely to occur during the Tunguska event: a) the meteoroid motion in the atmosphere at approximately constant speed down to 15 km; b) strong deceleration in the atmosphere at 5–10 km (airblast), electromagnetic and ultraviolet impulse – enhancement of TL in irradiated minerals below the trajectory– during about 1 s; c) ultraviolet and thermal radiation from the fireball – burnt trees and fire ignition at the epicentre, enhancement of TL by UV, annealing action of the heat flux on stored TL in the minerals – to about 20 s from the airblast; d) shock waves reach the ground – felled trees, lifting and spreading of dust, partial extermination of the fire at ground zero – in about 5 s after the airblast, during some minutes; e) precipitation of meteoroid condensed material, electrification of falling particles, growth of the fire, settling of large dust particles – during some minutes; f) development of the plume, electric discharges to the ground – influence on soil magnetization and thermoluminescence in a zone below the trajectory, – from half to some minutes after the airblast; g) global discharge, ballistic plume strikes the ionosphere – perturbations in the ionospheric current system, regional distortion of the geomagnetic field – in several minutes after the blast.

6) The problem considered here is far from providing an exhaustive solution as indicated in the paper title. Much research remains to be done to shed light on the nature of the geological anomalies and some physical aspects of large bolide airbursts.

Further field investigations are needed to acquire better statistics for the distribution of soil samples with the abnormal TL level. The magnetic anomaly also deserves closer examination. Additional fieldwork would be necessary to study the distribution of anomalous samples and depth dependence of soil magnetization. Laboratory experiments could be also done to investigate the response of the minerals to irradiation by hard UV spectrum, the mechanisms of soil magnetization, and the magnetohydrodynamic effect of the magnetic field generation. Our knowledge of meteoroid electrification needs improvements through both theoretical and experimental studies. And, at last, numerical simulation of the hydrodynamic flow in the atmosphere and of the magnetohydrodynamic effect is necessary to get more reliable data on the values of the electric currents generated by moving plasma and by a possible global discharge.

7) From a geological viewpoint, TL and magnetic anomalies could serve as markers of large airblasts caused by meteoroids hundreds of meters in size. Collisions with such bodies are fairly frequent in terms of the geological time scale, but it is difficult to identify the consequences of these airblasts because of the absence of craters. For example, the mysterious event presumably caused by an airblast in northern Syria more than four thousand years ago (Courty 1998) could be studied for magnetic and TL properties of the minerals. Meteoroids larger than the Tunguska bolide, also highly dispersed, come closer to the ground and could have left stronger magnetic and TL fingerprints.

References

Aitken MJ (1985) Thermoluminescence dating. Academic Press, London, 359 pp

Allibone TE, Schonland BEJ (1934) Development of the spark discharge. Nature 134: 736–737

Anderson R, Bjornsson S, Blanchard DC, Gathman S, Hughes J, Jonasson S, Moore CB, Survilas HJ, Vonnegut B (1965) Electricity in volcanic clouds. Science 148: 1179–1189

Anfinogenov JF, Fast VG (1985) A bright bolide at the South of Siberia. Zemlya i Vselennaya 3: 72–75 (in Russian)

Angino EE (1964) The effects of non-hydrostatic pressures on radiation-damage thermoluminescence. Geochim Cosmochim Acta 28: 381–388

Antonov-Romanovskiy VV (1966) Kinetika fotolyuminestsentsii kristalloforov (Kinetics of crystallophor photoluminescence). Nauka, Moscow (in Russian)

Arseniev VV, Mishin GI, Serov YuL, Yavor IP (1989) Electric charges on bodies moving at supersonic speeds in air. Zh Tekhnich Fiz 59: 122–126 (in Russian)

Astapovich IS (1958) Meteornye yavleniya v yamosfere Zemli (Meteoric phenomena in the Earth atmosphere). Fizmatgiz, Moscow (in Russian)

Bazelyan EM, Raizer YuP (1997) Spark discharge. CRC Press, Boca Raton, Florida, 294 pp

Beech M, Foschini L (1999) A space charge model for electrophonic bursters. Astron Astrophys Lett 345: L27–L31

Ben-Menahem A (1975) Source parameters of the Siberian explosion on June 30 1908 from analysis and synthesis of seismic signals at four stations. Phys Earth Planet Int 11: 1–35

Bering EA, Few AA, Benbrook JR (1998) The global electric circuit. Phys Today 51: 24–30

Bidyukov BF (1988) Thermoluminescent analysis of soils of the Tunguska fall area. In: Aktualnye voprosy meteoritiki v Sibiri. Nauka, Siberian branch, Novosibirsk, pp 96–104 (in Russian)

Bidyukov BF, Krasavchikov VO, Razum VA (1990) Thermoluminescent anomalies of soils of the Tunguska fall area. In: Sledy kosmicheskih vozdeystviy na Zemlyu. Nauka, Siberian branch, Novosibirsk, pp 88–108 (in Russian)

Bidyukov BF (1997) Thermoluminescent anomalies in a zone of the Tunguska phenomenon influence. Tungusskiy Vestnik KSE 5: 26–33 (in Russian)

Boslough MBE, Crawford DA (1997) Shoemaker-Levy 9 and plume-forming collisions on Earth. In: Remo JL (ed) Near-Earth objects. New York Academy of Sciences, New York, pp 236–282

Boyarkina AP, Sidoras SD (1974) Paleomagnetic studies in the area of Tunguskian meteorite falling. Geologiya i Geofizika 3: 79–84 (in Russian)

Boyarkina AP, Goldine VD, Sidoras SD (1980) About territorial structure of the remanent magnetization vector of soils in the area of the Tunguska meteorite fall. In: Vzaimodeystvie meteoritnogo vesshestva s Zemlyoy. Nauka, Siberian branch, Novosibirsk, pp 163–168 (in Russian)

Bronshten VA (1983) A magnetohydrodynamic mechanism for generating radio waves by bright fireballs. Solar System Research 17: 70–74

Bronshten VA (1991a) Electrical and electromagnetic phenomena associated with meteor flight. Solar System Research 25: 93–104

Bronshten VA (1991b) Nature of the anomalous sky luminescence connected with the Tunguska event. Solar System Research 25: 369–380

Bronshten VA (1999) Trajectory and orbit of the Tunguska meteorite revisited. Meteoritics Planet Sci 34: A137–A143.

Churchill DR, Armstrong BH, Johnston RR, Muller KG (1966) Absorption coefficients of heated air: a tabulation to 24 000 °K. Journal of Quantitative Spectroscopy and Radiative Transfer 6: 371–442

Chyba CF, Thomas PJ, Zahnle KJ (1993) The 1908 Tunguska explosion: atmospheric disruption of a stony asteroid. Nature 361: 40–44

Courty M-A (1998) Causes and effects of the 2350 BC Middle East anomaly evidenced by micro-debris fallout, surface combustion and soil explosion. In: Peiser BJ, Palmer T, Bailey ME (eds) Natural catastrophes during bronze age civilisation: archaeological, geological, astronomical and cultural perspectives. British Archaeological Reports – S728, Archaeopress, Oxford, pp 93–108

Crawford DA, Schultz PH (1988) Laboratory observations of impact-generated magnetic fields. Nature 336: 50–52

Crawford DA, Schultz PH (1991) Laboratory investigations of impact-generated plasma. J Geophys Res 96: 18807–18817

Crawford DA, Schultz PH (1993) The production and evolution of impact-generated magnetic fields. International Journal of Impact Engineering 14: 205–216

Crawford DA, Schultz PH (1999) Electromagnetic properties of impact-generated plasma, vapor and debris. International Journal of Impact Engineering 23: 169–180

Fabiano EB, Peddie NW, Zunde AK (1983) A magnetic field of the Earth, 1980 (map). US geological survey, Denver, USA

Fast VG (1967) Statistical analysis of the Tunguska flattened area. In: Problema Tungusskogo meteorita, iss 2. Tomsk Univ Press, Tomsk, pp 40–61 (in Russian)

Fast NP, Fast VG (1976) On possible influence of the Tunguska meteorite on rainfall of summer of 1908. In: Voprosy meteoritiki. Tomsk Univ Press, Tomsk, pp 132–143 (in Russian)

Fast VG, Boyarkina AP, Baklanov MV (1967) Destruction caused by the Tunguska meteorite shock wave. In: Problema Tungusskogo meteorita, iss 2. Tomsk Univ Press, Tomsk, pp 62–104 (in Russian)

Fast VG, Barannik AP, Razin SA (1976) On the field of directions of felled trees in the area of the Tunguska meteorite fall. In: Voprosy meteoritiki. Tomsk Univ Press, Tomsk, pp 39–52 (in Russian)

Fast VG, Fast NP, Golenberg NA (1983) Catalogue of trees felled by the Tunguska meteorite. In: Meteoritnyje i meteornyje issledovanija. Nauka, Novosibirsk, pp 24–74 (in Russian)

Florensky KP (1962) The problem of space dust and modern state of the Tunguska meteorite studies. Geokhimiya 23: 3–29 (in Russian)

Fragoulis DV, Readhead ML (1991) Feldspar inclusions and the anomalous fading and enhancement of thermoluminescence in quartz grains. Nuclear Tracks and Radiation Measurements 18: 291–296

Godfrey-Smith DI (1994) Thermal effect in the optically simulated luminescence of quartz and mixed feldspars from sediments. J Phys D: Appl Phys 27: 1737–1746

Guterman IG (ed) (1979) Novyy aeroklimaticheskiy spravochnik svobodnoy atmosfery nad SSSR (New aeroclimatic handbook on the free atmosphere over USSR), vol 2, Harakteristiki vetra i geopotenstiala. Meteoizdat, Moscow (in Russian)

Guterman IG (ed) (1986) Novyy aeroklimaticheskiy spravochnik pogranichnogo sloya atmosfery nad SSSR (New aeroclimatic handbook on the atmospheric boundary layer over USSR), vol 2, Statisticheskie harakteristiki vetra. Meteoizdat, Moscow (in Russian)

ter Haseborg JL, Trinks H (1980) Electric charging and discharging process of moving projectiles. IEEE Trans Aerospace Electr Sys, AES-16, 2: 227–231

Hills JG, Goda MP (1993) The fragmentation of small asteroids in the atmosphere. Astron J 105: 1114–1144

Hramov AN, Sholpo LE (1967) Paleomagnetizm. Nedra, Leningrad (in Russian)

Hunt JA, Palmer R, Penney W (1960) Atmosphere waves caused by large explosions. Phyl Trans Roy Soc Lond 252: 275–315

Ivanov KG (1961) Geomagnetic phenomena observed at Irkutsk magnetic observatory after the Tunguska meteorite explosion. Meteoritika 21: 46–48 (in Russian)

Ivanov KG (1964) Geomagnetic effect of the Tunguska catastrophe. Meteoritika 24: 141–151 (in Russian)

Ivanov KG (1967) On the nature of the Tunguska fall influence on the upper ionosphere, geomagnetic field and night glows. Geomagnetizm i Aeronomiya 7: 1031–1035 (in Russian)

Ivliev AI, Badyukov DD, Kashkarov LL (1995) The study of thermoluminescence in shock-loaded samples: 1. Oligoclase. Geokhimiya 9: 1367–1377 (in Russian)

Ivliev AI, Badyukov DD, Kashkarov LL (1996) Investigations of thermoluminescence in experimentally shocked samples: 2. Quartz. Geokhimiya 10: 1010–1018 (in Russian)

Kamra AK (1973) Experimental study of the electrification produced by dispersion of dust into air. J Appl Phys 44: 125–131

Kardash AV (1984) On magnetic declination in the area of the Tunguska meteorite fall. In: Dolgov YuA (ed) Meteoritnye issledovaniya v Sibiri. Nauka, Novosibirsk, pp 77–80

Keay SCL (1980) Anomalous sounds from the entry of meteor fireballs. Science 210: 11–15

Keay CSL (1992) Electrophonic sounds from large meteor fireballs. Meteoritics 27: 144–148

Keay CSL, Ceplecha Z (1994) Rate of observation of electrophonic meteor fireballs. J Geophys Res 99: 13163–13165

Keller GV (1982) Electrical properties of rocks and minerals. In: Carmichael RS (ed) Handbook of physical properties of rocks, vol 1. CRC Press, Boca Raton, Florida, pp 217–293

Kirichenko LV (1975) On verification of hypothesis of the Tunguska meteorite "nuclear explosion" by solid radioactivity at a trace of explosion products precipitation. In: Problemy meteoritiki. Nauka, Siberian branch, Novosibirsk, pp 88–101 (in Russian)

Kolesnikov EM, Lavrukhina AK, Fisenko AV (1975) A novel method of verification of hypothesis related to annihilation and thermonuclear nature of 1908 Tunguska explosion. In: Problemy meteoritiki. Nauka, Siberian branch, Novosibirsk, pp 102–110 (in Russian)

Kolesnikov EM, Kolesnikova NV, Boettger T (1998) Isotopic anomaly in peat nitrogen is a probable trace of acid rains caused by 1908 Tunguska bolide. Planet Space Sci 46: 163–167

Kolesnikov EM, Boettger T, Kolesnikova NV (1999) Finding of probable Tunguska cosmic body material: isotopic anomalies of carbon and hydrogen in peat. Planet Space Sci 47: 905–916

Korobeinikov VP, Chushkin PI, Shurshalov LV (1976) Mathematical model and computation of the Tunguska meteorite explosion. Acta Astronaut 3: 615–621

Korobeinikov VP, Chushkin PI, Shurshalov LV (1982) Interaction between large cosmic bodies and the atmosphere. Acta Astronaut 9: 641–652

Korobeinikov VP, Shurshalov LV, Vlasov VI, Semenov IV (1998) Complex modelling of the Tunguska catastrophe. Planet Space Sci 46: 231–244

Krinov EL (1966) Giant meteorites. Pergamon Press, Oxford, 397 p

Lee P (1965) Photodissosiation and photoionization of oxygen (O_2) as inferred from measured absorption coefficients. J Opt Soc Am 45: 703–709

Longo G, Serra R, Secchini S, Galli M (1994) Search for microremnants of the Tunguska cosmic body. Planet Space Sci 42: 163–177

Lvov YuA, Vasilyev NV (1976) Radiation burn of trees in the area of the Tunguska meteorite fall. In: Voprosy meteoritiki. Tomsk Univ Press, Tomsk, pp 53–57 (in Russian)

Lyne JE, Tauber M, Fought R (1996) An analytical model of the atmospheric entry of the large meteors and its application to the Tunguska event. J Geophys Res 101: 23207–23212

MacGorman DR, Rust WD (1998) The electrical nature of storms. Oxford Univ Press, New York, Oxford, 432 pp

Marfunin AS (1975) Spektroskopiya, lyuminestsenstiya i radiatsionnye tsentry v mineralah (Spectroscopy, luminescence and radiative centres in minerals). Nedra, Moscow, 326 pp (in Russian)

McKeever SWS (1985) Thermoluminescence of solids. Cambridge Univ Press, Cambridge, 390 pp

Medlin WL (1963) Thermoluminescence in quartz. J Chem Phys 38: 1132–1143

Melosh HJ, Schneider NM, Zahnle KJ, Latham D (1990) Ignition of global wildfires at the Cretaceous/Tertiary boundary. Nature 343: 251–254

Molotskiy MI (1977) Ionno-electronic mechanism of mechanic emission. Fizika Tverdogo Tela 19: 642–644 (in Russian)

Mühleisen R (1977) The global circuit and its parameters. In: Electrical processes in atmospheres. Dr Dietrich Steinkopff, Darmstadt, pp 467–476

Nemtchinov IV, Losseva TV, Merkin VG (1999a) Estimate of geomagnetic effect after the fall of the Tunguska meteoroid. In: Zettser YuI (Ed) Fizicheskie protsessy v geosferah. Institute for Dynamics of Geospheres, Russian Academy of Sciences, Moscow (in Russian)

Nemtchinov IV (1999b) Estimates of nonequilibrium effects on plasma composition and conductivity in the wake. Report, Institute for Dynamics of Geospheres, Russian Academy of Sciences, Moscow

Neumayer G (1891) Atlas des Erdmagnetismus. In: Berghaus H (Ed) Berghaus' Physikalischer Atlas, Part IV, Justus Perthes, Gotha.

Nevskiy AP (1978) Phenomenon of positive stabilised electric charge and the effect of electric discharge explosion of large meteorites in planetary atmospheres. Solar System Research 12: 173–180

Ogawa T (1995) Lightning currents. In: Handbook on atmospheric electrodynamics, vol 1. CRC Press, Boca Raton, Florida, pp 93–136

Ovchinnikov VM, Pasechnik IP (1988) Earthquake caused by bolide explosion in the basin of Chulym river. Dokl Akad Nauk SSSR 299: 595–598 (in Russian)

Pasechnik IP (1976) Evaluation of parameters of the Tunguska meteorite explosion by seismic and microbarographic data. In: Kosmicheskoe vesshestvo na Zemle. Nauka, Siberian branch, Novosibirsk, pp 24–54 (in Russian)

Pasechnik IP (1986) Refinement of time of the 30 June 1908 Tunguska explosion using seismic data. In: Kosmicheskoe vesshestvo i Zemlya. Nauka, Siberian branch, Novosibirsk, pp 62–69 (in Russian)

Petrova GN (1961) Various laboratory methods for determination of geomagnetic stability of rocks. Isvestiya Akad Nauk SSSR, Ser Geofiz 11: 1585–1598 (in Russian)

Pilyugin NN, Baulin NN (1993) Measurement of electric charges formed on bodies and in their wakes during hypersonic motion. Solar System Research 27: 558–573

Plekhanov GF (1964) Some results of works by a complex amateur expedition on study of the Tunguska meteorite problem. Meteoritika 24: 170–176 (in Russian)

Plekhanov GF, Kovalevskiy AF, Zhuravlyov VK, Vasilyev NV (1961) On effect of the Tunguska meteorite explosion on the geomagnetic field. Geologiya i Geofizika 6: 94–96 (in Russian)

Putyatin BV (1980) The radiation action on the Earth during the large meteorite body flight in the atmosphere. Dokl Akad Nauk SSSR, 252: 318–320 (in Russian)

Ramseyer K, Aldahan A, Collini B, Landstrom O (1992) Petrological modifications in granitic rocks from the Siljan impact structure: evidence from cathodoluminescence. Tectonophysics 216: 195–204

Rasmussen KL, Olsen HJF, Gwozdz R, Kolesnikov EM (1999) Evidence for a very high carbon/iridium-ratio in the Tunguska impactor. Meteoritics Planet Sci 34: 891–895.

Romanov GS, Stankevich YuA, Stanchits LK, Stepanov KL (1993) Thermodynamic properties, spectral and average absorption coefficients for multicomponent gases in a wide range of parameters. Preprint no 2, Lycov Institute of Heat and Mass Exchange, Minsk

Rulenko OP, Klimin NN, Diakonova IN, Kirianov VYu (1986) Studies of electrification in clouds created by dispersion of volcanic ash. Vulkanologiya i Seismologiya 5: 17–29 (in Russian)

Sapronov NL (1986) Drevnie vulkanicheskie struktury na yuge Tungusskoy sineklizy (Ancient volcanic structures at the south of the Tunguska syneclise). Nauka, Moscow (in Russian)

Sekanina Z (1983) The Tunguska event: no cometary signature in evidence. Astron J 88: 1382–1414

Serra R, Cecchini S, Galli M, Longo G (1994) Experimental hints on the fragmentation of the Tunguska cosmic body. Planet Space Sci 42: 777–783

Shakhovets SA, Shlyukov AI (1987) Thermoluminescence dating of the deposits of the Lower Volga. In: Novye dannye po geokhronologii chetvertichnogo perioda (New data in Quaternary geochronology). Nauka, Moscow, pp 197–204 (in Russian)

Sidoras SD, Boyarkina AP (1976) About results of paleomagnetic studies in the area of the Tunguska meteorite fall. In: Voprosy meteoritiki. Tomsk Univ Press, Tomsk, pp 64–73 (in Russian)

Souslov IM (1967) Interviewing of eyewitnesses of the Tunguska catastrophe in 1926. In: Problema Tungusskogo meteorita, iss 2. Tomsk Univ Press, Tomsk, pp 21–30 (in Russian)

Svetsov VV (1996) Total ablation of the debris from the 1908 Tunguska explosion. Nature 383: 697–699

Svetsov VV (1998) Could the Tunguska debris survive the terminal flare? Planet Space Sci 46: 261–268

Teterin MA (ed) (1939) Mirovaya karta magnitnyh skloneniy (dlya epohi 1935 goda) (World map of magnetic declination for epoch of 1935). Moscow (in Russian)

Turco RP, Toon OB, Park C, Whitten RC, Pollac JB, Noerdlinger P (1982) An analysis of the physical, chemical, optical and historical impacts of the 1908 Tunguska meteor fall. Icarus 50: 1–52

Vasilenko BV, Dyomin DV, Zhuravlyov VK (1967) Thermoluminescent analysis of rocks from the Tunguska fall area. In: Problema Tungusskogo meteorita, iss 2. Tomsk Univ Press, Tomsk, pp 227–231 (in Russian)

Vasilyev NV, Zhuravlyov VK, Zhuravlyova RK, Kovalevsky AF, Plekhanov GF (1965) Nochnye svetyashiesya oblaka i opticheskie anomalii, svyazannye s padeniem Tungusskogo meteorita (Noctilucent clouds and optical anomalies connected to the Tunguska meteorite fall). Nauka, Moscow (in Russian)

Vasilyev NV, Zhuravlyov VK, Dyomin DV, Ammosov AD, Batissheva AI (1976) On some anomalous effects associated with the Tunguska meteorite fall. In: Kosmicheskoe vesshestvo na Zemle. Nauka, Siberian branch, Novosibirsk, pp 71–87 (in Russian)

Vasilyev NV, Kovalevskiy AF, Razin SA, Epiktetova LE (1981) Pokazaniya ochevidtsev Tungusskogo padeniya (Reports of eyewitnesses of the Tunguska fall). VINITI (All Union Institute of Science Information), Deponent N 5350-81, Moscow (in Russian)

Vasilyev NV (1988) History of study of Tunguska meteorite problem. In: Aktualnye voprosy meteoritiki v Sibiri. Nauka, Siberian branch, Novosibirsk, pp 3–31 (in Russian)

Vasilyev NV (1998) The Tunguska meteorite problem today. Planet Space Sci 46: 129–150

Vaz JE, Kemmey PJ, Levy PV (1968) The effects of ultraviolet light illumination on the thermoluminescence of calcite. In: McDougall DJ (ed) Thermoluminescence of geological materials. Academic Press, London, New York, pp 111–123

Vlasov VK, Karpov NA, Kulikov OA (1979) Limits of TL-method application for dating of newest deposits. Vestnik Moskovskogo Gosudarstvennogo Universiteta, Ser 5, Geografiya 4: 56–64 (in Russian)

Vorobyov VA, Demin DV (1976) New results of studies of heat affection on larch trees in the Tunguska meteorite fall area. In: Voprosy meteoritiki. Tomsk Univ Press, Tomsk, pp 58–63 (in Russian)

Wilson LN (1967) Far wake behavior of hypersonic spheres. AIAA Journal 5: 1238–1244

Zotkin IT (1966) Trajectory and orbit of the Tunguska meteorite. Meteoritika 27: 109–118 (in Russian)

Zotkin IT, Tsikulin MA (1966) Modelling of the Tunguska meteorite explosion. Dokl Akad Nauk SSSR 167: 59–62 (in Russian)

Zotkin IT, Chigorin AN (1991) Radiant of the Tunguska meteorite from visual observations. Solar System Research 25: 459–464

MIX
Papier aus verantwortungsvollen Quellen
Paper from responsible sources
FSC® C105338

If you have any concerns about our products,
you can contact us on
ProductSafety@springernature.com

In case Publisher is established outside the EU,
the EU authorized representative is:
Springer Nature Customer Service Center GmbH
Europaplatz 3, 69115 Heidelberg, Germany

Printed by Libri Plureos GmbH
in Hamburg, Germany